T0201725

Scientific Foundations of Engineering

Providing an overview of the foundations of engineering from a fundamental physical perspective, this book reinforces the basic scientific and mathematical principles which underpin a range of engineering disciplines.

It covers the basics of quantum physics as well as some key topics in chemistry, making it a valuable resource for both students and professionals looking to gain a more coherent and interdisciplinary understanding of engineering systems. Throughout, the focus is on common features of physical systems (such as mechanical and electronic resonance), showing how the same underlying principles apply to different disciplines.

Problems are provided at the end of each chapter, including conceptual questions and examples to demonstrate the practical application of fundamental scientific principles. These include real-world examples which are solvable using computational packages such as MATLAB.

Stephen W. McKnight is a Professor in the Electrical and Computer Engineering Department of the College of Engineering at Northeastern University, and is currently the education thrust leader for two NSF and DHS research centers based there.

Christos Zahopoulos is an Associate Professor in the College of Engineering and the Graduate School of Education at Northeastern University. He is also the founder and executive director of the university's Center for STEM Education and has been actively involved in initiating and implementing numerous STEM (Science, Technology, Engineering, and Mathematics) education programs and partnerships focused on improving teaching and learning.

Scientific Foundations of Engineering

STEPHEN W. MCKNIGHT

Northeastern University

CHRISTOS ZAHOPOULOS

Northeastern University

CAMBRIDGE
UNIVERSITY PRESS

University Printing House, Cambridge CB2 8BS, United Kingdom

Cambridge University Press is part of the University of Cambridge.

It furthers the University's mission by disseminating knowledge in the pursuit of education, learning and research at the highest international levels of excellence.

www.cambridge.org
Information on this title: www.cambridge.org/9781107035850

© Cambridge University Press 2015

This publication is in copyright. Subject to statutory exception
and to the provisions of relevant collective licensing agreements,
no reproduction of any part may take place without the written
permission of Cambridge University Press.

First published 2015

Printed in the United Kingdom by TJ International Ltd. Padstow Cornwall

A catalog record for this publication is available from the British Library

Library of Congress Cataloging in Publication data
McKnight, Stephen W.
Scientific foundations of engineering / Stephen W. McKnight, Northeastern University, Christos Zahopoulos, Northeastern University.
    pages   cm
Includes bibliographical references.
ISBN 978-1-107-03585-0 (Hardback)
1. Physics. 2. Engineering. 3. Technology. I. Zahopoulos, Christos. II. Title.
QC23.2.M385 2015
530–dc23   2015008860

ISBN 978-1-107-03585-0 Hardback

Additional resources for this publication at www.cambridge.org/9781107035850

Cambridge University Press has no responsibility for the persistence or accuracy of URLs for external or third-party internet websites referred to in this publication, and does not guarantee that any content on such websites is, or will remain, accurate or appropriate.

Contents

Introduction

This book is based on notes prepared for a graduate course "Scientific Foundations of Engineering" in the Gordon Engineering Leadership Program at Northeastern University. The elevator speech on why such a course is needed goes as follows: (1) most engineering students take all of their basic science courses during freshman year, (2) they don't like those freshman courses very much, and (3) they forget the material as quickly as they can and concentrate on the specifics of electrical engineering, or mechanical engineering, or other engineering discipline where their interests and enthusiasm lies.

This summary may be unfair to some engineering students, but most engineering students (and their professors) at least grudgingly admit that it isn't terribly far off. And, in general, this approach serves the students well through their undergraduate education process and in their industrial careers – as long as they remain specialized in their specific engineering discipline. However, consider the case where an electrical engineer is leading a multidisciplinary project. One day a mechanical engineer who reports to her walks into her office and says, "Boss, this isn't going to work – we can't get the heat out!" A conventionally trained electrical engineer isn't likely to be able to frame a single substantive question about the problem. She hasn't studied heat transport or thermodynamics since the freshman year (if at all!), and likely has forgotten anything she ever knew about the subject. The ability to frame questions which put fundamental boundaries on the problem, "What is the power load? How hot will the device get? How much blackbody emission is there at that temperature? What is the thermal conductivity of the substrate?", will not only enable an engineering leader to quickly frame the gravity of the problem, but will undoubtedly earn her a reputation as someone with whom you want to have done your homework carefully before making rash statements about engineering limits!

Science curricula often employ a "spiral curriculum" model. In physics for example, mechanics, thermal physics, and electromagnetics are surveyed in freshman year, revisited in specialized courses in sophomore or junior year, and often studied again from a quantum statistical viewpoint in a senior class. Engineering education more often selects a "breadth" coverage of the vast range of engineering applications, rather than the depth of understanding that the spiral curriculum seeks to impart.

Among the topics that the typical quick pass through scientific fundamentals causes to be neglected or skipped in most engineering educations is the entire field of quantum science. Despite the critical and growing importance of nanoscale quantum science on nearly every electronic device we use or carry on us, I suspect that most engineering professors would be surprised by how few engineering students or professional engineers can give even a rudimentary description of the source of semiconductor band gaps, the distinction between metals and insulators and semiconductors, or how a p-n junction or micro-mechanical device works.

This lack of depth of understanding of fundamental scientific principles and lack of any formal instruction in the science of quantum systems is what we intended the "Scientific Foundations of Engineering" course in the Gordon Engineering Leadership Program to address. Our model was the confidence with which a well-trained scientist can approach an unfamiliar problem and quickly understand the fundamental principles and make "back-of-the-envelope" calculations about how large an effect each of these fundamentals may play in the problem. Even a little bit of this ability to understand the basic physical laws underlying a phenomenon, to place a problem in context, and to estimate the size of various influences on a system can give an engineering leader an ability to direct a project through proof-of-principle experiments and make-or-break decision points to an operating marketable product. Our goal in the Gordon program is to develop engineers with gravitas to whom no area of engineering is outside of the sphere of understanding in the "here be dragons" area of the unknown, and developing knowledge of scientific fundamentals in the "Scientific Foundations" course is part of that goal.

In looking for a book to teach the "Scientific Foundations" course, however, the authors found freshman survey texts that did not make use of advanced mathematics and were intended for readers who were new to the study of science and engineering, and highly theoretical books for professional physicists, which lacked examples meaningful to engineers, but no book suitable for the intended audience. In this book, we have combined a unified treatment of classical and quantum physics with a wealth of worked examples with an engineering flavor. We typically begin each chapter with a question about physical phenomena that engineers may know and have wondered about. This helps put the treatment of the physics in a context of what the implications of the theory are and why anyone would want to know about it.

In Chapters 1 and 2, we address kinematics and dynamics in the broadest framework to understand the limits of reaching for the "d-equals-one-half-a-t-squared" solution, and how this low-order solution can be used iteratively to find solutions even outside the limits of constant acceleration problems. In these chapters and throughout the book we consider examples primarily in Cartesian x–y–z space. While many real problems are best addressed in cylindrical or spherical coordinates, the concepts are fundamentally the same and we have opted for conceptual clarity over computational completeness whenever possible.

In Chapter 3, we discuss rotational motion and use the generalization of the scalar mass in linear motion into moment of inertia in rotational motion as a way to introduce the concept of tensor quantities. The linear motion concepts of force, momentum, and energy are similarly generalized into torque, angular momentum, and rotational kinetic energy with the fundamental principles of conservation of momentum (always!) and conservation of mechanical energy (in ideal systems that do not generate heat) reinforced.

Chapter 4 introduces rotation matrices to deal with cases where tensor quantities are not aligned with the principal axes. As an example, the generalization of an isotropic mass into an effective mass tensor for electrons in anisotropic materials is considered.

With the fundamental mechanical laws established, the role of materials is introduced in the fundamental property of elasticity in Chapter 5. Shear and compressive stress and strain are shown to be elements of the 3×3 second-order stress and strain tensors, making elasticity that links stress and strain a fourth-order elastic tensor which can be reduced by symmetries to a 6×6 symmetric tensor with 21 independent material parameters. This is further reduced in cubic materials to an elastic tensor with three parameters: the Young's modulus, the shear modulus, and the Poisson ratio.

In Chapter 6, simple harmonic motion of a system with a linear restoring force and a velocity-dependent loss mechanism is discussed with and without a harmonic driving force. The use of Euler's relation to describe oscillatory motion with amplitude and phase by a complex exponential is introduced, allowing the inclusion of a small imaginary part to model the loss mechanism. The identical form and solution of an LCR electrical circuit with the spring–mass system is shown with complex exponentials representing the charge and current, leading to the electrical engineering method of complex impedance.

In Chapter 7, harmonic motion in time is extended to systems with spatial coupling, creating the phenomenon of waves in mechanical systems: one-dimensional waves in strings and three-dimensional sound waves in fluids. By continuing the complex exponential notation for the wave oscillation we can model lossy media, interference in films or from multiple sources, and diffraction phenomena with a complex exponential wave form with a complex wave vector k.

Chapters 8, 9, 10, and 11 provide an introduction to quantum physics and chemistry. In Chapter 8, we discuss the historical origins of quantum theory, why it was such a radical departure from classical physics, why it became necessary to accept such a totally different approach to understand the world, and why the "quantum picture" continues to be anti-intuitive and difficult to accept. In Chapter 9, we examine the postulates of quantum mechanics and how we can "shut up and calculate" everything that is determinable for quantum systems, including tunneling, and low-dimensional quantum systems, such as 1D and 2D quantum wells. In Chapter 10, we extend the quantum analysis to real systems from quantum dots to the hydrogen atom and touch on the chemistry of the "spdf" quantum states and hybridized outer orbitals. In Chapter 11, we look at electrons in extended lattices and band states with both direct and indirect band gaps, and how these are reflected in the electronic properties of metals, insulators, and semiconductors and optical interactions, such as in light-emitting diodes.

In Chapter 12, we look at thermal physics from the perspective of random equipartition of particles into energy states, extending into thermal transport, thermal equilibrium, heat capacity, and thermodynamics. In Chapter 13, we look at thermal effects through the mathematics of quantum statistics: Maxwell–Boltzmann statistics for classical distinguishable particles, Fermi–Dirac statistics for quantum indistinguishable particles that obey the exclusion principle ("fermions") and Bose–Einstein statistics for quantum particles such as photons that do not have any restrictions on the occupancy of a single quantum state ("bosons"). The differences in the occupation of states for the three different statistical models is demonstrated though a simple "thought experiment" and the different statistical models are developed in the examples of blackbody radiation (photon/boson statistics) and semiconductor occupancy and p-n junctions (electron/fermion statistics).

In Chapters 14, 15, and 16, we look at electromagnetic theory, effects, and materials. Chapter 14 develops Maxwell's equations in the mathematics of vector calculus ("div, grad, curl"), using Gauss's and Stokes' integral relations to move from the four Maxwell's equations and the electromagnetic Lorentz force equation to understand a wide variety of electromagnetic effects: Faraday effect, electrical generation, eddy currents, transformers, and electromagnetic motors. Starting from the four Maxwell's equations, Chapter 15 develops the electromagnetic wave equation and applies it to examine wave propagation in uniform, lossy, and anisotropic materials. The electromagnetic boundary conditions are derived and applied to plane-wave reflection at surfaces for both normal and non-normal incidence, and to interference effects in thin films. In Chapter 16, electromagnetic wave propagation in materials is studied from the viewpoint of the constitutive relations and the material-dependent tensor permittivity, conductivity, and permeability. Physical models are used to develop examples including the plasma edge in gases, semiconductors, and metals, Lorentzian oscillators in the infrared properties in polar crystals, quantum absorption in transparent gases, and ferromagnetic resonance in magnetic materials.

Chapter 17 provides an introduction to the physics of fluids, using mechanical concepts for static fluid effects, such as buoyancy, and vector calculus from electromagnetics to develop the continuity equation, Euler equation, Bernoulli equation, and Navier–Stokes equation for moving fluids. The transition from laminar to turbulent flow with increasing Reynolds number, still one of the most important unsolved scientific problems, is presented.

Throughout, we have emphasized the connections between concepts and phenomena in different fields and the similarity of the mathematics used to describe them: how a spring–mass harmonic oscillator described by a complex exponential leads to familiar expressions in AC electrical circuits, in infrared properties in crystals with optical phonons, and in quantum-energy-level absorption lines in gases and transparent solids. Or how anisotropic materials effects in elasticity, electrical permittivity, or magnetic resonance can be described by tensor properties similar to a general description of rotational motion with a tensor moment of inertia. The ability of a scientist or engineer to apply models and concepts from their area of specialization to new phenomena in different limits can serve to demystify unfamiliar technologies and allow them to apply their knowledge to novel systems.

The authors would like to express their gratitude to Professors George Adams and John Cipolla for their assistance in Chapters 5 and 17, and to the classes of students from the Gordon Engineering Leadership Program who, with their questions and comments, have helped move this project from a set of cryptic notes to a text which we trust will be comprehensible and useful to engineering and science students and professionals.

1   Kinematics and vectors

The foundation of much of engineering concerns the location of stationary and moving objects in three dimensions, and it is critical to have the mathematical tools to quantify the position of those objects and predict their movement. In fact, although we are surrounded by moving objects every day, relatively few people have a conceptual framework and vocabulary to describe motion. In addition, simple three-dimensional problems such as finding the volume of a parallelepiped defined by its three edges are hard to visualize and calculate. In this chapter we will develop the mathematical framework for motion, for describing the three-dimensional position of objects, and for calculating their non-uniform motion.

1.1   Kinematics

Kinematics is the mathematical description of motion, of how position changes with time. In one dimension, motion is completely described by the function f where

$$x = f(t). \tag{1.1}$$

In general, $f(t)$ is a complicated function. Even what we might consider to be simple motion, someone running at a more or less constant speed for example, is likely to be quite complicated if we look in detail. And the motion of the hand of that runner, as she moves it back in forth in stride, is likely to be extremely complex and unlikely to be described by any closed-form mathematical expression. How can one describe motion, especially if it is complex? One could ask, for example, if there is any mathematical way to describe the one-dimensional motion of the graph shown in Figure 1.1.

Commonly in science and engineering, we can only make progress if we apply some level of approximation to the problem. Appropriately approximated, the problem may be conceptually straightforward and mathematically tractable. One mathematical technique that is convenient for approximation is expansion of the function describing the motion in a *Taylor series*. The Taylor series expansion of a function $x(t)$ around a particular time t_o is:

$$x = x_o + \frac{dx}{dt}\bigg|_{t=t_o}(t - t_o) + \frac{1}{2!}\frac{d^2x}{dt^2}\bigg|_{t=t_o}(t - t_o)^2 + \frac{1}{3!}\frac{d^3x}{dt^3}\bigg|_{t=t_o}(t - t_o)^3 + \dots \tag{1.2}$$

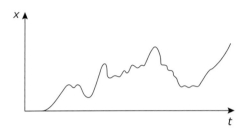

Figure 1.1 Position vs. time for complex one-dimensional motion.

Question 1.1

What would the general expression for the nth term of the series be?

A Taylor series allows us to predict the position at time Δt later if we know the initial position and all orders of rates of change of the function at the initial position. Mathematically this series is expected to converge because of the $1/n!$ prefactor – the factorial function $n!$ exceeds even the exponential function e^n for large n. Practically, given an appropriate time scale where the motion is reasonably smooth, the higher-order derivatives tend to be progressively smaller, and we can approximate the motion with a few terms. Letting $t_o \to 0$, which is equivalent to starting our clock at $t = t_o$ or setting $t = \Delta t = (t - t_o)$, we find an series expansion for $x(t)$ as below, where we have indicated the usual names given to the time derivatives of distance: $velocity = v = dx/dt$, $acceleration = a = dv/dt = d^2x/dt^2$, and $jerk = da/dt = d^3x/dt^3$.

$$x = x_o + \underbrace{\frac{dx}{dt}\bigg|_{t=t_o}}_{\substack{\uparrow \\ v \\ (\text{velocity})}} t + \underbrace{\frac{1}{2!}\frac{d^2x}{dt^2}\bigg|_{t=t_o}}_{\substack{\uparrow \\ a \\ (\text{acceleration})}} t^2 + \underbrace{\frac{1}{3!}\frac{d^3x}{dt^3}\bigg|_{t=t_o}}_{\substack{\uparrow \\ da/dt \\ (\text{jerk})}} t^3 + \frac{1}{4!}\frac{d^4x}{dt^4}\bigg|_{t=t_o} t^4 + \ldots \tag{1.3}$$

Note that there is no reason to truncate this series at any term except the expectation that the higher-order derivatives will be negligible compared to $1/n!$ since higher-order derivatives correspond to rapidly changing forces. Nevertheless, freshman physics books are so filled with examples of constant acceleration that it is not surprising that engineers often assume that distance is always found from

$$x = x_o + v_o t + \frac{1}{2}at^2. \tag{1.4}$$

In fact, for constant acceleration, when $a = $ constant and we have

$$0 = \frac{da}{dt} = \frac{d^2x}{dt^2} = \frac{d^3x}{dt^3} = \frac{d^4x}{dt^4} \quad (\ldots \text{and all higher orders}),$$

we can recover our familiar freshman physics equations by integration:

$$a = \frac{dv}{dt} = \frac{d^2x}{dt^2} = \text{constant} \Rightarrow$$

(1) $\quad v = \int a\,dt = at + v_o$

$\qquad\qquad\qquad\;\;$└─── Constant of integration = initial velocity

(2) $\quad x = \int v\,dt = \int (at + v_o)\,dt = \frac{1}{2}a\,t^2 + v_o\,t + x_o$ $\qquad\qquad\qquad$ (1.5)

(3) $\quad$ Eliminating t by algebra from (1) and (2) $\left(t = \dfrac{v - v_o}{a}\right)$

$$\text{gives}\quad v^2 = v_o^2 + 2a(x - x_o)$$

$$\text{or}\qquad v = \sqrt{v_o^2 + 2a(x - x_o)}.$$

In the more difficult (and more common) case where the acceleration is not constant, but depends on t or v or x, the equations for constant acceleration are not accurate. In this case, we can solve the problem numerically by taking steps in time or distance that are small enough that the acceleration can be considered as constant within the time interval Δt considered. (For Δt small, $\Delta t \gg \Delta t^2 \gg \Delta t^3$ and terms beyond constant acceleration are negligible.) The new position, velocity, and acceleration are calculated at the end of the small interval Δt, and then the next Δt interval is solved with these new initial values.

Such an iterative solution for a problem with non-constant acceleration can be diagramed as in Figure 1.2.

Question 1.2

Consider the case of a car slowing down and coming to a stop at a traffic light. Can this be represented with a constant acceleration? Why or why not?

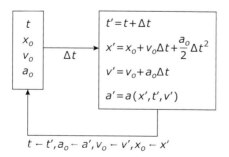

Figure 1.2 Block diagram of iterative solution for non-constant acceleration.

Iterative solutions are the basis of numerical differential equations solvers such as the second-and-third-order and third-and-fourth-order Runge–Kutta methods that are the basis of the MATLAB differential equation solvers `ode23` and `ode45`.

Example 1.1 If the effects of air resistance are ignored, a falling object near the Earth's surface experiences a constant acceleration due to the force of gravity $g = 9.8$ m/s^2 downward (toward the center of the Earth). This is a good approximation at low velocities for dense heavy objects – a bowling ball, for example – but is completely inadequate to describe a falling piece of paper. The effect of air resistance is to create a *non-uniform* acceleration that decreases as the square of the velocity

$$a = -9.8 - \gamma|v|v$$

where the constant γ depends on the mass and shape of the object and the air density. The air resistance component of the acceleration in the second term is the product of number of air molecules that the object collides with in a given time interval (proportional to the velocity) multiplied by the retarding force due to a collision with an air molecule which also increases with the velocity as we will see in Chapter 2, giving the velocity-squared dependence. The minus sign in the equation indicates that the gravitational acceleration is downward, and the absolute value of the velocity multiplied by the velocity guarantees that the air friction opposes the motion: the acceleration is more negative when the velocity is up (positive) and less negative when the velocity is down (negative). The MATLAB solution for an object with $\gamma = 0.05$ projected directly upward with an initial velocity of 30 m/s from $t = 0$ to $t = 4$ s is shown in the figure below (solid line) compared to the no-friction result from Eq. (1.4) (dashed):

```
>> type air_friction.m
function [ Ydot ] = air_friction(t, Y )
%AIR_FRICTION Function  to  find  acceleration  and  velocity
   with  air  friction
% Input vector Y=[y, v_y] ; output vector Ydot= dY/dt=[v_y, a_y}
v_y=Y(2);
a_y=-9.8-0.05*abs(v_y)*v_y;
Ydot=[v_y, a_y]';
end


>> [ t,y]=ode23(@air_friction, [0,4] , [0,30] );
>> plot(t, y(:,1))
>> hold on
>> tnf=[0:.01:4] ;
>> ynf=30*tnf - 0.5* 9.8* tnf.^2;
>> plot(tnf, ynf, '--')
>> ylabel('Height (m)'); xlabel('Time (s)')
```

Continued

Example 1.1 (*cont.*)

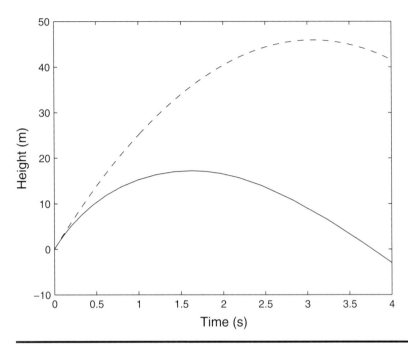

1.2 Vectors

A vector is an ordered group of n numbers that follow particular rules under the operations of addition, subtraction, and multiplication. The individual components of the vector are added or subtracted independently and not mixed when the vectors are added or subtracted, and multiplication by a scalar number multiplies each component equally. A vector equation represents n independent scalar equations.

The mathematical properties of vectors make them appropriate to describe displacements in two- ($n = 2$) or three- ($n = 3$) dimensional space, if the components of the vector represent distances in orthogonal directions. Addition of two vectors represents the effect of two successive displacements of an object, and subtraction of two vectors represents the change in position of an object. It is often useful to define a unit vector in the direction of a given vector $\vec{V}$. A vector of length 1 in the direction of $\vec{V}$ is given by $\hat{V} = \vec{V}/|\vec{V}|$, where $|\vec{V}|$ indicates the magnitude (length) of the vector. By simple geometry we find the length of a vector $\vec{V} = [x, y, z]$ is given by $|\vec{V}| = \sqrt{x^2 + y^2 + z^2}$.

Question 1.3

Why does the $\hat{V}$ vector have length=1? What is the unit vector in the direction [2, 1]?

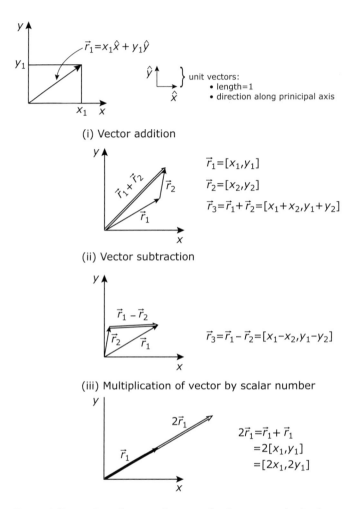

Figure 1.3 Expression of a vector in terms of *unit vectors* and rules for vector addition, subtraction, and multiplication by a scalar, illustrated for two dimensions. (Generalization to three dimensions is straightforward: $\vec{r}_3 = \vec{r}_1 + \vec{r}_2 = [x_1 + x_2, y_1 + y_2, z_1 + z_2]$, etc.)

The rules for vector addition, subtraction, and multiplication by a scalar number are summarized in Figure 1.3 for vectors $\vec{r}_1 = x_1\,\hat{x} + y_1\,\hat{y}$ and $\vec{r}_2 = x_2\,\hat{x} + y_2\,\hat{y}$ indicating positions in two dimensions.

1.3   Vector kinematics

Since position in two or three dimensions can be represented by vectors, motion which is the displacement of position in a time interval Δt can also be represented by vectors:

$$\vec{v} = \frac{\Delta \vec{r}}{\Delta t} = \frac{(\vec{r}_2 - \vec{r}_1)}{\Delta t} \tag{1.6}$$

$$\vec{a} = \frac{\Delta \vec{v}}{\Delta t} = \frac{(\vec{v}_2 - \vec{v}_1)}{\Delta t}. \tag{1.7}$$

Question 1.4

In Figure 1.3, knowing that scalar multiplication makes $(-1)\vec{r}_2 = -\vec{r}_2$ a vector of the same length but opposite direction as $\vec{r}_2$, can you show that $\vec{r}_1 + (-\vec{r}_2)$ when they are oriented tip-to-tail as in (i) gives a resultant vector from the tail of $\vec{r}_1$ to the tip of $(-\vec{r}_2)$ which is the same as the vector $\vec{r}_1 - \vec{r}_2$ in (ii)?

In terms of vector position, velocity, and acceleration, two- or three-dimensional motion can be represented through vector equations. For example, for constant acceleration we have the vector equivalent of the equation we saw previously

$$\vec{r} = \vec{r}_o + \vec{v}\,(t - t_o) + \frac{1}{2}\vec{a}\,(t - t_o)^2. \tag{1.8}$$

This one equation represents two (for 2D) or three (for 3D) independent scalar equations associated with the x-, y-, and z-components of the vector equation:

$$x = x_o + v_{ox}(t - t_o) + \frac{1}{2}a_x(t - t_o)^2$$

$$y = y_o + v_{oy}(t - t_o) + \frac{1}{2}a_y(t - t_o)^2$$

$$z = z_o + v_{oz}(t - t_o) + \frac{1}{2}a_z(t - t_o)^2.$$

1.4   Vector multiplication operations (dot and cross products)

In addition to the multiplication by a scalar we have defined above, there are two vector multiplication operations that we will be using, the dot product (also referred to as the *scalar product* – not to be confused with multiplication by a scalar number already discussed), and the cross product (also referred to as the *vector product*).

1.4.1   Dot (scalar) product

The *dot product* of two vectors is the application of the matrix operation of *inner product* to two- or three-dimensional vectors:

$$\vec{A} \cdot \vec{B} = A_x B_x + A_y B_y + A_z B_z. \tag{1.9}$$

Figure 1.4 The angle θ between vectors $\vec{A}$ and $\vec{B}$.

Note that the dot product of two vectors results in a scalar: the similar components are multiplied and added to get a single number (hence the name scalar product). Since the components are just scalar numbers that commute, the dot product also commutes:

$$\vec{A} \cdot \vec{B} = \vec{B} \cdot \vec{A}. \tag{1.10}$$

Using the Law of Cosines, it can be demonstrated that the dot product of two vectors is geometrically given by the product of the length of the two vectors times the cosine of the angle between them shown in Figure 1.4

$$\vec{A} \cdot \vec{B} = |\vec{A}||\vec{B}| \cos \theta. \tag{1.11}$$

Question 1.5

Show from the Law of Cosines $|\vec{C}|^2 = |\vec{A} - \vec{B}|^2 = |\vec{A}|^2 + |\vec{B}|^2 - 2|\vec{A}||\vec{B}| \cos \theta$ and the component definition of the dot product in Eq. (1.9) that the Eq. (1.11) relation is correct.

The physical meaning of the dot product is that it gives the product of the length of $\vec{A}$ times the length of the perpendicular projection of $\vec{B}$ in the direction of $\vec{A}$ (or equivalently product of the length of $\vec{B}$ times the length of the perpendicular projection of $\vec{A}$ in the direction of $\vec{B}$) as shown in Figure 1.5. Thus, the projection of $\vec{B}$ in the direction of $\vec{A}$ is equal to $\vec{B} \cdot \hat{A} = \vec{B} \cdot \vec{A}/|\vec{A}|$, and the projection of $\vec{A}$ in the direction of $\vec{B}$ is equal to $\vec{A} \cdot \vec{B}/|\vec{B}|$, for example.

Since the angle between any vector and itself is 0 and cos(0)=1, a corollary of the physical definition in Eq. (1.11) is that the magnitude (length) of any vector is the square root of its dot product with itself: $|\vec{A}| = \sqrt{\vec{A} \cdot \vec{A}}$. This result is also apparent from applying the Pythagorean Theorem to the component definition of the dot product

$$\sqrt{\vec{A} \cdot \vec{A}} = \sqrt{A_x A_x + A_y A_y + A_z A_z} = \sqrt{A_x^2 + A_y^2 + A_z^2} = |\vec{A}|. \tag{1.12}$$

The *unit vectors* of length = 1 and directions along perpendicular Cartesian axes are: $\hat{x} = [1\,0\,0]$, $\hat{y} = [0\,1\,0]$, and $\hat{z} = [0\,0\,1]$. Using component notation it is easy to show that taking the dot product of the unit vectors with any vector $\vec{A}$ gives the components of $\vec{A}$:

$$\vec{A} \cdot \hat{x} = [A_x \ A_y \ A_z] \cdot [1 \ 0 \ 0] = A_x$$
$$\vec{A} \cdot \hat{y} = [A_x \ A_y \ A_z] \cdot [0 \ 1 \ 0] = A_y$$
$$\vec{A} \cdot \hat{z} = [A_x \ A_y \ A_z] \cdot [0 \ 0 \ 1] = A_z.$$

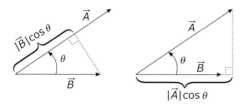

Figure 1.5 The projection of vector $\vec{B}$ in the direction of $\vec{A}$ (left) and of $\vec{A}$ in the direction of $\vec{B}$ (right).

It is also easy to show that the **dot products of the unit vectors $\hat{x}$, $\hat{y}$, and $\hat{z}$ with themselves = (the length of the vector)^2 = 1** and **dot products of any two different unit vectors = 0,** because the angle between them is 90° and cos(90°) = 0:

$$\hat{x} \cdot \hat{x} = 1 \qquad \hat{x} \cdot \hat{y} = \hat{y} \cdot \hat{x} = 0$$
$$\hat{y} \cdot \hat{y} = 1 \qquad \hat{x} \cdot \hat{z} = \hat{z} \cdot \hat{x} = 0$$
$$\hat{z} \cdot \hat{z} = 1 \qquad \hat{y} \cdot \hat{z} = \hat{z} \cdot \hat{y} = 0.$$

1.4.2   Vector product (cross product) of two vectors

The second multiplication operation for vectors is the *vector product* or *cross product* which is defined for three-dimensional vectors in component notation as the following matrix determinant:

$$\vec{A} \times \vec{B} = \begin{vmatrix} \hat{x} & \hat{y} & \hat{z} \\ A_x & A_y & A_z \\ B_x & B_y & B_z \end{vmatrix} = \hat{x}(A_yB_z - A_zB_y) - \hat{y}(A_xB_z - A_zB_x) + \hat{z}(A_xB_y - A_yB_x).$$

(1.13)

Note that the cross product of two vectors is a vector: its direction is perpendicular to the plane defined by $\vec{A}$ and $\vec{B}$. We should point out that the cross product, unlike the dot product, is not commutative. Since if we interchange any two rows in a determinant we change the sign, we have $\vec{A} \times \vec{B} = -\vec{B} \times \vec{A}$.

The geometrical interpretation of the cross product is that the magnitude of the cross product of two vectors is equal to the product of the magnitudes of the two vectors times the sine of the angle between the vectors: $|\vec{A} \times \vec{B}| = |\vec{A}||\vec{B}| \sin\theta$. The direction is perpendicular to the plane of $\vec{A}$ and $\vec{B}$ in the direction (since there are two directions perpendicular to a plane) given by the *right-hand rule*: curl the fingers of your right hand in the direction of the smallest angle from $\vec{A}$ to $\vec{B}$ – in the direction of the angle arrow in Figure 1.6 – and your thumb will point in the direction of the cross product. This geometrical interpretation of the cross product is illustrated in Figure 1.6. A physical interpretation of the cross product is that the magnitude of the cross product is equal to the area of the parallelogram defined by the two vectors, as shown in Figure 1.7.

It is easy to show by the component equation in Eq. (1.12) that $\hat{x} \times \hat{y} = \hat{z}$, $\hat{y} \times \hat{z} = \hat{x}$, and $\hat{z} \times \hat{x} = \hat{y}$. Of course, the cross product of any vector (including a unit vector) with itself is 0, since the angle between any vector and itself is 0 and sin(0) = 0.

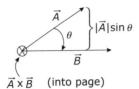

$\vec{A} \times \vec{B}$ (into page)

Figure 1.6 Cross product of vectors $\vec{A}$ and $\vec{B}$.

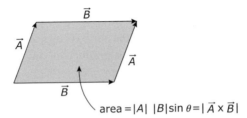

area $= |A|\,|B|\sin\theta = |\vec{A} \times \vec{B}|$

Figure 1.7 The area of the parallelogram defined by $\vec{A}$ and $\vec{B}$ is given by the magnitude of the curl $\vec{A} \times \vec{B}$.

There are two common vector identities for the cross product involving three vectors. First, the cross product of three vectors can be written

$$\vec{A} \times (\vec{B} \times \vec{C}) = (\vec{A} \cdot \vec{C})\vec{B} - (\vec{A} \cdot \vec{B})\vec{C}. \tag{1.14}$$

This is known as the *vector triple product* because the result is a vector.

The dot product of a vector and a cross product is known, since it results in a scalar quantity, as the *scalar triple product*

$$\vec{A} \cdot (\vec{B} \times \vec{C}) = \vec{C} \cdot (\vec{A} \times \vec{B}) = \vec{B} \cdot (\vec{C} \times \vec{A}). \tag{1.15}$$

Although the dot product is commutative, the order and grouping of the vectors in Eq. (1.13) and Eq. (1.14) is important because the cross product anti-commutes.

The scalar triple product has a physical interpretation: the magnitude of the scalar triple product gives the volume of the parallelepiped defined by the three vectors, as shown in Figure 1.8.

Vector algebra – adding, subtracting, taking the dot and cross products – is made much easier by using a vector processing software such as MATLAB. For example, the following MATLAB code gives the volume of a parallelepiped defined by three vectors:

```
>> A=[2 2 1];
>> B=[0.5 0.75 1.5];
>> C=[−2 4 −2];
>>
>> Volume = dot(cross(A,B), C)
Volume =
  −15.5000
```

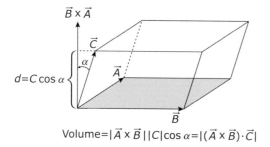

$$\text{Volume}=|\vec{A}\times\vec{B}||C|\cos\alpha=|(\vec{A}\times\vec{B})\cdot\vec{C}|$$

Figure 1.8 The volume of a parallelepiped defined by three vectors $\vec{A}$, $\vec{B}$, and $\vec{C}$ is given by the magnitude of $(\vec{A}\times\vec{B})\cdot\vec{C}$.

1.5 Curvilinear motion: velocity and acceleration on a curved path

We have defined acceleration as the change in velocity over the change in time $\vec{a}=\frac{d\vec{v}}{dt}$. Since velocity is a vector, its *direction* can also change, not just its *magnitude*. A description of motion on a curved path – as opposed to motion in a straight line – needs to take account of both the change in magnitude and the change in direction.

Consider a point object moving in a curved path A–B as shown in Figure 1.9. The velocity $\vec{v}(t)$ at any time t is *tangential* to the curve in the direction given by the *tangential unit vector* $\hat{e}_t$, $\vec{v}(t) = v\,\hat{e}_t$. We can also define at any point on the path of motion a *normal unit vector* $\hat{e}_n$, which is perpendicular to $\hat{e}_t$ in the plane of the motion. At a time Δt later the velocity is changed to $\vec{v}(t + \Delta t)$, which has a different magnitude from $|\vec{v}(t)|$ and whose direction is displaced from $\vec{v}(t)$ by the angle $\Delta\theta$ as shown in Figure 1.10.

The change in velocity $\Delta\vec{v} = \vec{v}(t + \Delta t) - \vec{v}(t)$ is given by the usual process of vector subtraction: put the tails of the vectors together and the vector $\Delta\vec{v}$ goes from the tip of $\vec{v}(t)$ to the tip of $\vec{v}(t + \Delta t)$ as shown in Figure 1.10. We can resolve the vector $\Delta\vec{v}$ into two components: one in the tangential direction Δv, due to the change in the magnitude of the velocity, and the other in the normal direction, due to the change in direction. In the limit $\Delta t \to 0$, the tangential component is $\Delta v = |\vec{v}(t + \Delta t)| - |\vec{v}(t)|$ and the perpendicular component is $(v + \Delta v)\tan\Delta\theta \cong v\Delta\theta$. We have neglected the term $\Delta v\Delta\theta$ from the length of the adjacent side in finding the magnitude of the opposite side because it is of order Δ^2, and the last equality is a result of the small angle approximation $\tan(\Delta\theta) \cong \Delta\theta$ when $\Delta\theta$ is small and expressed in radians.

Using these results, we can write

$$\Delta\vec{v} = \vec{v}(t + \Delta t) - \vec{v}(t) = \Delta v\,\hat{e}_t + v\Delta\theta\,\hat{e}_n \tag{1.16}$$

and the acceleration of the point object is

$$\vec{a} = \lim_{\Delta t\to 0}\frac{\Delta\vec{v}}{\Delta t} = \frac{dv}{dt}\,\hat{e}_t + v\frac{d\theta}{dt}\,\hat{e}_n \tag{1.17}$$

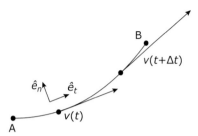

Figure 1.9 Object in curvilinear motion experiences acceleration in the tangential and normal directions.

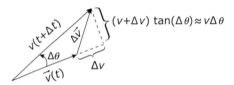

Figure 1.10 Resolving the vector $\Delta \vec{v} = \vec{v}(t + \Delta t) - \vec{v}(t)$ from Figure 1.9 into a tangential and normal component.

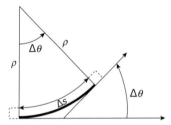

Figure 1.11 The arc length Δs can be expressed in terms of the angular deviation $\Delta \theta$ and the instantaneous radius of curvature ρ.

where the normal unit vector $\hat{\rho}_n$ is perpendicular to the tangent toward the *instantaneous concave* side of the curvilinear motion.

If we extend lines perpendicular to two adjacent segments of the particle path as shown in Figure 1.11, we can define an *instantaneous radius of curvature ρ*. Since the lines are perpendicular to the adjacent segments, the angle between them is the same as the $\Delta \theta$ that defines the deviation of $\vec{v}$ $(t + \Delta t)$ and $\vec{v}$ (t). From the arc length theorem, the distance along the segment $\Delta s = \rho \Delta \theta$. The particle velocity is

$$v = \frac{\Delta s}{\Delta t} = \rho \frac{\Delta \theta}{\Delta t} \quad \Rightarrow \quad \frac{d\theta}{dt} = \frac{v}{\rho} \tag{1.18}$$

from which we find the normal component of the acceleration

$$a_n = v \frac{d\theta}{dt} = \frac{v^2}{\rho} = \rho \omega^2, \tag{1.19}$$

where the *angular velocity* $\omega = \frac{d\theta}{dt} = \frac{v}{\rho}$. For circular motion where the radius of curvature is not changing, $\rho = R = $ constant, and we have the expressions for velocity and acceleration

$$\vec{v}_{cir} = \frac{ds}{dt}\hat{e}_t = \frac{Rd\theta}{dt}\hat{e}_t = R\omega\hat{e}_t$$

$$\vec{a}_{cir} = R\frac{d\omega}{dt}\hat{e}_t + \frac{v^2}{R}\hat{e}_n = R\alpha\hat{e}_t + R\omega^2\hat{e}_n \qquad \text{(circular motion)} \qquad (1.20)$$

where we have introduced the derivative of the angular velocity, the *angular acceleration* $\alpha = \frac{d\omega}{dt} = \frac{d^2\theta}{dt^2}$.

Chapter 1 problems

1. A person jumps straight up. Do they spend more time in the air while in the top half or bottom half of the total height? What is the ratio of the two times?

2. Two identical balls are released simultaneously from the same point S. One follows path "a" and the other path "b." Which one will reach point F first? Explain your answer. (Assume that the ball that follows path "b" is always in contact with the ground.)

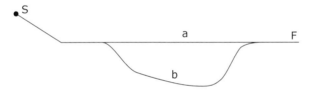

3. Describe the one-dimensional motion of the objects represented by the following graphs:

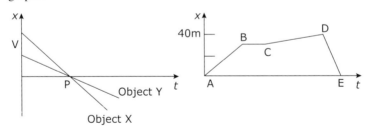

   a. What does point P represent?

   b. Which object in graph 1 covered a larger distance?

   c. Which object in graph 1 has a larger net displacement?

   d. What is the net displacement of the object in graph 2?

   e. What is the total distance covered by the object in graph 2?

4. A car is moving in the track shown below (AB = 100 m, $r = 50$ m, and DF = 10 m) and its tangential acceleration is given by $a_t = 2t^2$. If its initial velocity at A is zero,

calculate the total acceleration (magnitude and direction) at point C. At which point is the car's magnitude of the acceleration maximum? What is the magnitude of the maximum value of the acceleration? Is this a realistic problem? Justify your answer.

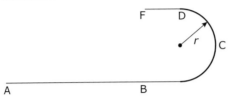

5. Given points A (0,0,0), B(6,0,0), C(5,0,3), and D(4,3,1) find the volume of the parallelepiped formed by vectors
   a. **AB, AC**, and **AD**

   b. **BA, BC**, and **BD**

6. Given two vectors $\vec{a} = x_1\,\hat{x} + y_1\,\hat{y} + z_1\,\hat{z}$ and $\vec{b} = x_2\,\hat{x} + y_2\,\hat{y} + z_2\,\hat{z}$, under what conditions are the vectors $\vec{a} + \vec{b}$ and $\vec{a} - \vec{b}$ perpendicular to each other?

7. A basketball player throws a ball at a 60° angle with the horizontal and at a height of 2.0 m. If the hoop is 5.0 m away and at a height of 3.0 m, what is the initial speed of the ball to make the shot?

8. A ball is thrown straight up. What is the magnitude and direction of its acceleration
   a. While it is going up?

   b. At the highest point?

   c. While it is falling down?

9. A ball is thrown straight up and, when it reaches the maximum height h, a second identical ball is thrown up with the same initial velocity as the first one. Where will these balls meet?

2 Newton's Laws, energy, and momentum

We spend most of our life in a world of macroscopic objects moving and interacting, yet many – perhaps most – people can't describe correctly the trajectory of an object thrown from a moving car. We are aware there are rules that govern motion and interactions, but what they are, when they apply, and how they can be applied to predict the motion of a pool cue ball after it collides with an object ball, for example, are a mystery to most people, judging by how often casual pool players "scratch" by pocketing the cue ball. Even engineers, who have all had at least a year-long course in physics, are often uneasy when they are asked what exactly Newton's Laws *mean* instead of just reciting the words. As a matter of fact, the vast majority of people are still Aristotelian thinkers in their understanding of the laws that govern motion. In this chapter we will revisit these fundamental concepts with the goal of increasing the level of comfort with the concepts of forces and motion.

2.1 Newton's Laws

The traditional formulation of Newton's Laws is something like these below:

> **First Law:** *An object at rest tends to remain at rest and an object in motion tends to remain in uniform motion (in a straight line at a constant speed) unless acted upon by a net force.*
> **Second Law:** *An object of mass m acted upon by a net force F experiences constant acceleration a according to the formula F = ma.*
> **Third Law:** *Every action has an equal and opposite reaction.*

These statements are a combination of definitions and empirical ("determined by observation") laws. For example, a force (F) is often described as a "push or pull that, if acting alone on a body, causes a constant acceleration." This is a circular definition that makes the Second Law a meaningless statement. The First Law also has conditions on it: you can't be in an accelerating or rotating system. If your system is accelerating or rotating you will experience "fictitious forces" such as the Coriolis force – an apparent force on a moving object to the right in the northern hemisphere and to the left in the southern hemisphere (looking along the direction of an object's motion) that is caused by the object trying to travel in a straight line in a rotating coordinate system. The effect of rotating or accelerating coordinate systems is so common as to be more the rule than the exception, such that the First Law is sometimes described as "objects in motion tend to remain in uniform motion – except when they don't!"

There are several ways of separating out the definitions from the empirical content of Newton's Laws. We will use the formulation of Robert Weinstock, as contained in his *American Journal of Physics* article, "Laws of Classical Motion: What's F? What's m? What's a?"[1] (The title is based on a conversation overheard between two freshmen students going into a physics midterm examination. One says to the other, "Sure I know that $F = ma$, but what's F? What's m? What's a?")

Weinstock says that the First Law is simply a definition of an inertial reference frame in which Newton's Laws hold. In an inertial reference frame objects remain at rest or in uniform velocity. An inertial reference frame is approximated by any system that is stationary or moving at constant velocity (but *not* accelerating!). We may have a hard time actually finding such a reference system, but we can find reference systems that are fairly good approximations. The room you are in is an approximation of an inertial reference frame if we neglect effects due to the rotation of the Earth, building tremors, etc. We can at least conceptually neglect such effects and consider mechanical behavior in their absence.

The Second Law, $F = ma$, is a combination of a definition and an empirical observation. A net "force" can be defined as a push or pull acting on an object that causes the object to undergo constant acceleration in our inertial reference frame. The empirical content of the Second Law is that if we determine that $\vec{F}_1$ and $\vec{F}_2$ cause accelerations $\vec{a}_1 = \vec{F}_1/m$ and $\vec{a}_2 = \vec{F}_2/m$ acting in isolation on an object, when they act together they cause an acceleration which is the **vector sum** of $\vec{a}_1$ and $\vec{a}_2$: forces add as vectors!

In cases where the mass changes, it can be demonstrated that a more fundamental expression of the Second Law is that the net force equals the time derivative of the *momentum*:

$$\vec{F} = \frac{d\vec{p}}{dt} \tag{2.1}$$

where the momentum of any object $\vec{p} = m\vec{v}$, the mass times the velocity.

Momentum is a foundational concept in relativity and quantum mechanics, so the expression of Newton's Laws in terms of momentum is considered more fundamental (it's also the way that Newton formulated the Laws). If the mass is constant, of course, $\frac{d\vec{p}}{dt} = m\frac{d\vec{v}}{dt} = m\vec{a}$ and you get back the familiar $\vec{F} = m\vec{a}$.

The Third Law is also a combination of definition and empirical content. It is, first, a definition of inertial mass: if we let two objects interact in our inertial reference frame (put a spring between them on a frictionless table, for example) the ratio of their observed accelerations is equal to the inverse ratio of their corresponding inertial masses: $m_2/m_1 = |a_1/a_2|$, for example. Since only the ratio of the masses is defined, we need to agree on a standard mass, and, in fact, there is (or used to be) a "standard

---

[1] Weinstock, Robert, "Laws of Classical Motion: What's F? What's m? What's a?", *American Journal of Physics* **29**(10), p. 698 (1961).

mass" in a climate-controlled vault in the International Bureau of Standards that is the standard kilogram. Periodically, secondary standard masses are compared to the standard mass by interacting through a mass balance, and comparison with the secondary masses is how we calibrate our masses.

Note that usually we compare masses by comparing how "heavy" they are – how the gravitational force of the Earth on them compares. In theory, there is no reason why the *gravitational mass* determined from the Law of Universal Gravitation $F = G\frac{m_1 m_2}{r^2}$ needs to be the same as the *inertial mass*. To illustrate the difference between inertial and gravitational mass, note that a brick floating in intra-stellar space with no gravitational force on it still has inertia – if you kick it you will still hurt your toe. The property of the brick that resists acceleration – and exerts a force back on your toe – is inertial mass. If the brick is then brought near another mass, it will have a gravitational force exerted on it which is proportional to its gravitational mass. There is no a priori reason why those two masses have to be the same, but in fact they appear to be identical as accurately as they can be measured. (If this apparent coincidence is interesting to you, you can learn about one proposed explanation by looking up *Mach's Principle*.) In any case, it is the inertial mass that is involved in the Third Law action/reaction pairs – the ratio of the accelerations of two objects (the brick and your foot, for example) is given by the inverse ratio of the corresponding inertial masses.

So, **if two objects interact**, the mass of the first times its acceleration equals the mass of the second times its acceleration: **the forces on the two objects are equal**. The empirical content of the *Third Law* is that the forces are in opposite directions: **they are not only equal but precisely opposite**. The force vectors are 180 degrees apart.

It is important when thinking about and applying the Third Law to remember that *the action/reaction pair forces act on different objects.* If the action force is the force of the foot on the brick, then the reaction force is the force of the brick on the foot. It is impossible to have an action force without a reaction force. If this seems confusing, try to push against someone who won't push back – you cannot exert a force on him if he doesn't push back on you. If we define $\vec{F}_{12}$ as the force on object 1 exerted by object 2, then the *Third Law* says $\vec{F}_{12} = -\vec{F}_{21}$: **forces always come in pairs and the force exerted by object 1 on object 2 is equal and opposite to the force exerted by object 2 on object 1**.

Question 2.1

If for every *action* there is an equal and opposite *reaction*, shouldn't these two forces always cancel each other and, as a result, have no motion? Or, how can a horse and a cart move if they both exert equal and opposite forces on each other?

Question 2.2

Suppose you apply a force on an object and it is accelerating. Is that object pushing back with an equal and opposite force? Justify your answer.

2.2   The Third Law and momentum conservation

Whenever we have two objects interacting, we can write using the Third Law and the Second Law:

$$\vec{F}_{12} = -\vec{F}_{21}$$

$$m_1\vec{a}_1 = -m_2\vec{a}_2$$

$$\frac{d}{dt}(m_1\vec{v}_1) = -\frac{d}{dt}(m_2\vec{v}_2)\left(\text{since } \vec{a}_1 = \frac{d\vec{v}_1}{dt}, \text{etc.}\right)$$

$$\Rightarrow \frac{d}{dt}(\vec{p}_1) = -\frac{d}{dt}(\vec{p}_2) \text{ where } \vec{p} = m\vec{v} \text{ (momentum!)}.$$

So in any interaction of two bodies,

$$\frac{d}{dt}(\vec{p}_1 + \vec{p}_2) = 0 \Rightarrow \text{momentum is conserved!} \tag{2.2}$$

While we have shown this for only two bodies interacting with their masses constant, in fact, by considering the Third Law forces between component pieces, it can be shown that it works even if the mass is time-varying (a rocket burning fuel, for example). It also works for systems of more than two bodies. This is one of the most fundamental results of physics: in any isolated system of bodies interacting with each other in any way, **momentum is conserved.** Since it follows just from the Third Law, it is more fundamental even than conservation of mechanical energy. Even in interactions in which mechanical energy is not conserved (usually due to conversion of some of it into thermal energy or heat), momentum is still conserved.

Example 2.1 Reaction rocket thrust

Consider a rocket in space which burns fuel and the exhaust is ejected out of the tail of the rocket at a velocity (relative to the rocket velocity) of V_{ex}. On a microscopic level the expanding molecules that hit the front wall of the combustion chamber bounce backward due to the force exerted on them by the front wall to reach a velocity of V_{ex} toward the tail of the rocket, and thus by the Third Law the molecules push back on the rocket with the same force forward. Since we know this process works in the vacuum of space, the propulsive force has nothing to do with the exhaust gases exerting a force on the atmosphere!

Solution: The rocket and its fuel are an isolated system with no external forces, so momentum is conserved. If at one instant the rocket plus fuel have a mass of M and a positive velocity v, after a time interval dt the rocket will have a velocity $v + dv$ and mass of $M - dM$ where dM is the mass of the fuel consumed. The mass dM of fuel has been ejected out of the back of the rocket at a velocity relative to the rocket of V_{ex}. In an inertial frame at rest with respect to the rocket, the exhaust gas has a velocity

Continued

Example 2.1 (*cont.*)

$V_{ex} - (v + dv)$ backwards. Setting the momentum before the interval dt equal to the momentum after dt, we have

$$Mv = (v + dv)(M - dM) - dM[V_{ex} - (v + dv)].$$

Expanding and canceling terms on both sides, we find

$$Mdv = -dM \, V_{ex} \Rightarrow Ma = -\frac{dM}{dt} V_{ex}$$

where we have explicitly focused on the (negative) change of mass of the spacecraft instead of the mass of the gas. The last equation, found by dividing by the time interval dt, tells us that the instantaneous force on the rocket $M\frac{dv}{dt} = Ma$, the *thrust*, is given by the exhaust velocity times the rate of fuel expenditure.

Solving the Mdv equation for dv and integrating, we find the change in velocity in any interval $\Delta t = t_f - t_i$

$$v_f - v_i = \int_{v_i}^{v_f} dv = -\int_{M_i}^{M_f} V_{ex} \frac{dM}{M} = V_{ex} \ln \frac{M_i}{M_f}.$$

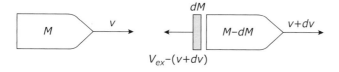

An important case of momentum conservation are collision problems, where two (or more) objects with given masses and velocities interact and it is desired to find the final velocities of the objects. Collision problems can be in one dimension (where motion is confined to a line), in two dimensions, where the velocities are in a single plane, or in three dimensions. There are two limiting cases when collision problems are particularly easy to solve using momentum conservation: perfectly elastic collisions and perfectly inelastic collisions.

1. Perfectly elastic collision: A perfectly elastic collision is one in which momentum *and* mechanical energy are both conserved in the collision. For example, if we know the masses and initial velocities of two colliding objects in two or three dimensions, we can predict the final velocities as a function of the angle that one of the particles moves after the collision.
2. Perfectly inelastic collision: A perfectly inelastic collision is one in which the bodies collide and stick together. Mechanical energy is not conserved in a perfectly inelastic collision, but we can still predict the final velocity and direction of the combined body if we know the masses and initial velocities of the two bodies before the collision, as in the example below.

Example 2.2 Perfectly inelastic collision

Two masses, M_1 with a mass of 10 kg and a velocity of 2 m/s $\hat{x}$ and M_2 with a mass of 5 kg and a velocity of -6 m/s $\hat{y}$, collide. After the collision the two masses stick together as illustrated. What is the velocity (magnitude and direction) of the motion of the combined masses after the collision?

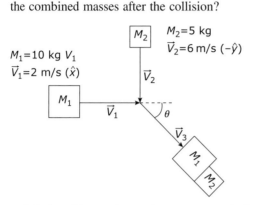

Solution: The total x- and y-momentum before the collision is given by:

$$p_{xi} = M_1 V_1 = (10\,\text{kg})(2\text{ m/s}) = 20\,\text{kg m/s}$$
$$p_{yi} = M_2 V_2 = (5\,\text{kg})(-6\text{ m/s}) = -30\,\text{kg m/s}.$$

Since there are no external forces on the system, the momentum doesn't change, and the momentum after the collision must be the same as the momentum before the collision. The momentum after the collision is just the combined mass times the final velocity. Breaking the final velocity up into its x- and y-components, we get:

$$p_{xf} = (M_1 + M_2)\, V_3 \cos\theta = 15\, V_3 \cos\theta$$
$$p_{yf} = (M_1 + M_2)\,(-V_3 \sin\theta) = -15\, V_3 \sin\theta.$$

We can set the x-component before the collision equal to the x-component after, and the y-component before equal to the y-component after. (Since momentum is a vector, if the momentum before equals the momentum after, then each component must be equal as well!)

$$p_{xi} = p_{xf} \Rightarrow 20\,\text{kg m/s} = 15\, V_3 \cos\theta \quad \text{①}$$
$$p_{yi} = p_{yf} \Rightarrow -30\,\text{kg m/s} = -15\, V_3 \sin\theta \quad \text{②}.$$

We now have two equations (① and ②) in two unknowns (V_3 and θ), so the problem is solved! The mathematics to get the actual answer is below:

(a)   Divide by ② by ①:

$$\frac{15\, V_3 \sin\theta}{15\, V_3 \cos\theta} = \tan\theta = \frac{-30}{20} = -1.5$$

$$\Rightarrow \theta = -56.31°.$$

Continued

Example 2.2 (*cont.*)

(b)   Plug $\theta = 56.31°$ back into ①

$$20 = 15\, V_3 \, \cos\left(-56.31°\right)$$

$$\Rightarrow V_3 = \frac{20}{15} \, \frac{1}{\cos\left(-56.31°\right)} = 2.404 \text{ m/s}.$$

2.3   Work–Energy Theorem

Work is defined as the dot product of force applied and the distance moved:

$$Work = W = \vec{F} \cdot \vec{d}. \tag{2.3}$$

If the force is changing during the movement, we would define work in terms of the integral of force times the distance moved:

$$W = \int_{\vec{r}_1}^{\vec{r}_2} \vec{F} \cdot d\vec{r}. \tag{2.4}$$

Why is the concept of work useful? Consider motion in one dimension (the result generalizes to three dimensions as well)

$$W = F_x \left(x - x_o\right) = m \, a_x \left(x - x_o\right). \tag{2.5}$$

Using

$$v^2 = v_o^2 + 2a_x \left(x - x_o\right) \Rightarrow a_x = \frac{v^2 - v_o^2}{2\left(x - x_o\right)}, \tag{2.6}$$

we find:

$$W = F_x \left(x - x_o\right) = m \left(\frac{v^2 - v_o^2}{2}\right) = \frac{1}{2} mv^2 - \frac{1}{2} mv_o^2 = \Delta KE \tag{2.7}$$

where $KE = Kinetic \ Energy = \frac{1}{2} mv^2$.

The significance of kinetic energy is that if the work done on a system is zero, then

$$\Delta KE = \frac{1}{2} mv^2 - \frac{1}{2} mv_o^2 = 0. \qquad \text{(no work done)} \tag{2.8}$$

So, kinetic energy is conserved if there is no work done on a system!

   Conservation laws are very useful in analyzing systems, because they put constraints on how a system evolves and allow you to predict the final state given the initial state. Just as momentum is a vector quantity that is conserved in an isolated system if there is no net external force, kinetic energy is a scalar quantity that is conserved if there is no work done on the system.

Even in cases where there is an external force doing work on the system, we can extend the concept of energy conservation if the work done by a force to move an object from point A to point B doesn't depend on the path taken. These types of fields, where the work done is independent of the path taken are called *conservative fields* and the forces generated by them are called *conservative forces*. In these cases, where we have a conservative field, we can define a *potential energy difference*

$$\Delta U = U_B - U_A = \int_A^B \vec{F}_C \cdot d\vec{r} \tag{2.9}$$

where $\vec{F}_C$ is the net force due to the conservative field. We can define $U = 0$ at any convenient point without loss of generality.

For example, since the gravitational force $\vec{F}_g = -mg\hat{y}$ is conservative, it doesn't matter if we lift a mass directly up or push it up an inclined plane; the work done depends only on the change in height above some reference plane

$$W = mg(h - h_o). \quad (Gravitational\ potential\ energy) \tag{2.10}$$

Similarly, the work done by an electric force moving a charge from one point to another in an electric field depends only on the *voltage* difference between the initial point and final point, and not on the path taken

$$W = \int_A^B q\vec{\varepsilon} \cdot d\vec{r} = q(\phi - \phi_o) \quad (Electrical\ potential\ energy) \tag{2.11}$$

where ϕ is the final voltage and ϕ_o is the initial voltage.

The sum of the kinetic and potential energies is the total mechanical energy, E:

$$E = KE + U. \tag{2.12}$$

From this, the general Work–Energy Theorem follows: the change in total mechanical energy between A and B is the work done by *non-conservative forces* in moving from A to B

$$\Delta E = W_{nc} = \int_A^B \vec{F}_{nc} \cdot d\vec{r}. \tag{2.13}$$

Non-conservative forces include friction and anything else that converts mechanical energy into other forms, such as heat, deformation, etc. **If there are no non-conservative forces, then mechanical energy – the sum of kinetic and potential energy – is conserved**. Kinetic energy can be converted into potential energy and potential energy into kinetic energy, but the total mechanical energy stays the same.

The classic problem in conservation of mechanical energy is the ski jump problem illustrated in Figure 2.1. A skier approaches the top of a frictionless ski jump with a velocity v_i – how fast is he going when he reaches the slope, a vertical distance of h meters below?

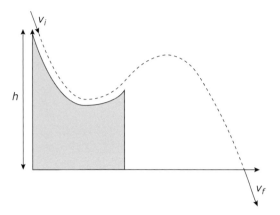

Figure 2.1 A skier approaches a frictionless ski jump with initial velocity $\vec{v}_i$ and hits the ground a vertical distance h below at velocity $\vec{v}_f$. If the magnitude of $\vec{v}_i$ is known, the magnitude of $\vec{v}_f$ can be found by conservation of energy.

$$E_i = \frac{1}{2}mv_i^2 + mgh; \qquad E_f = \frac{1}{2}mv_f^2$$

$$E_f = E_i \Rightarrow \frac{1}{2}mv_f^2 = \frac{1}{2}mv_i^2 + mgh \qquad (2.14)$$

$$\Rightarrow v_f = \sqrt{v_i^2 + 2gh}.$$

Assuming that the skier has negligible frictional losses, the result for the magnitude of the final velocity doesn't depend on the skier's mass, the shape of the ski jump, or the trajectory of the skier. Conservation of energy tells you that the final velocity magnitude is the same, regardless of path.

Question 2.3

(a) If a golf ball is thrown straight down with the same speed (magnitude of velocity) as that of the skier at height h, how will their speeds compare when they reach $h = 0$? (b) What if the ball is thrown straight up with the same speed and from the same height as before?

The time it takes for the skier to reach the ground, however, *does* depend on the path. Consider, for example, a skier descending from a height h down a constant slope of angle θ as shown in Figure 2.2. The force of the skier's weight $m\vec{g}$ directed vertically downward can be resolved into a component perpendicular to the slope $\vec{F}_\perp$ (canceled by the normal force $\vec{N}$ exerted by the slope on the skier), and a component parallel to the slope $\vec{F}_\parallel$ with $|\vec{F}_\parallel| = mg \sin\theta$, which is the net force on the skier (see Fig. 2.2). The acceleration is $g \sin\theta$ and the total distance down the slope is $h/\sin\theta$. Using

$$d = \frac{h}{\sin\theta} = \frac{1}{2}at^2 = \frac{1}{2}g\sin\theta\, t^2 \qquad (2.15)$$

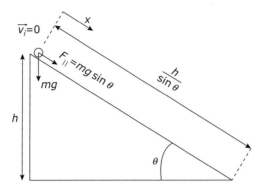

Figure 2.2 An object descends an inclined plane with an inclination angle θ.

we find $t = \frac{1}{\sin\theta}\sqrt{\frac{2h}{g}}$, which depends on the details of the path since it depends on the angle θ.

2.4   Normal forces and frictional forces

When an object sits on a surface, we understand that the force of the Earth's gravity on the object does not disappear. Since the body is not accelerating, however, there must be an equal force that the surface exerts upward on the object. This so-called *normal force* exists only when the object is in contact with the surface, increases to match the total applied force if the object is pressed harder into the surface, and disappears when the object is lifted out of contact with the surface. How does the surface know how much force to apply? Where does this force come from?

The source of the normal force is in the elastic interactions between the molecules of the surface. As shown in Figure 2.3, on a microscopic level the molecules of the surface are displaced by the force applied by the weight of the object or any other force exerted on the surface. To a good approximation the intermolecular elastic forces are proportional to the displacement of the atoms. The elastic force exerted by the atoms opposes the displacement $\Delta\vec{x}$ and is given by

$$\vec{F}_e = -k\Delta\vec{x} \tag{2.16}$$

where the elastic constant k depends on the material of the surface and the area over which the force is applied (the area of the object resting on the surface). Since the force and the displacement are at an angle of $180°$, the work done by the normal force is

$$W = \int \vec{F} \cdot d\vec{x} = \int_0^{x_f} (kx)dx \cos 180° = -\frac{1}{2}kx_f^2. \tag{2.17}$$

The work done is path-independent because if the surface is displaced to a distance $x_f + x_1$ greater than x_f and then allowed to relax to the final displacement x_f the work done is the same as the direct displacement to x_f as given in Eq. (2.16).

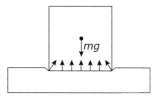

Figure 2.3 An object resting on a surface causes a microscopic displacement of the surface molecules. The elastic restoring force on these molecules creates the normal force which cancels the object's weight.

Question 2.4

What work is done by the normal force if the surface is displaced from 0 to $x_f + x_1$? Is this work positive or negative? What work is done by the normal force if the surface is allowed to relax from $x_f + x_1$ to x_f? Is this work positive or negative? Show that adding these two gives the same result as Eq. (2.17).

Since the work done by the normal force is thus path-independent, the elastic force is conservative and we can define an elastic potential energy

$$PE_{el} = \frac{1}{2}kx^2. \qquad (2.18)$$

So when you place an object on a surface, the weight of the object does work on the surface converting gravitational potential energy into the elastic potential energy of the deformed surface. When the object is lifted from the surface, the normal force does work on the object equal to the amount of work done by the object on the surface, transferring the elastic potential energy back into gravitational potential energy.

The two characteristics of the normal force are

1. The normal force is always *perpendicular to the surface*. This comes from the fact that any lateral forces on the left in Figure 2.3 are canceled by the lateral forces on the right and deduced from the observation that an object resting on a perfectly flat surface with no friction will not accelerate either to the right or left.
2. The normal force is always equal to the applied force (assuming that the surface remains intact!).

Frictional forces are forces on an object on a surface that are *parallel* to the surface. Frictional forces are caused either by irregularities in the surfaces that mechanically interfere with the relative motion or by weak intermolecular bonding between atoms on the surfaces. The characteristics of frictional forces are

1. The frictional forces are always *parallel to the surface.*
2. They act in a direction which is opposite to the motion of an object relative to the surface (*kinetic friction*), or opposite to the applied parallel force if the object is not in motion (*static friction*).

3.   The frictional forces are proportional to the normal force on the object, and, to the first approximation, independent of the area of contact between the object and the surface.

4.   Generally static friction is greater than kinetic friction: it takes a greater force parallel to the surface to start an object moving on a surface with friction than to keep it in motion.

These characteristics are captured in the equation for frictional force

$$F_f = \mu N \qquad (2.19)$$

where F_f is the force of friction, N is the normal force on the object, and μ is the *coefficient of friction* for the object moving over the surface. If the object is in motion the *coefficient of kinetic friction* μ_k is used, while if the object is not in motion the *coefficient of static friction* μ_s is the appropriate parameter, with $\mu_s > \mu_k$.

The area of the interface does not enter into this equation which implies that to first approximation the frictional force does not depend on which surface of an object is in contact with the surface. Thus the frictional force for an object of mass M is the same in configuration (a) and configuration (b) in Figure 2.4.

In general the total force exerted by a surface on an object is a combination of the normal force and the frictional force and can be directed at any angle from the surface. In a statics problem, the direction of the frictional force may not be immediately obvious, since the static direction of the frictional force depends on (is opposite to) the direction of the applied force.

We also note that the work done by the frictional force depends on the distance traveled, so the frictional work is different going between position A and B by path D_1 and path D_2 in Figure 2.5. Thus the frictional force is *non-conservative* and does not

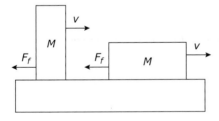

Figure 2.4 The frictional force depends on the normal force but, to first approximation, not on the contact area. Thus the object has the same frictional force (opposite to the direction of the velocity) if (a) it is on its end as on the left, or (b) on its side as on the right.

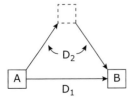

Figure 2.5 Top view of object being moved from position A to position B by two paths D_1 and D_2. The work done by friction depends on the distance moved and is not the same for each path.

conserve energy. In fact the work done by the frictional force is converted into heat and lost from the mechanical system.

Example 2.3 A 10 kg box sits on a surface that is lifted up on one side to make an inclined plane 15 m long. The coefficient of static friction is $\mu_s = 0.4$ and the coefficient of kinetic friction is $\mu_k = 0.25$. What is the angle of the inclined plane when the box begins to slide? If the plane is held at that angle, how long does it take the box to slide down the plane? Does this depend on the mass of the box?

Solution: The forces on the box are the normal force N perpendicular to the surface, the weight $W = mg$ vertically downward, and the force of friction F_f directed parallel to the plane. We can resolve the weight into a component parallel to the plane $W_\parallel = W \sin \theta$

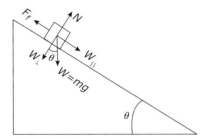

and a component perpendicular to the plane $W_\perp = W \cos \theta$ as shown in the figure.

Since there is no acceleration perpendicular to the plane, we can conclude that the perpendicular forces add to zero

$$W_\perp = mg \, \cos \theta = N.$$

Before the box begins to slide, the force of friction is directed up the plane in the opposite direction to the parallel force acting on the box, and the box will begin to slide when the static force of friction is equal to $W_\parallel$

$$W_\parallel = mg \, \sin \theta = F_f = \mu_s N = \mu_s mg \, \cos \theta$$

which yields $\tan \theta = \mu_s \Rightarrow \theta = \tan^{-1} 0.4 = 21.8°$.

Once the box is sliding down the plane, the frictional force is due to kinetic friction which is in the opposite direction to the velocity, i.e. the frictional force is up the plane. The normal force remains equal to the component of the weight perpendicular to the plane and the total force on the block is parallel to the plane

$$F_{tot} = W_\parallel - \mu_k N = mg \, \sin 21.8° - 0.25 \, mg \, \cos 21.8°.$$

The acceleration is thus

$$a = {}^{F_{tot}}\!/_m = (9.8)(\sin 21.8° - 0.25 \cos 21.8°) = 1.365 \, \text{m/s}^2$$

and the time to slide down the 15 m plane is

$$t = \sqrt{\frac{2d}{a}} = \sqrt{\frac{2 \cdot 15}{1.365}} = 4.69 \text{ s}$$

which does not depend on the mass of the box.

2.5   Conservation of momentum and energy: elastic collisions

Using the concept of kinetic energy, we can now solve the problem of *elastic* collisions where there are no net external forces and so both momentum and kinetic energy are conserved. The algebra can be a little complicated because the v^2 term in the kinetic energy can lead to a quadratic equation, but the concept is simple. As an example, consider a collision in one dimension between a large object (a baseball, for example) and a much smaller object (such as an air molecule) which leads to the v^2 damping of the acceleration of an object affected by air friction that we saw in Chapter 1. Assume that the molecule is at rest before the collision and the baseball is moving with an initial velocity V_i. The picture before and after the collision are shown in Figure 2.6.

Since momentum and energy are conserved we can set $P_i = P_f$ and $KE_i = KE_f$. Assuming that $M \gg m$ so that $\Delta V \ll V_i$:

$$\text{Conservation of } KE \Rightarrow \frac{1}{2}MV_i^2 = \frac{1}{2}M(V_i^2 - 2V_i\,\Delta V + \Delta V^2) + \frac{1}{2}mv^2$$

$$(\text{neglect!})$$

$$\Rightarrow \frac{1}{2}mv^2 = MV_i\,\Delta V. \qquad ①$$

$$\text{Conservation of momentum} \Rightarrow MV_i = M(V_i - \Delta V) + mv$$

$$\Rightarrow M\Delta V = mv. \qquad ②$$

Substituting $\Delta V = \frac{m}{M}v$ from ② into ①, we find

$$\frac{1}{2}mv^2 = MV_i\frac{m}{M}v \;\Rightarrow\; v = 2V_i.$$

By Newton's Third Law, you know that the force of the molecule on the ball is equal and opposite to the force of the ball on the molecule. If you know that N collisions take place in a time interval dt you can find the force from the change in momentum (Δp)

$$F = \frac{d}{dt}(N\Delta p) = Nm\frac{\Delta v}{dt} = Nm\frac{(2V_i - 0)}{dt}. \qquad (2.20)$$

Since the number of molecular collisions in dt is also proportional to the initial velocity of the object, $N = \rho_n A V_i\, dt$, where ρ_n is the "number density" of molecules in #/m^3, A is

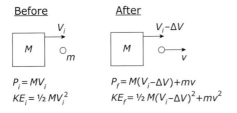

Before   After

V_i $V_i - \Delta V$

| M | O m | | M | O→ v |

$P_i = MV_i$ $P_f = M(V_i - \Delta V) + mv$
$KE_i = \frac{1}{2}MV_i^2$ $KE_f = \frac{1}{2}M(V_i - \Delta V)^2 + mv^2$

Figure 2.6 A large object of mass M and initial velocity V_i collides with a smaller object (mass m). The motions are all confined to the $\hat{x}$ direction; energy and momentum are conserved in one dimension.

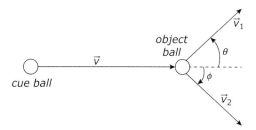

Figure 2.7 A cue ball with initial velocity $v\,\hat{x}$ collides with an object ball at rest. After the collision the object ball travels at an angle of ϕ while the cue ball is defected an angle of θ from the original path.

the cross-section area in m^2, and $V_i\,dt$ is how far the object moves in the time interval dt, we immediately get the result that the force of the air on the moving object is proportional to the square of the object's velocity. In terms of ρ_o, the mass density of the atmosphere in kg/m^3, which is related to the number density and the molecule mass by $\rho_o = m\rho_n$, we find

$$F = 2\rho_o A V_i^2.$$
(2.21)

This result neglects the effect of the shape of the object (a "streamlined" object can convert much of the momentum change of the molecule into a direction perpendicular to the object's motion) and the fact that the collision is not entirely elastic because both the object and the molecule can absorb energy in internal processes and convert mechanical energy to heat. But the V_i^2 dependence is characteristic of fluid friction in non-viscous media.

A classic elastic collision problem in two dimensions is the pool cue-ball problem as shown in Figure 2.7. If you want to make a "cut shot" and strike the cue ball to hit the object ball (initially at rest) into a pocket at an angle of ϕ, what direction θ does the cue ball travel after the collision? We assume that the balls slide across the table with minimal friction, so we can neglect the rotational motion, and we assume that kinetic energy and momentum are both conserved.

The momentum and kinetic energy before is all from the cue ball:

$$\vec{p}_i = m_1 v\,\hat{x}; \quad KE_i = \frac{1}{2}m_1 v^2.$$

The momentum and kinetic energy afterward is the sum of the cue ball and the object ball:

$$\vec{p}_f = m_1 v_1 \,(\cos\theta\,\hat{x} + \sin\theta\,\hat{y}) + m_2 v_2 (\cos\phi\,\hat{x} - \sin\phi\,\hat{y})$$

$$KE_f = \frac{1}{2} m_1 v_1^2 + \frac{1}{2} m_2 v_2^2.$$

Separately equating the $\hat{x}$ momentum components, $\hat{y}$ momentum components, and kinetic energy before and after the collision and canceling out the masses (since the cue ball and object ball have the same mass), we end up with the following three equations:

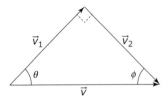

Figure 2.8 Relationship of initial cue ball velocity to final velocities of cue and object ball implied by momentum conservation (vector addition) and conservation of kinetic energy.

$$v = v_1 \cos\theta + v_2 \cos\phi \qquad ①$$
$$0 = v_1 \sin\theta - v_2 \sin\phi \qquad ②$$
$$v^2 = v_1{}^2 + v_2{}^2. \qquad ③$$

Assuming you know the initial velocity v, you still have three equations in four unknowns: v_1, v_2, θ, and ϕ. This is because there are many angles ϕ that the object ball can go – there are essentially an infinite number of "cut shots" that can be made. If we select one angle ϕ that we want the object ball to go and the cue ball initial velocity v, we can then solve the three equations for the other three unknowns (v_1, v_2, and θ) numerically, using the MATLAB function `fzero`, for example, as shown in Example 2.4.

However in this problem, because the masses are equal, there is a trick that allows us to find ϕ immediately if we know θ. We note that momentum must be conserved, $\vec{p}_i = \vec{p}_f = \vec{p}_1 + \vec{p}_2$, which, since the masses cancel out, implies $\vec{v} = \vec{v}_1 + \vec{v}_2$. By the rules of vector addition, this means that the vectors must make a triangle as shown in Figure 2.8.

Further, by the third equation, the Pythagorean Theorem says that the triangle must be a right triangle and the angle opposite to v must be a right angle. Therefore, the angle between v_1 and v_2 – the direction of travel of the object ball and the cue ball after the collision – must be 90°! Knowing this fact can significantly improve your pool game; you will sometimes see a pool player lining up his pool cue from the object ball to the pocket he intends to cut the ball into. He can then judge where the cue ball, moving off perpendicular to the line of the object ball from the point of collision, will go. If the perpendicular path will cause the cue ball to drop into another pocket (a "scratch," which causes the shooter to lose his turn), the player will need to do something to modify the path of the cue ball after the collision. This usually involves putting some spin on the cue ball: overspin, or even just rolling the cue ball slowly without slipping, will decrease the angle between the object ball and the cue ball after collision. Putting backspin on the cue ball (by chalking up your cue and hitting the cue ball sharply below center), can increase the angle between the object ball and cue ball. The topic of rotational dynamics of solid objects is the topic of the next chapter.

Example 2.4 Elastic collision with unequal masses – MATLAB solution. The collision dynamics of two objects of unequal mass with one object initially at rest can be found numerically. Consider an alpha particle (a subatomic particle consisting of two protons

Continued

Example 2.4 (*cont.*)

bound with two neutrons) with a mass approximately four times the proton mass $m_\alpha \cong 4m_p = 4 \times 1.67 \times 10^{-27}$ kg with energy of 1000 eV scattering off C^{12} atoms with a mass $M_{C^{12}} \cong 12m_p$. Find the recoil angle ϕ and velocity v_2 of the C^{12} atoms as a function of the scattering angle θ of the alpha particle.

Solution: Using the notations of Figure 2.7, the equation of y-momentum conservation can be solved for v_1 in terms of v_2

$$v_1 = \frac{m_2\, v_2\, \sin\phi}{m_1\, \sin\theta}.$$

Substituting this into the equation for x-momentum conservation, we find an expression for v_2 in terms of θ and ϕ

$$v_2 = \frac{m_1 v}{m_2 \cos\phi + m_2 \cos\theta\, \sin\phi / \sin\theta} = \frac{m_1 v}{m_2 \cos\phi + m_2 \sin\phi / \tan\theta}.$$

If we define the differences between the initial and final kinetic energies as

$$K(\phi) = \frac{1}{2}\left(m_1 v^2 - m_1 v_1^2 - m_2 v_2^2\right)$$

the physically correct solution for the C^{12} scattering angle ϕ for a given initial velocity v and alpha particle scattering angle θ is given when $K(\phi) = 0$. Given an initial guess ϕ_o, the value of a parameter ϕ that gives the zero of a function can be found from the MATLAB function fzero as below:

```
global m1 m2 v theta
m1=4*1.67e-27; m2=12*1.67e-27; q=1.602e-19;
v=sqrt(2*1000*q/m1);      % Initial velocity
phi_0=60;                 % Initial estimate of phi for FZERO
thetaA=1:1:179;
for n=1:179
    theta=thetaA(n);
    phi(n)=fzero(@collision,phi_0);
    v2(n)=m1*v/(m2*sind(phi(n))/tand(theta) + m2*cosd(phi(n)));
    phi_0=phi(n);
end
plot(thetaA,phi,'-',thetaA,v2/1000,'-')
xlabel('\Theta (degrees)')
ylabel('\phi (degrees), v_2 (km/s) ')

function [K] = collision(phi)
global m1 m2 v theta
A=m2*sind(phi)/tand(theta);
```

Continued

Example 2.4 (*cont.*)

```
B=m2*cosd(phi);
v2=m1*v/(A+B);
v1=(m2/m1)*v2*sind(phi)/sind(theta);
K=0.5*(m1*v^2-m1*v1^2-m2*v2^2);
end
```

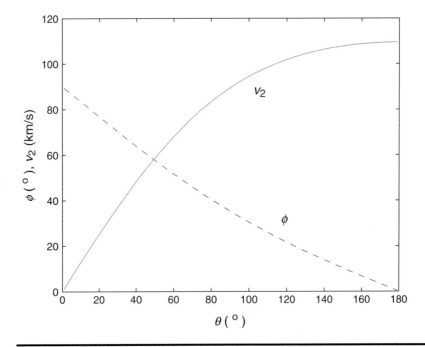

Chapter 2 problems

1. If you were to jump from a 3-m height, would you rather land on sand or concrete?
   a. Justify your answer qualitatively.

   b. Make the least number of assumptions and give an approximate quantitative answer.

2. A 1-N apple falls on your head from a height of h. Assume that, when it hits your head, it deforms by d and does not bounce back.
   a. Neglecting air friction, calculate the average force your head feels during the collision. Is it 1 N? Explain.

   b. Do the calculations for a height of 1 m and a deformation of 0.5 mm.

3. If, instead of an apple (see Problem 2), a 1-N stainless steel ball falls on your head. If your head deforms by 0.1 mm and the ball does not bounce back, calculate the

average force your head feels. What happens if the ball bounces back? Will the force on your head be larger or smaller?

4. When designing equipment, passing the "drop" test is essential. What are some of the factors one should take into account to make sure that the new product passes this test? [Here only focus on aspects that are relevant to this chapter.]

5. What is the force you will feel if you jump from 10 cm and do not bend your knees? Make some reasonable assumptions and come up with an estimate.

6. A book is sitting on a table. Draw free body diagrams showing all the forces on each object and identify the so called "action–reaction" pairs.

7. When a large SUV collides with a small car, Newton's Third Law says that the forces are equal between the objects of the collision. Why are the occupants of the small car in more danger?

8. A bird sits on a perch in a closed cage that is resting on a scale. The bird leaves the perch and flutters in the air. Does the scale change its reading?

9. A small car pushes a large car. Draw free body diagrams showing all the forces on each object. Identify the pair forces.

10. Consider the set up shown below. If the pulleys are ideal (massless and frictionless)

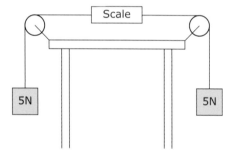

a. what does the scale show?

b. Justify your answer.

11. A 20 g bullet fired with an initial velocity V_m becomes embedded in a 1.6 kg block which is resting on a surface with a coefficient of friction $\mu = 0.6$. The block moves $\Delta x = 12.5$ cm before coming to rest. What was the initial velocity of the bullet?

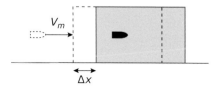

12. The air frictional force on an object is conventionally given by $F_{af} = \frac{1}{2}C_D A \rho v^2$ where A in the cross-section area of the object, ρ is the density of air (approximately 1.22 kg/m³ near sea level), v is the object's velocity, and C_D is the object's *drag coefficient* which depends on the shape and texture of the object. Find an expression for the *terminal velocity* of an object, the maximum velocity the object reaches if it is dropped from a great distance, in terms of the mass m, the acceleration of gravity g, and the parameters from the drag force. Given that a baseball has mass of 0.145 kg and a radius of 0.0366 m and has a terminal velocity around 95 miles/hour, what is the baseball's drag coefficient? How can a skydiver "catch up" with other falling divers to make group formations?

13. Modify the program from Example 2.4 to calculate the scattering of a 1000 eV C¹² atom off a stationary alpha particle (interchange m_1 and m_2). Replace the "for" loop with commands to solve for an initial velocity $v = 2 \cdot 1000 \cdot 1.6 \times 10^{-19}/(12 \cdot 1.67 \times 10^{-27})$ and scattering angles of $\theta = 10, 12, 14, 16,$ and 18°. What results do you get? Do they make sense? Now try for $\theta = 20°$. What happens? Can you make sense out of this?

14. Describe the trajectory as seen by a stationary observer of a plastic water bottle thrown from a moving car moving at a constant velocity. (A practice to be greatly discouraged!) Draw sequential drawing to show how the car moves and bottle moves and explain the forces involved. Does it make a difference if the bottle is filled with water or is empty? Explain why or why not. How does your picture change if the car is decelerating by braking?

3 Rotational motion

While we have been discussing Newton's Laws so far, mass – resistance to acceleration – has been considered as a *scalar* quantity. No matter which direction you apply a force to a point object, its mass is the same. This simple formalism breaks down when we consider rotational motion. If we have a long cylinder, it matters very much whether we rotate it about an axis down the center of the cylinder or an axis which is perpendicular to the cylinder! This will require us to extend the concept of resistance to acceleration to depend on the direction of the rotation.

3.1 Rotational motion and moment of inertia

Consider a mass attached to a string rotating in a circle one complete rotation each second, as shown in Figure 3.1. The speed, the magnitude of the velocity, is constant. Geometry gives us that the arc length s is related to the radius and the angle that the arc transects by $s = \theta r$ where θ is the angle in radians as shown in Figure 3.2. (Note if θ is 2π radians – 360 degrees – this gives us arc length = circumference = $2\pi r$, and we recover our familiar relation that circumference = $\pi \times$ diameter.) The speed is the time derivative of the distance traveled (the arc length): $|v| = ds/dt = r\, d\theta/dt = r\omega$ where $\omega = d\theta/dt$ is the *angular velocity* in radians per second.

At each moment, the ball has a velocity which is tangential to the circle. Since the *direction* of the velocity is constantly changing – even if the speed is constant – the ball is undergoing constant acceleration. This *centripetal acceleration* is caused by the force exerted on the string to pull the ball into a circular path. As illustrated in Figure 3.3, when the angle changes by a small $d\theta$, the change in velocity $d\vec{v}$ is directed toward the center of the circle. Bisecting the angle to make two right triangles, we find the magnitude of the velocity change

$$\frac{dv}{2} = v \sin \frac{d\theta}{2} \approx v \frac{d\theta}{2} \tag{3.1}$$

where the last equality comes from the small angle approximation where the sine of a small angle is nearly equal to the angle in radians. This gives $dv = v\, d\theta$. Dividing both sides by dt, we find that the magnitude of the centripetal acceleration of an object moving in a circular path at a constant speed is

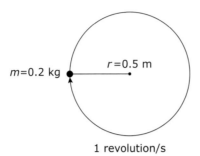

1 revolution/s

Figure 3.1 Circular rotational motion of a mass with $m = 0.2$ kg, circular radius $r = 0.5$ m, and at a rate of 1 revolution/s $= 2\pi r$ m/s.

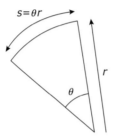

Figure 3.2 Relation of arc length to the included angle (in radians).

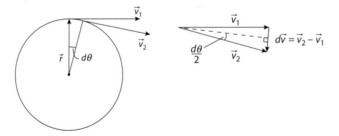

Figure 3.3 Change of velocity vector for particle rotating in circular motion at constant velocity. By the rules of vector subtraction the change in velocity $d\vec{v}$ is a vector extending from the tip of $\vec{v}_1$ to the tip of $\vec{v}_2$.

$$a = \frac{dv}{dt} = v\frac{d\theta}{dt} = v\omega = \frac{v^2}{r}. \qquad (3.2)$$

The last equality follows from the relation $\omega = v/r$ that we found above. The direction of the acceleration for an object moving in a circle at a constant speed is toward the center of the circle.

Question 3.1

How does this equation compare with Eq. (1.20)? In what limit are the two equations the same? Are the direction and magnitude of $\vec{a}_{cir}$ from Eq. (1.20) consistent with Eq. (3.2)?

The forces to sustain the centripetal acceleration of the mass elements that make up a rotating solid body come from the elastic cohesive strength of the materials, and they are invisible to the observer (unless they exceed the material strength of the body – resulting in a catastrophic fracture!). The forces are directed toward the center of the rotation, perpendicular to the motion of each mass particle of the object. If $\vec{F}_c$ is the centripetal force and $d\vec{s}$ is the distance moved, the work done by the centripetal force $W = \vec{F}_c \cdot d\vec{s} = |\vec{F}_c||d\vec{s}| \cos(90°) = 0$. Thus a rotating object has a rotational energy that is constant unless acted upon by an outside force.

Kinetic energy, $\frac{1}{2}mv^2$, depends only on the speed of an object. For the mass on a string above, we have

$$KE = \frac{1}{2}mv^2 = \frac{1}{2}m(r\omega)^2 = \frac{1}{2}mr^2\omega^2 = \frac{1}{2}I\omega^2 \tag{3.3}$$

where we have introduced the symbol $I = mr^2$ for the *moment of inertia* of the ball. So for a string length of 0.5 m and a ball mass of 0.2 kg rotating at one revolution per second ($\omega = 2\pi$ radians/s), the kinetic energy would be $KE = \frac{1}{2}(0.2)(0.5)^2(2\pi)^2 = 0.987$ joules.

3.2   Moment of inertia of a solid body

To find the kinetic energy of a rotating solid body, we observe that the angular velocity of each particle of mass in the body is rotating at the same angular velocity ω (they each make one rotation – 2π radians – in the same time interval). We can define the rotational kinetic energy of the body to be

$$KE = \frac{1}{2}I\omega^2 \tag{3.4}$$

if we define the moment of inertia by summing up mr^2 for each particle of the body

$$I = \sum_i m_i r_i^2 = \int r^2 \, dm \tag{3.5}$$

where dm is an infinitesimal mass element and r is the distance of each element from the axis of rotation. For example, a uniform disk of mass M, thickness h, and radius R rotating around a perpendicular axis through the center of the disk, as shown in Figure 3.4, would have $dm = \rho r h \, d\theta \, dr$ where ρ is the mass density, the total mass divided by the volume, $\rho = M/(\pi R^2 h)$.

We find the moment of inertia of the disk around its axis by integrating over the whole disk, for dr from 0 to R and $d\theta$ from 0 to 2π as shown in Figure 3.4.

$$I = \int r^2 \, dm = \iint r^2 \, \rho h r \, d\theta \, dr$$

$$= \iint r^3 \left(\frac{M}{\pi R^2 h}\right) h \, d\theta \, dr = 2\pi \frac{R^4}{4} \left(\frac{M}{\pi R^2 h}\right) h = M\frac{R^2}{2}.$$

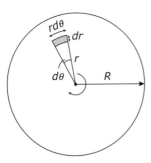

Figure 3.4 Calculation of moment of inertia I by integration of $dm = pdv = p\,h\,da = p\,h\,rd\theta dr$.

The kinetic energy of a disk of mass 5 kg and radius 0.3 m rotating at 1 revolution per second is therefore $KE = \frac{1}{2}\left(5\frac{(0.3)^2}{2}\right)(2\pi)^2 = 4.44$ J.

3.3   Center of mass (COM) and parallel axis theorem

If we know the moment of inertia of a solid object around an axis passing through its *center of mass (COM)*, it is easy to find the moment of inertia around any axis parallel to the COM axis by the *parallel axis theorem*. The COM of a body or system of masses is the point that moves as if all the system's mass were concentrated at that point and as if all external forces were concentrated there. For example, a baseball bat thrown up with some rotation describes a complicated trajectory, but if we concentrate on one particular point at the center of the bat near the thick end (the COM), we will find that that point follows a parabolic trajectory of a point mass with an initial velocity acted on by a gravitational force.

The coordinates of the COM of a solid body are found by taking the mass-weighted average of the (x, y, z) positions of all the mass elements dm in the body. For an object with a constant mass density $\rho = dm/dV$ this becomes the same as a volume-weighted average:

$$x_{COM} = \frac{1}{M}\int x\,dm = \frac{1}{V}\int x\,dV$$

$$y_{COM} = \frac{1}{M}\int y\,dm = \frac{1}{V}\int y\,dV \tag{3.6}$$

$$z_{COM} = \frac{1}{M}\int z\,dm = \frac{1}{V}\int z\,dV.$$

For the disk whose moment of inertia we calculated in Section 3.2, we find

$$x_{COM} = \frac{1}{\pi R^2 h} \iiint (r \cos\theta) r d\theta dr dz = \frac{1}{\pi R^2 h} \left(\frac{R^3}{3} h \sin\theta \right) \Big|_{\theta=0}^{2\pi} = 0$$

$$y_{COM} = \frac{1}{\pi R^2 h} \iiint (r \sin\theta) r d\theta dr dz = \frac{-1}{\pi R^2 h} \left(\frac{R^3}{3} h \cos\theta \right) \Big|_{\theta=0}^{2\pi} = 0$$

$$z_{COM} = \frac{1}{\pi R^2 h} \iiint z r d\theta dr dz = \frac{1}{\pi R^2 h} \left(\frac{R^2}{2} 2\pi \frac{z^2}{2} \right) \Big|_{z=0}^{h} = \frac{h}{2}$$

giving the result that we expected that the COM is at $(0,0,h/2)$, at the center of the disk, and the moment of inertia around the COM is $I_{COM} = MR^2/2$ as we calculated.

Example 3.1 Find the center of mass of a cone with height $h = 2$ m and a circular base of radius $R = 0.5$ m.

Solution: The COM will clearly be along the line of rotational symmetry that passes from the center of the base through the point of the cone. The volume of a slice of the cone of thickness dz is $dV = \pi[r(z)]^2 dz$ where the radius of the cone as a function of position z measured *from the tip of the cone* is $r(z) = \frac{z}{h} R$. Integrating dV over the height of the cone from 0 to h gives us the volume of the cone that may be a dim memory from high school geometry

$$V = \int_0^h \pi \left(\frac{zR}{h} \right)^2 dz = \left(\frac{\pi R^2}{h^2} \right) \frac{z^3}{3} \Big|_0^h = \frac{1}{3} \pi R^2 h.$$

The COM is then found from

$$z_{COM} = \frac{3}{\pi R^2 h} \int_0^h z \pi \left[\frac{z}{h} R \right]^2 dz = \frac{3}{h^3} \left(\frac{z^4}{4} \right) \Big|_0^h = \frac{3}{4} h,$$

three-quarters of the way from the tip to the base. The COM is thus along the center line of the cone 0.5 m from the base.

Since the motion of an object may be considered to be concentrated at the COM, a solid object of mass M rotating around an axis parallel to and a distance d from an axis passing through the COM has a contribution due to the point mass of $I_{point\ mass} = Md^2$. In addition, however, on each rotation around the parallel axis the object completes one complete rotation around its COM. So the total moment of inertia around the parallel axis is

$$I = I_{COM} + Md^2 \qquad \text{(Parallel axis theorem)} \qquad (3.7)$$

where I_{COM} is the moment of inertia of the object rotating around the axis passing through the COM. So, for example, the disk we considered in Section 3.2, if rotating around an axis on the edge of the disk instead of around its center, has a total moment of inertia given by the parallel axis theorem with the parallel axis separated by a distance $d = R$ from the axis through the COM

$$I = M\frac{R^2}{2} + MR^2 = \frac{3}{2}MR^2.$$

Thus the kinetic energy of a 5 kg disk 0.3 m in radius rotating at 1 revolution/s around an axis on the edge of the disk is $KE = \frac{1}{2}\left(\frac{3}{2}5(0.3)^2\right)(2\pi)^2 = 13.32\,\text{J}$.

Question 3.2

Can you show that the kinetic energy of the above disk rotating at one revolution/s around an axis on its edge is the sum of the 4.44 J for the rotation around its COM plus $\frac{1}{2}Mv^2$, the kinetic energy of the disk in its rotational motion assuming that its entire mass is concentrated at its COM?

3.4   Tensor moments of inertia

A point mass on a string would have the same rotational kinetic energy rotating around the x-axis, the y-axis, or the z-axis. This clearly is not true for all solid objects, however. Consider, for example, a long cylinder as shown in Figure 3.5. Clearly the cylinder will have more energy rotating at the same rate around the x-axis than it would have rotating around the z-axis because the mass near the end of the cylinder would be moving a much greater distance per revolution (and thus have a greater velocity) rotating around the x-axis than if it were rotating around the z-axis.

The moment of inertia of a non-spherical object must therefore have a moment of inertia that depends on the axis of rotation. A mathematical object that has the quality

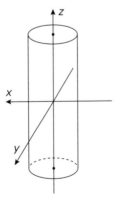

Figure 3.5 The rotational energy of the cylinder is larger if it rotates around the x-axis or y-axis than if it rotates around the z-axis.

that its value depends on direction is a *tensor*, and the moment of inertia of an object, expressed in its *principal axes coordinates* (as discussed later), can be written as a tensor

$$I = \begin{bmatrix} I_{xx} & 0 & 0 \\ 0 & I_{yy} & 0 \\ 0 & 0 & I_{zz} \end{bmatrix}. \tag{3.8}$$

Example 3.2 Find the components of the moment of inertia tensor in the coordinate system of Figure 3.5 for a cylinder of radius R and height h.

The I_{zz} component of I is the same as we calculated for rotation of a disk around its axis $I_{zz} = \frac{MR^2}{2}$. The moments around the x-axis and y-axis, I_{xx} and I_{yy}, are equal and can be found by integration

$$I_{xx} = \int r_x^2 dm.$$

Here the mass element $dm = \rho dV = \rho r d\theta dr dz$, $\rho = \frac{M}{\pi R^2 h}$ is the mass density of the cylinder, and $r_x = \sqrt{z^2 + y^2}$ is the distance of the mass element from the x-axis. In terms of the distance r from the cylinder axis and the angle θ around the cylinder axis $x = r \cos \theta$ and $y = r \sin \theta$. Thus

$$I_{xx} = \left(\frac{M}{\pi R^2 h} \right) \iiint (z^2 + r^2 \sin^2 \theta) r d\theta dr dz$$

$$= \left(\frac{M}{\pi R^2 h} \right) \left(\iiint z^2 r d\theta dr dz + \iiint r^3 \sin^2 \theta d\theta dr dz \right)$$

where the integrals are over the whole cylinder from $z = -h/2$ to $z = +h/2$, $r = 0$ to $r = R$, and the cylinder axial angle θ from $\theta = 0$ to $\theta = 2\pi$. The first integral gives

$$\left(\frac{M}{\pi R^2 h} \right) \iiint z^2 r d\theta dr dz = \left(\frac{M}{\pi R^2 h} \right) \left[2\pi \frac{R^2}{2} \left(\frac{(h/2)^3}{3} - \frac{(-h/2)^3}{3} \right) \right] = \frac{Mh^2}{12}.$$

The second integral is

$$\left(\frac{M}{\pi R^2 h} \right) \iiint r^3 \sin^2 \theta d\theta dr dz = \left(\frac{M}{\pi R^2 h} \right) \left[h \frac{R^4}{4} \left(\frac{\theta}{2} - \frac{\sin \theta \cos \theta}{2} \right) \Big|_0^{2\pi} \right] = \frac{MR^2}{4}.$$

Thus $I_{xx} = I_{yy} = \frac{Mh^2}{12} + \frac{MR^2}{4}$, and the moment of inertia tensor is

$$I = \begin{bmatrix} \dfrac{Mh^2}{12} + \dfrac{MR^2}{4} & 0 & 0 \\ 0 & \dfrac{Mh^2}{12} + \dfrac{MR^2}{4} & 0 \\ 0 & 0 & \dfrac{MR^2}{2} \end{bmatrix}.$$

In terms of the moment of inertia tensor, I, the kinetic energy can be written

$$KE = \frac{1}{2}\vec{\omega}^T I \vec{\omega} \tag{3.9}$$

where $\vec{\omega}$ is the angular velocity vector with magnitude equal to the rotation rate in radians/s and direction perpendicular to the plane of rotation, given by the right-hand rule. (If the fingers of your right hand curl in the direction of the rotation, your thumb will be in the direction of the perpendicular direction that corresponds to a positive angular velocity.) The superscript "T" refers to the transpose of the vector.

For the cylinder and the coordinate system shown above, we have $I_{xx} = I_{yy}$ because of the symmetry of rotation around the x- and y-axis. I_{zz} will be less, representing the smaller moment of inertia of the cylinder rotating around the z-axis. For rotation around the z-axis, we have

$$\vec{\omega} = \begin{bmatrix} 0 \\ 0 \\ \omega_z \end{bmatrix} \quad \text{and} \quad \vec{\omega}^T = \begin{bmatrix} 0 & 0 & \omega_z \end{bmatrix}. \tag{3.10}$$

From the rules of matrix multiplication, the kinetic energy is thus

$$KE = \frac{1}{2}\begin{bmatrix} 0 & 0 & \omega_z \end{bmatrix}\begin{bmatrix} I_{xx} & 0 & 0 \\ 0 & I_{yy} & 0 \\ 0 & 0 & I_{zz} \end{bmatrix}\begin{bmatrix} 0 \\ 0 \\ \omega_z \end{bmatrix} = \frac{1}{2}\begin{bmatrix} 0 & 0 & \omega_z \end{bmatrix}\begin{bmatrix} 0 \\ 0 \\ I_{zz}\,\omega_z \end{bmatrix} = \frac{1}{2}I_{zz}\,\omega_z^2 \tag{3.11}$$

in agreement with our previous result for kinetic energy if $\omega = \omega_z$ and the moment of inertia $I = I_{zz}$.

Question 3.3

What would you find for the kinetic energy of an object with a moment of inertia as in Eq. (3.8) that has an angular velocity of ω_y around the y-axis and an angular velocity of ω_x around the x-axis? Can an object have different angular velocities around different axes? What does the motion look like if $\omega_y = 2\omega_x$? If $\omega_x = 2\omega_y$?

3.5   Torque and angular momentum

For linear motion, we have seen that the quantity *linear momentum,* $\vec{p} = m\vec{v}$, has a crucial role in Newton's Laws give us

$$\frac{d}{dt}\vec{p} = \frac{d}{dt}m\vec{v} = m\frac{d}{dt}\vec{v} = m\vec{a} = \vec{F}_{ext} \tag{3.12}$$

where $\vec{F}_{ext}$ is the external force applied to the object. It follows that **if there is no external applied force, then $\frac{d}{dt}\vec{p} = 0$ and linear momentum is conserved.**

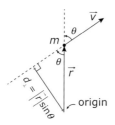

Figure 3.6 The angular momentum of a particle around an origin can be expressed in terms of the perpendicular distance $d_\perp$ to the line of the velocity.

(We assumed that the mass is a constant above, but the conclusion that momentum is conserved in the absence of external forces holds as well if the mass depends on time.)

In rotational motion, we can define a quantity *angular momentum*, $\vec{L}$, as below

$$\vec{L} = \vec{r} \times \vec{p} \tag{3.13}$$

where $\vec{p}$ is the linear momentum of a particle of mass m, and $\vec{r}$ is the position vector from any selected origin to the particle. By the definition of the cross product, the magnitude of $\vec{L}$ is the magnitude of the momentum times $d_\perp$, the perpendicular distance from the line of the momentum to the origin (see Figure 3.6)

$$|\vec{L}| = |\vec{r} \times \vec{p}| = |\vec{r}| \sin\theta |\vec{p}| = d_\perp |\vec{p}|. \tag{3.14}$$

The value of the angular momentum clearly depends on where we select the origin to be. For a rotating object, it usually makes sense to define it with respect to the axis of rotation, but we will see some examples later where other choices of the origin make it easier to understand the problem.

The significance of the angular momentum becomes apparent when we look at the time derivative of $\vec{L}$

$$\frac{d}{dt}\vec{L} = \frac{d}{dt}(\vec{r} \times m\vec{v}) = \frac{d}{dt}\vec{r} \times m\vec{v} + \vec{r} \times \frac{d}{dt}(m\vec{v}) = \vec{v} \times m\vec{v} + \vec{r} \times \vec{F} \tag{3.15}$$

where we have used the chain rule to expand the derivative and noted that the time derivative of the position vector $\vec{r}$ is the velocity $\vec{v}$ (regardless of where we select our origin) and, from our discussion of linear momentum above, the time derivative of the linear momentum $\vec{p}$ is the external force applied to the object, $\vec{F}$. Since the cross product of any two parallel vectors is zero, $\vec{v} \times m\vec{v} = 0$, and we have

$$\frac{d\vec{L}}{dt} = \vec{r} \times \vec{F} = \vec{\tau}, \tag{3.16}$$

where the *torque* is defined as $\vec{\tau} = \vec{r} \times \vec{F}$ is the "turning force" that is tending to rotate the mass around the origin. Similar to the situation with linear momentum where the momentum doesn't change if there is no external applied force, **if the external torque is zero, then $\frac{d\vec{L}}{dt} = 0$, and** *angular momentum is conserved!*

This is true of angular momentum defined around any origin. If we are considering a solid body rotating around an axis, the angular momentum around the axis of rotation is particularly simple to define in terms of the angular velocity ω. The velocity of each element of mass dm in the rotating object is perpendicular to the position vector from the axis of rotation

$$|d\vec{L}| = |\vec{r}| \, dm \, |\vec{v}| = r^2 \, dm \, \omega \tag{3.17}$$

where we have used the relation $v = \omega r$. Since the angular velocity of each element of mass in the solid body is the same, we have

$$|\vec{L}| = \int r^2 \, dm \, \omega = I \, \omega \tag{3.18}$$

using our previous definition of the moment of inertia I. Taking the direction of the $\vec{L}$ to be in the same direction as $\vec{\omega}$, perpendicular to the direction of rotation as given by the right-hand rule, we can write the vector equation for the angular momentum of a rotating solid body in terms of the (tensor) moment of inertia

$$\vec{L} = I \, \vec{\omega}. \tag{3.19}$$

if a torque is applied around an axis of rotation of a solid object with moment of inertia I, we have the following relation

$$\vec{\tau} = \frac{d}{dt}\vec{L} = I\frac{d}{dt}\vec{\omega} = I\vec{\alpha}. \tag{3.20}$$

We have defined $\vec{\alpha}$, the time derivative of the angular velocity, as the *angular acceleration* in units of radians/s^2. Corresponding to our definition of angular velocity, the magnitude of the acceleration of any mass element in terms of the angular acceleration and the distance from the axis of rotation is $a = r\alpha$.

Example 3.3 A 2.0 kg mass is attached to a rope wrapped around the outside of a 5 kg disk of radius $= 0.3$ m as shown in the figure below. The disk is free to rotate around an axis through its center. The mass is released from rest and drops under the force of gravity, causing the disk to rotate. What is the acceleration a of the mass? What is the angular velocity of the disk after the mass falls 1 meter?

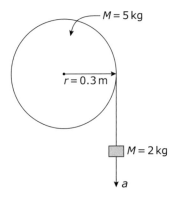

Continued

Example 3.3 (*cont.*)

Solution: As found before, the moment of inertia of the disk rotating around an axis through its center is $I = \frac{1}{2} M R^2$. Free body diagrams of the mass and the rotating disk are shown below.

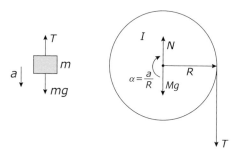

T is the tension in the rope which pulls up on the mass and down on the edge of the disk. N is the normal force exerted on the disk by the axle. Because the mass is connected by the rope to the edge of the disk, the acceleration of the mass is the same as the acceleration of the edge of the disk. The angular acceleration of the disk α is related to the acceleration of the edge by $\alpha = a/R$.

Applying Newton's Second Law to the mass yields

$$ma = mg - T. \qquad ①$$

Since the disk has no net up/down movement, the net vertical force is zero, and the normal force up must be equal and opposite to the tension plus the weight of the disk: $N = -Mg + T$. Calculating torques around the COM of the disk, we find $\vec{\tau} = \vec{r} \times \vec{T}$ since both the normal force and the weight act on the COM and therefore do not contribute to the torque. Applying the torque equation we find

$$|\tau| = |\vec{r} \times \vec{T}| = T d_\perp = TR = I |\vec{\alpha}| = I \frac{a}{R}. \qquad ②$$

Solving the second equation for $T = I a/R^2$ and substituting into the first equation, we find

$$ma = mg - I \frac{a}{R^2} = mg - \frac{1}{2} MR^2 \frac{a}{R^2}$$

$$\Rightarrow a = \frac{g}{(1 + \frac{M}{2m})} = \frac{9.8}{(1 + \frac{5}{4})} = 4.36 \, \text{m/s}^2.$$

To find the angular velocity after the mass falls 1 m, we could use $v^2 = v_o^2 - 2ax$ to find $v = 2.95$ m/s^2 yielding $\omega = \frac{v}{R} = 9.84$ radians/s. Alternately, conservation of energy gives:

Continued

Example 3.3 (*cont.*)

$$\Delta KE = \frac{1}{2} mv^2 + \frac{1}{2} I \omega^2 = -\Delta PE = -mg\Delta h$$

$$\frac{1}{2} m(\omega R)^2 + \frac{1}{2} \left(\frac{MR^2}{2} \right) \omega^2 = -mg\Delta h \Rightarrow \omega = \sqrt{\frac{2\,g}{R^2 \left(1 + \frac{M}{2m} \right)}} = 9.84 \text{ radians/s.}$$

3.6   Torques acting on inertial tensor

Since the moment of inertia I is a tensor, the equation $\vec{\tau} = I\,\vec{\alpha}$ implies that the angular acceleration $\vec{\alpha}$ of a rigid body does not have to be in the same direction as the torque $\vec{\tau}$. Consider a cylinder floating weightlessly in Earth orbit acted on by a force exerted on the lower edge as shown in Figure 3.7. The appropriate origin to take torques around is the *center of mass* (COM) in the center of the cylinder, since that is the point that a weightless cylinder will rotate around. From the definition of the cross product, the direction of the torque $\vec{\tau} = \vec{r} \times \vec{F}$ is perpendicular to the force $\vec{F}$, which is into the page and perpendicular to the position vector from the COM to the point where the force is applied, as shown in the figure. Because the cylinder is easier to rotate around the cylindrical axis **A** than around the transverse axis **B**, the diagonal elements of the moment of inertia are not the same. The angular acceleration is found from Equation (3.20) by multiplying both sides by the inverse moment of inertia tensor, I^{-1}, defined as the tensor that when multiplying I it yields the *identity matrix*

$$I^{-1}\vec{\tau} = I^{-1} I\,\vec{\alpha} = \begin{bmatrix} 1 & 0 & 0 \\ 0 & 1 & 0 \\ 0 & 0 & 1 \end{bmatrix} \vec{\alpha} = \vec{\alpha} \qquad (3.21)$$

because any vector multiplied by the identity matrix is unchanged.

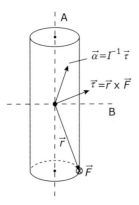

Figure 3.7 Torque around center of mass and direction of angular acceleration for weightless cylinder with force applied to bottom edge of cylinder in tangential direction.

Question 3.4

Show by direct matrix multiplication that the tensor moment of inertia

$$
I = \begin{bmatrix} I_{xx} & 0 & 0 \\ 0 & I_{yy} & 0 \\ 0 & 0 & I_{zz} \end{bmatrix} \text{ has an inverse } I^{-1} = \begin{bmatrix} \dfrac{1}{I_{xx}} & 0 & 0 \\ 0 & \dfrac{1}{I_{yy}} & 0 \\ 0 & 0 & \dfrac{1}{I_{zz}} \end{bmatrix}.
$$

(For tensors that are not in their principal axes system – and therefore not diagonal – the inverse can be found with the MATLAB `inv(I)` command, for example.)

Since for a tall cylinder as illustrated in Figure 3.7 $I_{xx} = I_{yy}$ is larger than I_{zz} by a factor of approximately $\frac{1}{6}\left(\frac{h}{R}\right)^2$, the angular acceleration $\vec{\alpha} = I^{-1}\vec{\tau} = \frac{1}{I_{xx}}\tau_x\hat{x} + \frac{1}{I_{yy}}\tau_y\hat{y} + \frac{1}{I_{zz}}\tau_z\hat{z}$ will have a larger z-component than the torque. As shown in the figure, the direction of $\vec{\alpha} = I^{-1}\vec{\tau}$ is in a different direction than $\vec{\tau}$, closer to the "easy axis" **A**.

Example 3.4 Tensor moment of inertia

Consider a 2500 kg cylindrical tank with height of 12 m and radius 3 m floating in space and being pushed on by a force couple of 250 N exerted in the $\hat{y}$- and $-\hat{y}$- directions at the top and bottom edge as shown in the figure. Find the magnitude and direction of the resulting angular acceleration.

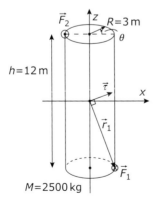

Solution: The force couple has no net force on the cylinder and, in its weightless environment, the cylinder will rotate around its center of mass. The principal axes of the moment of inertia tensor are the x-, y-, and z-axis as shown (the y-axis is directed into the paper).

Continued

Example 3.4 (*cont.*)

From Example 3.2, the moment of inertia tensor is

$$
I = \begin{bmatrix} \dfrac{Mh^2}{12} + \dfrac{MR^2}{4} & 0 & 0 \\ 0 & \dfrac{Mh^2}{12} + \dfrac{MR^2}{4} & 0 \\ 0 & 0 & \dfrac{MR^2}{2} \end{bmatrix} = \begin{bmatrix} 35625 & 0 & 0 \\ 0 & 35625 & 0 \\ 0 & 0 & 11250 \end{bmatrix} \text{ kg m}^2.
$$

The total torque is the sum of the torques of $\vec{F}_1$ and $\vec{F}_2$

$$
\begin{aligned}
\vec{\tau} &= \vec{r}_1 \times \vec{F}_1 + \vec{r}_2 \times \vec{F}_2 \\
&= (3\hat{x} - 4\hat{z}) \times (250\hat{y}) + (-3\hat{x} + 4\hat{z}) \times (-250\hat{y}) \\
&= 2000\hat{x} + 0\hat{y} + 1500\hat{z}.
\end{aligned}
$$

Operating on both sides of the equation $\vec{\tau} = I\vec{\alpha}$ with the inverse of the moment of inertia tensor, we find

$$
\vec{\alpha} = I^{-1}\vec{\tau} = \begin{bmatrix} 0.281e^{-4} & 0 & 0 \\ 0 & 0.281e^{-4} & 0 \\ 0 & 0 & 0.889e^{-4} \end{bmatrix} \begin{bmatrix} 2000 \\ 0 \\ 1500 \end{bmatrix} = \begin{bmatrix} 0.056 \\ 0 \\ 0.133 \end{bmatrix} \text{ rad/s}^2.
$$

As predicted, the angular acceleration is predominantly around the z-axis despite the fact that the torque is greater around the x-axis.

Example 3.5 Kepler's Laws

Isaac Newton developed his theory of gravitation to explain Johannes Kepler's Laws of Planetary Motion, based on Kepler's analysis of the planetary observations of Tycho Brahe. Kepler's Laws are:

1. The planetary orbits are ellipses with the Sun at one focus.
2. A line between the planet and the Sun sweeps out equal areas in equal times.
3. The square of the period of the planet's orbit is proportional to the cube of the semi-major axis (one-half of the longest diameter) of the ellipse.

All of these laws can be calculated as a consequence of Newton's Gravitational Law: $F = G\frac{m_1 m_2}{r^2}(-\hat{r})$. In particular, Kepler's Second Law is just an expression of the conservation of angular momentum. Since the gravitational force is parallel to the radius vector $\vec{r}$ the torque $\vec{r} \times \vec{F} = 0$, and the angular momentum doesn't change as the planet orbits around the Sun. As shown in the figure, the planet moves faster when it is closer to the Sun. The angular momenta at r_1 and r_2 are given by

Continued

Example 3.5 (*cont.*)

$$|\vec{L}_1| = |\vec{r}_1 \times m\vec{v}_1| = r_1 m \frac{\Delta s_1}{\Delta t}$$

$$|\vec{L}_2| = |\vec{r}_2 \times m\vec{v}_2| = r_2 m \frac{\Delta s_2}{\Delta t}.$$

Since the angular momentum is the same at 1 and 2, for the same time interval Δt we have

$$r_1 \Delta s_1 = 2 A_1 = r_2 \Delta s_2 = 2 A_2$$

showing that $A_1 = A_2$ and the area swept out by the radius vector is the same!

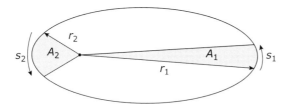

The Third Law can be easily proven in the limit in which the orbit is circular. The centripetal force, $mr\omega^2$, is equal to the gravitational force $G\frac{Mm}{r^2}$. Setting these two equal and noting that the orbital period is given by $T = \frac{2\pi}{\omega}$, we find that $T^2 = \frac{(2\pi)^2 r^3}{MG}$ confirming Kepler's Third Law.

3.7 Torques around an arbitrary axis: statics

In the examples so far, we have measured the torques and angular acceleration around the axis of rotation, but our equation for rotational motion $\vec{\tau} = \frac{d\vec{L}}{dt} = I\vec{\alpha}$ doesn't require that the torques and the angular velocity or acceleration be taken around an axis of rotation. In statics problems where the angular acceleration and velocity are zero, the total torque around any point is zero. The problem can often be simplified by taking the torques around a point that causes forces that are unknown to drop out of the problem.

Example 3.6 A 10 m long 100 N beam, with one end attached to a wall, supports a 500 N weight at the end with a cable attached at the end and attached to the wall at a 45° angle as shown in the figure. Find the tension in the cable.

Continued

Example 3.6 (*cont.*)

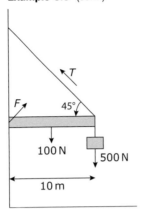

Solution: In a statics problem, since the objects are motionless and not accelerating either linearly or angularly, the sum of the forces in the x- and y-directions, and the sum of the torques around any point, must all be zero. In this case, taking the torques around the center of mass gives unknown torque both from the tension in the cable $\vec{T}$ and from the force that the wall exerts on the beam $\vec{F}$. Both the magnitude and the direction of $\vec{F}$ are unknown, so the torque equation will have three unknowns: the tension, the magnitude of $\vec{F}$, and the direction of $\vec{F}$. With the aid of the equations for the vanishing of the net x- and y-forces, a solution can be obtained. But if, instead, the torques are taken around the point where the beam meets the wall, then the torque due to $\vec{F}$ is zero since the force passes through the point where the torques are calculated and the perpendicular distance is zero. The torque equation has just one unknown

$$-100\,\text{N}\cdot 5\,\text{m} - 500\,\text{N}\cdot 10\,\text{m} + T\,\sin 45\cdot 10\,\text{m} = 0$$

with the solution, $T = 778$ N.

Another example of taking torques other than around the axis of rotation is below:

Question 3.5

Consider pulling on the string of a yo-yo which is sitting at rest on a horizontal surface. If we select the point of contact between the yo-yo and the table as the point to take the torques around, the only torque is due to the tension of string since the normal and the frictional force, as well as the weight, pass through the point of contact.

In the three cases below, use the right-hand rule on $\vec{\tau} = \vec{r} \times \vec{F}$ to find the direction of the torque on the yo-yo due to the tension on the string calculated around the point of contact. What direction would the angular acceleration $\vec{\alpha} = \vec{r} \times \vec{a}$ (measured with respect to the same origin at the point of contact) be if the yo-yo begins to move toward the right? What if it begins to move to the left? Comparing the

Continued

Question 3.5 (*cont.*)

direction of the torque to the direction of the angular acceleration, what direction will the yo-yo move in each of the three cases? Try this out with a yo-yo to convince yourself that your predictions are correct.

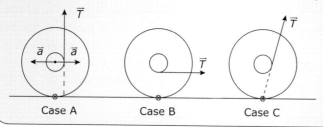

Case A   Case B   Case C

Chapter 3 problems

1. Two solid spheres are released from the same height and are rolling down the same inclined plane. Which one will reach the bottom first if

   a. they have same diameter, but different masses?

   b. they have the same mass, but different diameters?

   c. they have different masses and different diameters?

2. What if, instead of two spheres (see Problem 1), we roll down a solid cylinder and a solid sphere?

3. A solid and a hollow sphere with equal masses are released from the same height and roll down an inclined plane. Which one will reach the bottom first? What happens if you do the same for a solid and a hollow cylinder?

4. Consider the same inclined plane as before and compare the linear acceleration of a rolling sphere with that of a sliding block (without friction) of the same mass.

5. Consider the Atwood's Machine shown below. If the pulley has a mass m_p and the frictional force between the axle (with diameter r) and the pulley is f, find the angular acceleration of the pulley and the tensions on the rope on each side of the pulley as a function of the two hanging masses, the mass of the pulley, and the frictional force. [Assume that the pulley is a disk.]

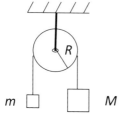

6. Demonstrate by integration that the center of mass of an 18 inch long $2'' \times 4''$ wood beam is in fact in the center of the beam.

7. Calculate the moment of inertia tensor for the beam in Problem 6 in the axis system with x parallel to the 2″ side, y parallel to the 4″ side and z parallel to the 18″ side. The beam weighs 3.5 lb.

8. The beam in Problem 7 is supported on its 2″ side by thin metal bars at the end of the beam and at a point 6″ from the other end of the beam. The metal bar at the beam end is removed suddenly. Calculate the angular acceleration of the beam. At some point the motion ceases to be simple rotation around the remaining metal bar. What are the conditions for when the beam ceases to rotate around the bar and what is its angular velocity when this happens?

9. A 75 kg man, approximated by a uniform cylinder of height 1.8 m and radius 0.15 m, holding two 5 kg masses at arm's length a distance of 1 meter from the center of his body, rotates on a frictionless turntable at 0.5 revolutions per second. Calculate the angular momentum of the man–weight system. If the man pulls the masses into his body, is there any torque exerted? What will his angular velocity be in revolutions/s with the weights against his body?

11. If the man in Problem 9, while rotating with the masses extended, drops the masses, describe the subsequent motions of the weights and of the man. What is the man's new angular momentum? Will his angular velocity change?

12. If the tension in the rope is 2 kN and the collar B is fixed to the 15 kg bar and the system is in equilibrium,

    a. Draw a free body diagram showing all forces on the bar AB.

    b. Find the reaction forces and torques at A.

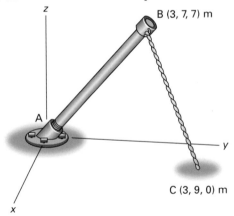

4   Rotation matrices

The form of tensor quantities, such as the moment of inertia I, depend on the coordinate system that is used to define them. For example, we have written an expression for the moment of inertia of a cylinder in the case where the cylindrical axis is the z-axis as $I_{zz} = MR^2/2$ and $I_{xx} = I_{yy} = Mh^2/12 + MR^2/4$. If the cylindrical axis instead is the y-axis, clearly we would have $I_{yy} = MR^2/2$ and $I_{xx} = I_{zz} = Mh^2/12 + MR^2/4$. If the coordinate system is not aligned with the cylinder's axis of symmetry, I is not diagonal and I_{xy}, I_{xz}, I_{yx}, I_{yz}, I_{zx}, and I_{zy} are not zero. Because the moment of inertia tensor is *symmetric* ($I_{xy} = I_{yx}$, $I_{xz} = I_{zx}$, and $I_{zy} = I_{yz}$) and the elements are real, there is always at least one coordinate system (the *principal axes system*) where I is diagonal. However, it is sometimes useful to select a coordinate system which is not aligned with the principal axes because, for example, we want one of the axes to align with a known torque or significant angular momentum axis. Since we will be studying many other tensor material properties – such as stress, strain, conductivity, permittivity, and or permeability – where these same considerations arise, it is important that we understand how a tensor quantity can be transformed into an arbitrary coordinate system by using *rotation matrices*.

4.1   Vectors in rotated coordinate systems

Consider a vector $[x_1, y_1]$ defined in terms of $\hat{x}$ and $\hat{y}$ axes that we want to describe in a new coordinate system with $\widehat{x'}$ and $\widehat{y'}$ axes that are rotated around the z-axis by an angle θ as shown in Figure 4.1.

Unit vectors (length $= 1$) in the $\widehat{x'}$ and $\widehat{y'}$ directions can be written as

$$\widehat{x'} = \cos\theta\,\hat{x} + \sin\theta\,\hat{y} \tag{4.1a}$$

$$\widehat{y'} = -\sin\theta\,\hat{x} + \cos\theta\,\hat{y}. \tag{4.1b}$$

Taking a dot prodect of the vector $[x_1, y_1]$ with $\widehat{x'}$ and $\widehat{y'}$, we find the projection onto the $\widehat{x'}$ and $\widehat{y'}$ axes

$$x_1' = [x_1\,\hat{x} + y_1\,\hat{y}]\cdot[\cos\theta\,\hat{x} + \sin\theta\,\hat{y}] = x_1\,\cos\theta + y_1\,\sin\theta \tag{4.2a}$$

$$y_1' = [x_1\,\hat{x} + y_1\,\hat{y}]\cdot[-\sin\theta\,\hat{x} + \cos\theta\,\hat{y}] = -x_1\,\sin\theta + y_1\,\cos\theta. \tag{4.2b}$$

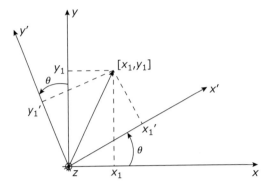

Figure 4.1 The primed axes are rotated counter-clockwise around the z-axis by an angle θ. The vector (x_1, y_1) will be described in the primed coordinates as (x_1', y_1').

If the vector has a component in the z-direction, the z_1' value is the equal to the z_1 value since the rotation around the z-axis doesn't change the projections in the $\hat{z} = \hat{z'}$ direction. These equations

$$x_1' = x_1 \cos\theta + y_1 \sin\theta$$
$$y_1' = -x_1 \sin\theta + y_1 \cos\theta$$
$$z_1' = z_1$$

can be expressed as a matrix equation

$$\begin{bmatrix} x_1' \\ y_1' \\ z_1' \end{bmatrix} = \begin{bmatrix} \cos\theta & \sin\theta & 0 \\ -\sin\theta & \cos\theta & 0 \\ 0 & 0 & 1 \end{bmatrix} \begin{bmatrix} x_1 \\ y_1 \\ z_1 \end{bmatrix}. \tag{4.3}$$

We see that multiplication of a vector by the *rotation matrix*

$$\boldsymbol{R}_z = \begin{bmatrix} \cos\theta & \sin\theta & 0 \\ -\sin\theta & \cos\theta & 0 \\ 0 & 0 & 1 \end{bmatrix} \tag{4.4}$$

corresponds to transformation of the vector into a new coordinate system rotated by an angle of θ around the z-axis. For example, if $\theta = \pi/2$ so that the y'-axis is the negative x-axis and the x'-axis is the y-axis we find

$$\begin{bmatrix} x_1' \\ y_1' \\ z_1' \end{bmatrix} = \begin{bmatrix} 0 & 1 & 0 \\ -1 & 0 & 0 \\ 0 & 0 & 1 \end{bmatrix} \begin{bmatrix} x_1 \\ y_1 \\ z_1 \end{bmatrix} = \begin{bmatrix} y_1 \\ -x_1 \\ z_1 \end{bmatrix}$$

as expected.

Question 4.1

What do you think the rotation matrix would be for rotation by an angle θ around the x-axis? For rotation around the y-axis?

4.2   Tensors in rotated coordinate systems

How does rotation of a system affect tensor properties? Consider a moment of inertia tensor I expressed in a particular coordinate system, with x-, y-, and z-axes directed in a particular direction in space. We can transform the tensor into a coordinate system with axes – call them x', y', and z' – rotated from the original using rotation matrices. In the unprimed coordinate system we have I defined by

$$\vec{\tau} = I\,\vec{\alpha}\ . \tag{4.5}$$

We can express the vectors and $\vec{\tau}$ and $\vec{\alpha}$ in the new coordinate system by using the rotation matrix R

$$R\,\vec{\tau} = I'R\,\vec{\alpha} \tag{4.6}$$

where I' is the rotation inertia tensor in the new coordinate system.

We can operate on both sides of this equation by the inverse of the rotation matrix R^{-1} defined by the relation: $R^{-1}\,R = \mathbb{I} = \begin{bmatrix} 1 & 0 & 0 \\ 0 & 1 & 0 \\ 0 & 0 & 1 \end{bmatrix}$. The product of R^{-1} and R is the *identity matrix* $\mathbb{I}$ that has the properties that $\mathbb{I}\ \vec{V} = \vec{V}$ and $\mathbb{I}\,M = M$ where $\vec{V}$ is any vector and M is any matrix. This gives us

$$R^{-1}\,R\ \vec{\tau} = \mathbb{I}\ \vec{\tau} = \vec{\tau} = \left[R^{-1}\,I'R \right]\vec{\alpha}\ . \tag{4.7}$$

Comparing this with our original equation $\vec{\tau} = I\,\vec{\alpha}$ we identify the matrix $[R^{-1}\,I'\,R] = I$. Operating on the left by R and the right by R^{-1}, we find the moment of inertia tensor rotated into the new $[x', y', z']$ coordinate system is $I' = [R\,I\,R^{-1}]$.

It is easy to demonstrate that for a rotation around the z-axis, where

$$R_z = \begin{bmatrix} \cos\theta & \sin\theta & 0 \\ -\sin\theta & \cos\theta & 0 \\ 0 & 0 & 1 \end{bmatrix}$$

the inverse of R is

$$R_z^{-1} = \begin{bmatrix} \cos\theta & -\sin\theta & 0 \\ \sin\theta & \cos\theta & 0 \\ 0 & 0 & 1 \end{bmatrix} = R_z^{\ T}. \tag{4.8}$$

The *inverse* of a rotation matrix is equal to the *transpose* matrix.

> **Question 4.2**
>
> Demonstrate that the transpose matrix given above is the inverse of R_z.

The rotation matrix for a rotation around the x-axis is given by

$$R_x = \begin{bmatrix} 1 & 0 & 0 \\ 0 & \cos\theta & \sin\theta \\ 0 & -\sin\theta & \cos\theta \end{bmatrix} \tag{4.9}$$

and the inverse matrix for rotation around the x-axis is

$$R_x^{-1} = R_x^T = \begin{bmatrix} 1 & 0 & 0 \\ 0 & \cos\theta & -\sin\theta \\ 0 & \sin\theta & \cos\theta \end{bmatrix}. \tag{4.10}$$

For rotation around the y-axis, we have

$$R_y = \begin{bmatrix} \cos\theta & 0 & \sin\theta \\ 0 & 1 & 0 \\ -\sin\theta & 0 & \cos\theta \end{bmatrix} \tag{4.11}$$

and

$$R_y^{-1} = R_y^T = \begin{bmatrix} \cos\theta & 0 & -\sin\theta \\ 0 & 1 & 0 \\ \sin\theta & 0 & \cos\theta \end{bmatrix}. \tag{4.12}$$

Any arbitrary rotation can be expressed as successive rotations around the x-axis, followed by a rotation around the (new) y-axis, followed by a rotation around the (new) z-axis. In each case the rotation acts on the tensor through left-multiplying by the rotation matrix and right-multiplying by its inverse (transpose)

$$I' = \begin{bmatrix} R & I & R^{-1} \end{bmatrix}. \tag{4.13}$$

Example 4.1 Consider an object rotating in an $(x–y–z)$ coordinate system with components of the angular velocity $\omega_x = 8$ rad/s, $\omega_y = 4$ rad/s, and $\omega_z = 0$. (a) Find a coordinate system $(x'–y'–z')$ where this motion is entirely rotation around the y' axis. What is the net angular velocity around the y' axis? (b) Repeat for the case where $\vec{\omega} = \begin{bmatrix} 8 \\ 4 \\ 3 \end{bmatrix}$.

Solution: (a) Rotation around the z-axis will not change the z-component of the angular velocity. The angular velocity in a coordinate system rotated by an angle ϕ around the z-axis will be given by

$$\vec{\omega}' = \begin{bmatrix} \cos\phi & \sin\phi & 0 \\ -\sin\phi & \cos\phi & 0 \\ 0 & 0 & 1 \end{bmatrix} \begin{bmatrix} 8 \\ 4 \\ 0 \end{bmatrix} = \begin{bmatrix} 8\cos\phi + 4\sin\phi \\ -8\sin\phi + 4\cos\phi \\ 0 \end{bmatrix}.$$

Continued

Example 4.1 (*cont.*)

If we pick an angle ϕ where $\omega'_x = 0$, we find $8\cos\phi + 4\sin\phi = 0 \Rightarrow \tan\phi = -2$. This

gives $\phi = -1.107$ radians with a rotation matrix $R_1 = \begin{bmatrix} 0.447 & -0.894 & 0 \\ 0.894 & 0.447 & 0 \\ 0 & 0 & 1 \end{bmatrix}$ and

$$\vec{\omega}' = R_1 \begin{bmatrix} 8 \\ 4 \\ 0 \end{bmatrix} = \begin{bmatrix} 0 \\ 8.9443 \\ 0 \end{bmatrix} \text{ with angular frequency of } |\vec{\omega}'| = 8.9443 \text{ rad/s.}$$

(b) With $\vec{\omega} = \begin{bmatrix} 8 \\ 4 \\ 3 \end{bmatrix}$ we can use the same R_1 to rotate -1.107 radians around the z-

axis giving $\vec{\omega}' = R_1 \begin{bmatrix} 8 \\ 4 \\ 3 \end{bmatrix} = \begin{bmatrix} 0 \\ 8.9443 \\ 3 \end{bmatrix}$. Next we rotate an angle of θ around the

x'-axis (since rotating around the z'-axis will not change the z' component!)
to give us the angular velocity in the $[x''-y''-z'']$ coordinate system $\vec{\omega}'' =$

$$\begin{bmatrix} 1 & 0 & 0 \\ 0 & \cos\theta & \sin\theta \\ 0 & -\sin\theta & \cos\theta \end{bmatrix} \begin{bmatrix} 0 \\ -\omega_x\sin\phi + \omega_y\cos\phi \\ \omega_z \end{bmatrix} = \begin{bmatrix} 0 \\ \cos\theta(-\omega_x\sin\phi + \omega_y\cos\phi) + \sin\theta\,\omega_z \\ -\sin\theta(-\omega_x\sin\phi + \omega_y\cos\phi) + \cos\theta\,\omega_z \end{bmatrix}.$$

Setting $\omega''_z = 0$ gives $\tan\theta = \frac{\omega_z}{(-\omega_x\sin\phi + \omega_y\cos\phi)} = \frac{3}{8.9443} \Rightarrow \theta = 0.324$ radians.
This gives

$$\vec{\omega}'' = R_2 R_1 \begin{bmatrix} 8 \\ 4 \\ 3 \end{bmatrix} = \begin{bmatrix} 1 & 0 & 0 \\ 0 & 0.948 & 0.318 \\ 0 & -0.318 & 0.948 \end{bmatrix} \begin{bmatrix} 0.447 & -0.894 & 0 \\ 0.894 & 0.447 & 0 \\ 0 & 0 & 1 \end{bmatrix} \begin{bmatrix} 8 \\ 4 \\ 3 \end{bmatrix}$$

$$= \begin{bmatrix} 0 \\ 9.434 \\ 0 \end{bmatrix} \text{rad/s.}$$

Note: The counter-clockwise angle of rotation around the z-axis ϕ and the angle of the subsequent counter-clockwise rotation around the x'-axis θ are two of the three *Euler angles* that describe the rotation of a solid body into any new body-centered coordinate system. The third Euler angle is a counter-clockwise rotation of ψ around the z'' axis.

Example 4.2 Find the angular momentum of a 2500 kg cylinder with radius $R = 3$ m and height $h = 12$ m rotating around a wire through the center of mass with an angular velocity $\omega = 0.5$ rad/s 30° from the axis of the cylinder, as in the figure.

Continued

Example 4.2 (*cont.*)

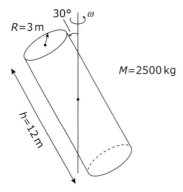

Solution: The principal axes of the cylinder are aligned with the cylinder axis in the *x–y–z* coordinate system shown in the next figure. In that coordinate system the moment of inertia is (as shown before)

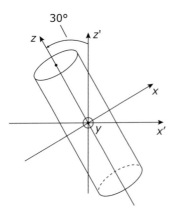

$$I = \begin{bmatrix} 35675 & 0 & 0 \\ 0 & 35675 & 0 \\ 0 & 0 & 11250 \end{bmatrix}.$$

To align the coordinate system with the axis of rotation at the instant shown, we would need to rotate the inertial tensor by 30° around the *y*-axis, which is directed into the page, to bring the *z*-axis into the *z'*-axis and *x*-axis into the *x'*-axis. The rotation matrix to rotate 30° around the *y*-axis is

$$R_y = \begin{bmatrix} \cos 30° & 0 & \sin 30° \\ 0 & 1 & 0 \\ -\sin 30° & 0 & \cos 30° \end{bmatrix}.$$

Continued

Example 4.2 (*cont.*)

The inertial tensor in the $x'-y'-z'$ coordinates is

$$I' = R_y I R_y^T = \begin{bmatrix} 0.866 & 0 & 0.500 \\ 0 & 1 & 0 \\ -0.500 & 0 & 0.866 \end{bmatrix} \begin{bmatrix} 35675 & 0 & 0 \\ 0 & 35675 & 0 \\ 0 & 0 & 11250 \end{bmatrix} \begin{bmatrix} 0.866 & 0 & -0.500 \\ 0 & 1 & 0 \\ 0.500 & 0 & 0.866 \end{bmatrix}.$$

Using MATLAB to do the matrix multiplication, we find

$$I' = \begin{bmatrix} 29569 & 0 & -10576 \\ 0 & 35675 & 0 \\ -10576 & 0 & 17356 \end{bmatrix}.$$

At the instant shown, the angular momentum is given by

$$\vec{L} = I' \, \vec{\omega} = I' \begin{bmatrix} 0 \\ 0 \\ 0.5 \end{bmatrix} = \begin{bmatrix} -5288 \\ 0 \\ 8678 \end{bmatrix} \text{ kg m}^2/\text{s}.$$

We note that the angular momentum is not parallel to the angular velocity $\vec{\omega}$ – in addition to the component in the z' direction, the angular momentum has a component in the x' direction which will change in time (and a component in the y' direction will appear) as the cylinder rotates around the z' axis. To keep the inertial tensor in the $x'-y'-z'$ coordinate space, the tensor needs to be rotated around the z' axis by an angle ωt **after** it is rotated around the y-axis. (Note that rotating around the z-axis **before** rotating around the y-axis will not change the inertial tensor because it is symmetric around the z-axis in its principal axis frame.)

Example 4.3 Find the moment of inertia as a function of time of the cylinder in Example 4.1.
  Solution: The rotation matrix for rotating an angle ωt around the z' axis is

$$R_{z'} = \begin{bmatrix} \cos \omega t & \sin \omega t & 0 \\ -\sin \omega t & \cos \omega t & 0 \\ 0 & 0 & 1 \end{bmatrix}.$$

Applying this rotation to the inertial tensor I' we find the moment of inertia as a function of time

$$I'(t) = R_{z'} I' R_{z'}^T$$
$$= \begin{bmatrix} 29569\cos^2\omega t + 35675\sin^2\omega t & 6102\cos\omega t \sin\omega t & -10576\cos\omega t \\ 6102\cos\omega t \sin\omega t & 29569\sin^2\omega t + 35675\cos^2\omega t & 10576\sin\omega t \\ -10576\cos\omega t & -10576\sin\omega t & 17356 \end{bmatrix}.$$

Continued

Example 4.3 (*cont.*)

This yields the angular momentum as a function of time as

$$\vec{L}(t) = \mathbf{I}'(t)\,\vec{\omega} = \begin{bmatrix} -5288\ \cos\ \omega t \\ 5288\ \sin\ \omega t \\ 8678 \end{bmatrix}.$$

Thus, the cylinder must have a torque on it due to the wire

$$\vec{\tau} = \frac{d\vec{L}}{dt} = \begin{bmatrix} 5288\ \omega\ \sin\ \omega t \\ 5288\ \omega\ \cos\ \omega t \\ 0 \end{bmatrix}.$$

If the torque is released, the rotation of the cylinder will change to maintain the angular momentum constant. For example, if the torque is released at $t = 0$, the angular momentum will be constant at $\vec{L} = \begin{bmatrix} -5288 \\ 0 \\ 8678 \end{bmatrix}$ kg m^2/s.

4.3   Semiconductor effective masses[1]

In free space particle masses are isotropic and can be represented by a scalar mass m. This is not the case for electrons in solids. As we shall see when we consider quantum theory, electrons have wave properties, and the interaction of these "matter waves" with the lattice results in an energy–momentum relation – the material "band structure" – which is far more complex than the simple free particle result $E = \frac{p^2}{2m}$ and depends on the direction of the momentum with respect to the crystal axes. As a result, when a force is applied to an electron in a crystal the electron acceleration is not necessarily parallel to the force, and we need to generalize the concept of mass to a tensor "effective mass" $\mathbf{m}$.

As an example, the energy bands in silicon – the allowed values of energy as a function of momentum – are shown in Figure 4.2 for energy E as a function of momentum in the x-direction p_x. At $T = 0$ K (absolute zero), all of the electron states below a value of energy known as the *Fermi energy* E_f are filled and all the states above E_f are empty. The bands below the Fermi energy are known as *valence bands*, while those above E_f are the *conduction bands*. If, as a result of heating or the addition of impurity atoms ("doping"), electrons are introduced into the conduction bands, they will tend to collect near the points of minimum energy in the bands into

---

[1] This section and the next involve concepts that are fully developed in Chapter 11. The sections can be considered optional and postponed for consideration until after Chapter 11, if desired.

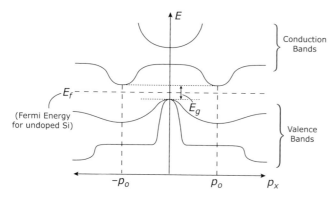

Figure 4.2 Energy vs. momentum for electrons in a silicon lattice. The states below E_f are usually filled with electrons and the states above E_f are usually empty. If the material is heated or doped with electron-rich impurities, electrons will collect in the lower energy states near $p = p_o$ and $p = -p_o$. The energy gap E_g is the energy separation between the top of the normally full valence bands and the bottom of the normally empty conduction bands.

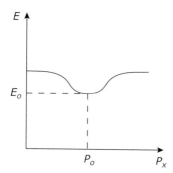

Figure 4.3 Expanded view of energy vs. momentum around one of the low energy points in the silicon conduction bands.

electron "valleys" or "pockets." In silicon, as shown in Figure 4.2, the minima of the conduction band are located at values of p_x which are not zero.

Question 4.3

What is the equation for energy in terms of momentum for a free particle of (scalar) mass m where the only energy is kinetic energy? What does the energy vs. momentum plot for a free particle of mass m look like? In three dimensions, what would a surface of constant energy look like for a free particle when plotted on (p_x, p_y, p_z) axes?

If we look at one conduction band minimum, as in Figure 4.3, we can expand the energy–momentum relation around p_o in a Taylor series as

$$E = E(p) = E_o + \frac{dE}{dp}\bigg]_{p=p_o} (p - p_o) + \frac{1}{2}\frac{d^2E}{dp^2}\bigg]_{p=p_o} (p - p_o)^2 + \dots \tag{4.14}$$

At a band minimum, we have the derivative of E with respect to p equal to zero, so the first-order term goes away, and we are left with

$$E = E_o + \frac{1}{2}\frac{d^2 E}{dp^2}\bigg]_{p_o} (p - p_o)^2 + \cdots \qquad (4.15)$$

Measuring the energy and momentum with respect to E_o and p_o, we can choose E_o and p_o both to be equal to zero, and we end up with

$$E = \frac{1}{2}\frac{d^2 E}{dp^2}\bigg]_{p_o} p^2 \qquad (4.16)$$

where we neglect the higher-order terms. Comparing this with the free particle equation for kinetic energy $E = \frac{p^2}{2m}$, it is natural to define an *effective mass* $m*$ by

$$m^* = \frac{1}{d^2 E/dp^2} \qquad (4.17)$$

where the second derivative is understood to be evaluated near the band minima. Since it is unlikely that the band curvature will be the same in all directions, the effective mass needs to be expressed as a tensor. In the *principal axis* coordinate system, with x, y, and z along symmetry directions in the crystal, we have

$$m^* = \begin{bmatrix} m_{xx} & 0 & 0 \\ 0 & m_{yy} & 0 \\ 0 & 0 & m_{zz} \end{bmatrix} \qquad (4.18)$$

where the tensor components relate to the band curvature along the different crystal directions:

$$m_{xx} = \frac{1}{d^2 E/dp_x^2}, \quad m_{yy} = \frac{1}{d^2 E/dp_y^2}, \quad m_{zz} = \frac{1}{d^2 E/dp_z^2}. \qquad (4.19)$$

Electrons in the conduction band of silicon, that has a crystal with cubic symmetry, are found in six electron pockets directed along the crystal [100], [010], and [001] axes. The effective mass tensor of each pocket is given by

$$m^*_{Si} = \begin{bmatrix} m_\perp & 0 & 0 \\ 0 & m_\perp & 0 \\ 0 & 0 & m_\| \end{bmatrix} \qquad (4.20)$$

where the mass parallel to the cubic axis (taken as the local z-axis in each case) $m_\| = 0.98\, m_o$ and the mass transverse to the axis is $m_\perp = 0.19\, m_o$. (The free electron mass m_o is 9.11×10^{-31} kg.)

4.4   Conductivity effective mass

To see what effect this effective mass has in calculating physically measurable quantities, let's consider the conductivity of a semiconductor, the measure of how much

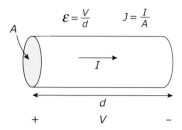

Figure 4.4 The relation of the current I and current density J and the voltage V and electric field $\vec{\mathcal{E}}$ in terms of sample dimensions d and A.

current flows in response to an electric field. The *microscopic form* of Ohm's Law relates the *current density J* to the *electric field $\mathcal{E}$*. Current density is defined as the current (in amperes $=$ coulombs/s) divided by the perpendicular cross-section area of the wire the current is traveling in. In Figure 4.4, for example, the current density in a wire of circular cross-section area is $\vec{J} = \frac{\vec{I}}{\pi r^2}$. The units of current density are amperes/m^2, and current density (and current) are vectors because they have a direction associated with them – the direction of current flow.

The microscopic form of Ohm's Law is given by

$$\vec{J} = \sigma \ \vec{\mathcal{E}} = \frac{1}{\rho} \ \vec{\mathcal{E}} \tag{4.21}$$

where σ is the *conductivity* of the material, measured in siemens/m $= 1/(\Omega \ \text{m})$. (The reciprocal of σ is the *resistivity ρ* measured in Ω m).

We can recover the usual form of Ohm's Law by observing that the *potential difference* or *voltage V* between two points is defined as the integral of the dot product of the electric field with the displacement along a path between two points, $V_{AB} = \int_B^A \vec{\mathcal{E}} \cdot d\vec{r}$. In the geometry of the figure with a current flowing through a wire of constant cross-section area A and length d, we have

$$J = \frac{I}{A} = \frac{1}{\rho} \mathcal{E} = \frac{1}{\rho} \frac{V}{d} \quad \Rightarrow I = \frac{V}{\rho \, d/A} = \frac{V}{R}.$$

The last equation is Ohm's Law giving the current in terms of the voltage difference and the *resistance R* in ohms. In terms of the length d and cross-section area A of the wire segment, $R = \rho \frac{d}{A} = \frac{1}{\sigma} \frac{d}{A}$. The resistivity ρ (or conductivity σ) are properties of the material, whereas the resistance also includes the geometry of the conductor.

Question 4.4

The conductivity of copper at 20 °C is $\sigma = 5.96 \times 10^7$ S/m. What would the resistance be of a 3-meter long copper wire 1mm in diameter?

The microscopic expression for conductivity is given by

$$\sigma = \frac{n\,q^2\tau}{m} \tag{4.22}$$

where n is the number of electrons per cubic meter, $q = 1.602 \times 10^{-19}$ coulomb is the charge on the electron, the *electron scattering time* τ is the average time between electron scattering events with impurities or thermal vibrations, and m is the electron mass.

Question 4.5

The conductivity of copper decreases when the temperature increases. Would you expect this from Eq. (4.22)? In a semiconductor such as silicon, on the other hand, the conductivity increases when the temperature increases. Why do you think this might be?

If we consider an electron in a semiconductor, the important parameters here are the electron density n, the electron scattering time τ, and the *conductivity effective mass m**. If we have several different electron pockets with different masses along the electric field direction, the currents from each pocket will add, and the total current will be

$$\vec{J}_T = \sum_i \vec{J}_i = \sum_i \sigma_i\,\vec{\mathcal{E}} = \sum_i \frac{n_i\,q^2\tau_i}{m^*_i}\,\vec{\mathcal{E}}. \tag{4.23}$$

The sum here is over the i pockets with n_i electrons with effective mass m^*_i and scattering time τ_i in each pocket. Assuming that the scattering time doesn't depend on the electron pocket, we can define the conductivity effective mass from the following equation:

$$\frac{1}{m^*} = \frac{1}{n} \sum_i \frac{n_i}{m^*_i}. \tag{4.24}$$

The effective mass is an important device parameter because it controls how fast the electron moves in response to an applied electric field – and thus the frequency response of the material. All other things being equal, a semiconductor with a small conductivity effective mass is thus a good material for high-frequency devices.

In general, the conductivity can be different in different directions. For example, in layered graphite, the conductivity is much higher along the layers than perpendicular to the layers. Thus, in general the conductivity should be represented as a tensor quantity.

$$\sigma = \sum_i \frac{n_i q^2\tau}{m_i} = \sum_i n_i\,m_i^{-1}\,q^2\,\tau \tag{4.25}$$

where m_i^{-1} is the inverse mass tensor. In the conductivity principal axes, the conductivity tensor is diagonal:

$$\sigma = \begin{bmatrix} \sigma_{xx} & 0 & 0 \\ 0 & \sigma_{yy} & 0 \\ 0 & 0 & \sigma_{zz} \end{bmatrix}. \tag{4.26}$$

In general, $\sigma_{xx} \neq \sigma_{yy} \neq \sigma_{zz}$, but for cubic crystals, like silicon, the conductivity principal axes are aligned with the cubic axes, and, since there is no difference in the crystal symmetry along the x-, y-, and z-directions, $\sigma_{xx} = \sigma_{yy} = \sigma_{zz} = \sigma$, and the conductivity is isotropic:

$$\boldsymbol{\sigma} = \sigma \begin{bmatrix} 1 & 0 & 0 \\ 0 & 1 & 0 \\ 0 & 0 & 1 \end{bmatrix} = \sigma \, \mathbb{I} \tag{4.27}$$

where $\mathbb{I}$ is the identity matrix. The conductivity may not be isotropic in non-cubic or stressed crystals where the directions are non-equivalent.

Since silicon is a cubic crystal, the total conductivity must be isotropic. However, this is not true for the conductivity of each pocket. Figure 4.5 shows the six electron pockets in silicon where we have sketched the constant energy surfaces in momentum space (p_x–p_y–p_z space). Each pocket is highly anisotropic with a much larger mass parallel to the pocket axis than perpendicular to the pocket axis. The momentum difference $p - p_o$ needs to be greater along the pocket axis than perpendicular to the pocket axis to have the same energy $E = \frac{p^2}{2\,m^*}$. In the "pocket" coordinate system with the z-direction along the direction of the pocket axis (as in pockets "a" in the diagram), we can describe the inverse effective mass tensor as

$$\boldsymbol{m}_a^{-1} = \begin{bmatrix} 1/m_\perp & 0 & 0 \\ 0 & 1/m_\perp & 0 \\ 0 & 0 & 1/m_\parallel \end{bmatrix} \tag{4.28}$$

where, in silicon, $m_\parallel = 0.98\,m_o$, $m_\perp = 0.19\,m_o$, and m_o is the free electron mass, 9.11×10^{-31} kg.

To find the total conductivity inverse mass, we need to multiply $\boldsymbol{m}_a^{-1}$, $\boldsymbol{m}_b^{-1}$, and $\boldsymbol{m}_c^{-1}$ by the fraction of electrons in each pocket and add them *in the same coordinate system*. To get the pockets in a common coordinate system we need to rotate the "c"

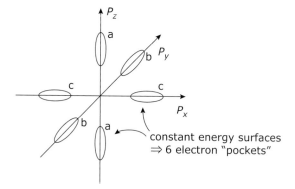

Figure 4.5 Constant energy surfaces in "momentum space" for electrons in the conduction band of silicon.

pockets 90° around the y-axis and the "b" pockets 90° around the x-axis. Since one third of the carriers are in each of the "a," "b," and "c" pockets, we will multiply each inverse mass by 1/3 and add them:

$$
m^{*-1} = \frac{1}{3} m_a^{-1} + \frac{1}{3}
\begin{bmatrix}
1 & 0 & 0 \\
0 & \cos 90 & \sin 90 \\
0 & -\sin 90 & \cos 90
\end{bmatrix}
\begin{bmatrix}
1/m_\perp & 0 & 0 \\
0 & 1/m_\perp & 0 \\
0 & 0 & 1/m_\parallel
\end{bmatrix}
\begin{bmatrix}
1 & 0 & 0 \\
0 & \cos 90 & -\sin 90 \\
0 & \sin 90 & \cos 90
\end{bmatrix}
$$

$$
+ \frac{1}{3}
\begin{bmatrix}
\cos 90 & 0 & \sin 90 \\
0 & 1 & 0 \\
-\sin 90 & 0 & \cos 90
\end{bmatrix}
\begin{bmatrix}
1/m_\perp & 0 & 0 \\
0 & 1/m_\perp & 0 \\
0 & 0 & 1/m_\parallel
\end{bmatrix}
\begin{bmatrix}
\cos 90 & 0 & -\sin 90 \\
0 & 1 & 0 \\
\sin 90 & 0 & \cos 90
\end{bmatrix}
$$

$$
= \frac{1}{3}
\begin{bmatrix}
2/m_\perp + 1/m_\parallel & 0 & 0 \\
0 & 2/m_\perp + 1/m_\parallel & 0 \\
0 & 0 & 2/m_\perp + 1/m_\parallel
\end{bmatrix}
= \frac{2/m_\perp + 1/m_\parallel}{3} \mathbb{I}.
$$

$$(4.29)$$

Thus, for silicon we recover an isotropic effective mass (as we must in a cubic crystal) with an effective mass value:

$$
m^* = \frac{3}{2/m_\perp + 1/m_\parallel} = 0.26 \, m_o. \tag{4.30}
$$

Chapter 4 problems

1. A cylinder has an inertial tensor $I = \begin{bmatrix} 9 & 0 & 0 \\ 0 & 9 & 0 \\ 0 & 0 & 6 \end{bmatrix}$ kg m^2 in the frame of reference with the z-direction along the axis of the cylinder. The cylinder is rotated around its x-axis so that the axis of the cylinder is $\alpha = 40°$ from vertical. (See figure: the x-axis is perpendicular to the page.)

   a. What is the inertial tensor of the cylinder in the coordinate system with the z-axis in the vertical direction?

   b. If the cylinder is then rotated around an axis in the vertical direction through its center of mass at 5 revolutions/s, what is the rotational kinetic energy of the cylinder?

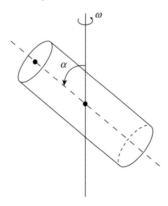

2. The stress tensor for pure shear stress along the y-axis in the x–y coordinate system

is $\sigma = \begin{pmatrix} 0 & F_{xy}/A & 0 \\ F_{xy}/A & 0 & 0 \\ 0 & 0 & 0 \end{pmatrix}$. The rotation matrix to rotate the tensor into an x'–y'

coordinate system at an angle of $45°$ to the x–y axes is given by

$$R_z = \begin{bmatrix} \cos 45° & \sin 45° & 0 \\ -\sin 45° & \cos 45° & 0 \\ 0 & 0 & 1 \end{bmatrix} = \begin{bmatrix} 1/\sqrt{2} & 1/\sqrt{2} & 0 \\ -1/\sqrt{2} & 1/\sqrt{2} & 0 \\ 0 & 0 & 1 \end{bmatrix}.$$

Find $\sigma' = R_z \, \sigma R_z^T$, the stress tensor in the new x'–y' coordinate system.

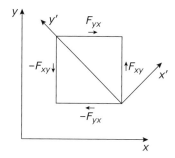

3. In electromagnetic theory the electric displacement vector $\vec{D}$ is related to the electric field vector $\vec{E}$ by $\vec{D} = \varepsilon \, \vec{E}$ by where the *dielectric permittivity* ε is a tensor quantity. Crystal calcite ($CaCO_3$) is an anisotropic dielectric material with uniaxial symmetry. The dielectric permittivity at optical frequencies ($\lambda = 560$ nm) is given by:

$$\epsilon = \epsilon_o \begin{bmatrix} 2.211 & 0 & 0 \\ 0 & 2.211 & 0 \\ 0 & 0 & 2.756 \end{bmatrix},$$

where the z-axis is along the symmetry axis of the crystal. Consider a crystal with the surface normal inclined at $20°$ from the x-axis toward the z-axis. Find the dielectric tensor in the coordinate system with the new x'-direction normal to the surface. For $\vec{E}$ parallel to the surface of this crystal, is $\vec{D}$ parallel to $\vec{E}$?

4. MATLAB Project: The electron bands in germanium have their minimum in the $[111]$ direction ($p_x = p_y = p_z$). There are effectively four electron pockets (eight half pockets) as shown in the figure below. The mass along the $[111]$ directions is $1.64 \, m_o$ and the mass perpendicular to $[111]$ is $0.082 \, m_o$ (where m_o is the free electron mass). So the effective mass tensor is given by:

$$m^* = \begin{bmatrix} 0.082 & 0 & 0 \\ 0 & 0.082 & 0 \\ 0 & 0 & 1.64 \end{bmatrix} m_o,$$

where the z-axis of the coordinate system is taken along the $[111]$ direction. Using MATLAB, apply appropriate rotations to the four inverse mass tensors to bring their z-axes parallel to the [001] axis as shown. Use the MATLAB `cross ()` cross product `dot ()` functions to find the crystal direction of the rotation axes perpendicular to the pocket z-axis and the crystal z-axis and the angle of rotation. Make sure that the rotated pockets have the same x–y axis directions (rotate around the new z if necessary) and add the transformed inverse effective mass tensors to find a net effective conductivity mass tensor in terms of m_o. Is the result what you expect? (Germanium is a cubic crystal with principal axes along the [100], [010], and [001] directions.)

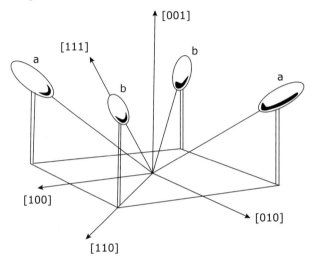

5 Materials properties: elasticity

Once fundamental scientific laws are understood, engineering technology depends on advances in materials – materials science – more than any other factor. Historically, ages have been named for the most advanced materials technology available: Stone Age, Bronze Age, Iron Age. If you look around the room you are in, you will recognize that our current age is the age of manufactured materials: plastics, ceramics, acrylics, polyester fibers, etc. Underlying the "Information Age" are the advances in electronic materials that have made possible the computers, displays, and communications equipment that enable information transfer and processing. Selecting materials with desirable properties and then processing to enhance these properties is now standard practice. The next generation of materials will include nanoscale manipulation and fabrication of materials unlike anything found in nature. These new *nanomaterials* can be predicted to have properties exceeding any currently available materials.

Materials are characterized by *response functions* that describe the materials' response to applied forces or fields. For example, the electromagnetic properties of materials can be characterized by the *electric permittivity* (electric polarization of materials in response to an applied electric field), *electrical conductivity* (electrical current in response to an applied electric field), and *magnetic permeability* (magnetic polarization in response to an applied magnetic field). One of the fundamental mechanical properties is the *elasticity*, which describes the deformation of the material in response to applied forces. The elasticity, as with other response functions, depends on the physics and chemistry at the atomic scale and on the structure of the material at a micro- and mesoscale, which is larger than the atoms but smaller than the macroscopic sample. For example, crystal structure, grain boundaries, and dislocations are key determining factors for elastic, plastic, and fracture properties. There are three elastic regimes: tensile/compressive elasticity, shear elasticity, and bulk elasticity. These are all special cases of a generalized elastic tensor, but we will consider them separately to begin.

5.1 Tensile/compressive elasticity

Tensile or compressive elasticity is the phenomenon where a force causes a material to stretch or compress in the same direction as the applied force as shown in Figure 5.1. In a typical tensile elasticity experiment, one side of the sample – usually in a rod form

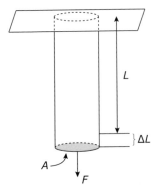

Figure 5.1 Elongation of a rod under tensile stress.

with a length L and cross-section area A – is fixed and force is applied to the other end while the change in length ΔL is measured. For relatively small applied force, in the *linear elastic range*, the change in length as a function of the force is linear. If the force is in a direction to stretch the rod ($L' = L + \Delta L$) the rod is in *tension*; if the force is in a direction to compress the rod ($L' = L - \Delta L$) the rod is in *compression*. For small ΔL, the slope for compression and tension are the same.

If the force is removed from the sample in the linear elastic range, it will return to its original length L. The elastic deformation force can be written as $\vec{F}_e = \left(E\frac{A}{L}\right)\Delta\vec{L}$ where E is the *Young's modulus*. In the linear elastic range where there is no permanent deformation and the change in length is reproducible, the work done to achieve any given ΔL does not depend on the path, and we can define an *elastic potential energy* in terms of the work to stretch or compress the material from length a to length b:

$$U_e = W_{ab} = \int_a^b \vec{F}_e \cdot d\vec{L'} = \int_0^{\Delta L} \left(E\frac{A}{L}\right)\Delta L \, d\Delta L = \frac{1}{2}\left(E\frac{A}{L}\right)\Delta L^2 \tag{5.1}$$

where we have set the initial extension $\Delta L = 0$. By including this elastic potential energy in the total energy the total mechanical energy $E = KE + U_e$ is conserved in the linear elastic region.

The elastic behavior of a material is sketched in Figure 5.2. If the force is increased beyond the *yield strength* of the material, the rod will undergo non-linear permanent deformation (*plastic deformation*). If the material has entered the plastic deformation region it will not return to its original length when the force is removed. Mechanical energy is not conserved in the plastic region – it is dissipated as permanent distortion and heat. Often the F vs. ΔL curve has a portion with a negative slope, as shown in Figure 5.2, and the length will continue to increase even if the force is kept constant. At some force, referred to as the *ultimate strength* of the material, the rod will break. Materials in which the yield strength is very close to the ultimate strength are called *brittle*. On the other hand, if there is a fairly large range of strain (or ΔL) where the material deforms without breaking, it is called *ductile*.

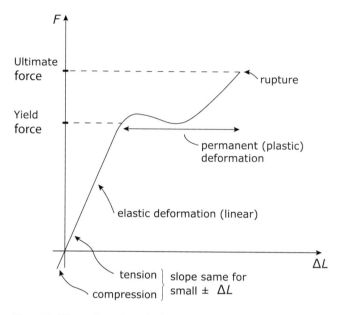

Figure 5.2 Elongation of a rod with tensile stress applied indicating the elastic deformation region and permanent deformation at higher applied force.

The amount by which a rod stretches or compresses depends on the length of the rod and the cross-section area as well as the material. To separate out these geometrical factors from the material properties, we can define the force divided by the rod cross-section area as the *stress* σ

$$\sigma = \frac{F}{A} \quad \left(\frac{\text{N}}{\text{m}^2}\right). \tag{5.2}$$

If we double the cross-section area of the rod, we need to double the force to get the same extension or compression.

Similarly, the deformation can be measured as the change in length divided by the unstressed length, a quantity known as *strain* ϵ

$$\epsilon = \frac{\Delta L}{L} \quad \text{(dimensionless)}. \tag{5.3}$$

The relation between stress and strain in the linear elasticity region is given by

$$\sigma = E\epsilon \tag{5.4}$$

where the constant of proportionality E (*Young's modulus*) is a property of the material. Since strain is dimensionless, the units of Young's modulus are the same as the units of stress, N/m^2. Since the strain is an effect for which stress is the cause, the inverse Young's modulus E^{-1} is sometimes used to express strain (the result) as a function of stress (the cause): $\epsilon = E^{-1}\sigma$.

Question 5.1

For a given applied force, is the elongation of a sample greater for a larger or a smaller Young's modulus E?

5.2 Shear elasticity

If a force is applied parallel to a surface, the stress can still be defined as the force divided by the area of the surface. In this case the deformation Δx will be parallel to the surface (perpendicular to the surface normal vector) and dependent on the length L of the object to the point where it is anchored on the side opposite to the surface where the force is applied as shown in Figure 5.3. In this case we can define the strain as

$$\epsilon = \frac{\Delta x}{L}. \tag{5.5}$$

The dimensionless quantity $\Delta x/L$ is the equal to the tangent of the deformation $\tan(\gamma)$, which for small deformation is the same as the angle γ. The proportionality between the shear stress F/A and the shear strain $\Delta x/L \cong \gamma$ is given by the *shear modulus G*, measured in units of N/m^2,

$$\frac{F}{A} = \sigma_{shear} = G\frac{\Delta x}{L} = G\gamma. \tag{5.6}$$

Example 5.1 A 20 cm long 4 cm square cross-section vertical steel rod supports a force of 50×10^3 N distributed uniformly across its upper surface directed at $15°$ from vertical. The bottom of the rod is rigidly fixed. If the Young's modulus of steel is $E = 200 \times 10^9$ N/m^2 and the shear modulus is $G = 80 \times 10^9$ N/m^2, find the elastic displacement of the upper surface of the rod, assuming that the horizontal displacement is entirely due to shear strain. (We will compare this to the displacement due to beam-bending in Problem 5.9.)

Solution: The total force is $\vec{F} = 50 \times 10^6 (\sin 15° \, \hat{x} - \cos 15° \, \hat{y})$ with the $\hat{y}$-direction vertical. The compression stress (the y-directed stress on the y-normal face) is $\sigma_{yy} = -\frac{50 \times 10^3 \cos 15°}{\pi(0.04)^2} = -30.2 \times 10^6$ N/m^2. The shear stress (the x-directed stress on the y-normal face) is $\sigma_{yx} = \frac{50 \times 10^3 \sin 15°}{\pi(0.04)^2} = 8.09 \times 10^6$ N/m^2. These stresses are less than the yield strength and ultimate strength of steel, 250 MPa and 400 MPa respectively.

The compressive strain is $\epsilon_{yy} = \sigma_{yy}/E = \frac{-30.2 \times 10^6}{200 \times 10^9} = -0.151 \times 10^{-3}$ which gives a displacement

$$\Delta y = (-0.151 \times 10^{-3})(0.20) = -30.2 \times 10^{-6} \text{ m} = -30.2 \text{ μm}.$$

The shear strain is $\epsilon_{yx} = \sigma_{yx}/G = \frac{8.09 \times 10^6}{80 \times 10^9} = 0.101 \times 10^{-3}$ yielding an x-displacement

$$\Delta x = (0.101 \times 10^{-3})(0.20) = 20.2 \times 10^{-6} \text{ m} = 20.2 \text{ μm}.$$

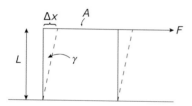

Figure 5.3 Shear strain of sample under the application of transverse (shear) stress.

The shear modulus characterizes the response to torsional torques that twist an object around an axis. The angular "twist" (in radians) of a cylindrical rod is given by

$$\theta = \frac{\tau L}{J_{zz} G} \tag{5.7}$$

in terms of the torque on the rod τ, the shear modulus G, the length of the rod L, and a geometrical factor J_{zz} which equals $\frac{1}{2} \pi r^4$ for a solid circular cylindrical rod of radius r. The shear modulus is also the parameter that describes the linear restoring force in helical springs which are formed so that a linear displacement is converted to a twisting force on the coils.

Example 5.2 Torsion pendulum

Consider a solid disk of radius $R = 4$ cm and mass $M = 0.3$ kg supported horizontally by a 25 cm long 0.5 mm diameter steel wire with a shear modulus $G = 8 \times 10^{10}$ N/m^2 connected to the center of the disk and the other end fixed. If the disk is rotating freely at an angular velocity $\omega = 6$ radians/s when $\theta = 0$, through what angle will the disk rotate before it comes to rest before beginning to rotate back in the other direction?

Solution: We will use the relation that the torque is equal to the time derivative of the angular momentum $\tau = \frac{d}{dt}(I\omega)$, where $I\omega$ is the angular momentum of the rotating disk in terms of its moment of inertia $I = \frac{MR^2}{2} = \frac{0.3(0.04)^2}{2} = 2.4 \times 10^{-4}$ kg m^2 and its angular velocity $\omega = \frac{d\theta}{dt}$. The torque exerted by the wire **which opposes the angular change** is given by $\tau = -\frac{J_{zz} G}{L}\theta = -\kappa\theta$ where the *torsional stiffness* $\kappa = \frac{J_{zz} G}{L} = \frac{\pi r^4 G}{2L} = \frac{\pi(0.25 \times 10^{-3})^4 (8 \times 10^{10})}{2(0.25)} = 0.02$ N m/radian. Relating the angular velocity to the angular deviation $\omega = \frac{d\theta}{dt}$ we find

$$\tau = -\kappa\theta = \frac{d}{dt}I\omega = I\frac{d^2\theta}{dt^2}.$$

In the next chapter we will be looking in detail at the solutions to the differential equations of the form

$$\frac{d^2\theta}{dt^2} = -\frac{\kappa}{I}\theta.$$

For this example we note only that a function $\theta(t)$ that yields $\left(-\frac{\kappa}{I}\right)\theta(t)$ when two time derivatives are taken and has $\theta(t) = 0$ at $t = 0$ is $\theta(t) = A \sin\left(\sqrt{\frac{\kappa}{I}}t\right)$. (You should

Continued

Example 5.2 (*cont.*)

demonstrate to your satisfaction that this $\theta(t)$ is a solution of the differential equation above!) From this solution, it follows that

$$\omega = \frac{d\theta}{dt} = A\sqrt{\frac{\kappa}{I}}\cos\left(\sqrt{\frac{\kappa}{I}}t\right).$$

At $\theta = 0$ when $t = 0$, we have $\omega(0) = A\sqrt{\frac{\kappa}{I}}\cos(0) = 6\,\mathrm{s}^{-1} \Rightarrow A = \frac{6}{\sqrt{\kappa/I}} = 2.10$. The maximum value of $\theta(t)$ when $\sqrt{\frac{\kappa}{I}}t = \frac{\pi}{2}$ is A, so the maximum angle of rotation is 2.10 radians $= 120.2°$.

5.3   Bulk elasticity (hydraulic stress)

Hydraulic stress is moderated through pressure in a fluid. Pressure is defined as force divided by area, $p = \frac{F}{A}$, which has the same form and dimension as stress (pressure is a negative compressive stress). By simple force considerations, as shown in Figure 5.4, the pressure on an infinitesimal fluid element can be shown to be the same in all directions in fluids that are at rest or moving at a constant velocity. A volume of material under hydraulic stress is thus acted upon by the same pressure, perpendicular to the surface, on all sides (assuming that the gravitational gradient due to the weight of the fluid is negligible). The response to the hydraulic stress can be defined as a dimensionless hydraulic strain, the change in volume divided by the initial volume $\Delta V/V$. The relation between the pressure change and the hydraulic strain is given by the *bulk modulus B*

$$\Delta p = -B\frac{\Delta V}{V}. \tag{5.8}$$

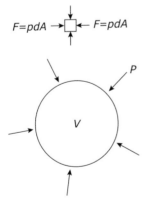

Figure 5.4 Equality of pressure on non-accelerating cubic volume and arbitrary shape under hydraulic stain.

> **Question 5.2**
>
> What are the units of the bulk modulus B? How do the units of B compare with the units of Young's modulus E and shear modulus G?

It might be expected that the bulk modulus B can be simply expressed in terms of the Young's modulus, E, since if we consider a cube with sides of length L under hydrostatic pressure, the force on each face is purely normal (no shear stress) and the volume under pressure V' is given by

$$V' = (L - \Delta L)^3 = L^3 \left(1 - \frac{\Delta L}{L}\right)^3 \cong L^3 \left(1 - 3\frac{\Delta L}{L}\right) = V_o \left(1 - 3\frac{\Delta L}{L}\right). \tag{5.9}$$

If we assume that the compression is given by the Young's modulus $\frac{\Delta L}{L} = \frac{\sigma}{E}$, we find $\frac{V' - V_o}{V_o} = \frac{\Delta V}{V} = 3\frac{\Delta L}{L} = 3\frac{\sigma}{E}$. Equating the stress σ to the pressure p, we would derive that $B = E/3$. This assumes, however, that the compression along the x-axis is uncorrelated with displacements along the y-, or z-axis, but we know that most materials when compressed tend to bulge out in the lateral direction. To derive the connections between the Young's modulus, the shear modulus, and the bulk modulus, we need to consider a generalized treatment of elasticity.

5.4 Unified treatment of elasticity

5.4.1 Stress tensor

To combine the concepts of tensile and shear elasticity, we begin by considering that the force on any surface of area A can be resolved into a component that is normal to the surface and a component in the plane of the surface. In a standard x–y–z coordinate system, we can define, for example, F_{xy} as the force on an x-normal face in the y-direction. Thus F_{xx}/A is the tensile elastic stress on an x-normal surface while F_{xy}/A is the shear stress in the y-direction on the same face. The stress is thus a tensor:

$$\sigma = \begin{bmatrix} \sigma_{xx} & \sigma_{xy} & \sigma_{xz} \\ \sigma_{yx} & \sigma_{yy} & \sigma_{yz} \\ \sigma_{zx} & \sigma_{zy} & \sigma_{zz} \end{bmatrix} \tag{5.10}$$

where the diagonal terms represent the normal (tensile or compressive) stress and the off-diagonal terms the shear stress.

For the stress to be applied without accelerating the object, the stressed object must be secured by forces that balance the applied forces, as shown in Figure 5.5(a). The $(-F_{xx})$ on the left side and the $(-F_{yx})$ force on the bottom are the balancing forces exerted by whatever holds the object in place. Moreover, since the object also needs to be secured against rotation, the shear forces on the perpendicular sides must also balance to make the torques sum to zero. As shown in Figure 5.5(b), this means that

$$F_{yx} = F_{xy}. \tag{5.11}$$

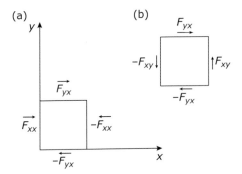

Figure 5.5 (a) Equal reaction forces on a non-moving object from the applied forces and the object holder. (b) Symmetric equal forces required by the condition that the total torque vanish.

Similarly, $F_{yz} = F_{zy}$ and $F_{zx} = F_{xz}$. Since $\sigma = F/A$, $\sigma_{xy} = \sigma_{yx}$, etc. so *the stress tensor is symmetric*

$$\sigma = \begin{bmatrix} \sigma_{xx} & \sigma_{xy} & \sigma_{xz} \\ \sigma_{xy} & \sigma_{yy} & \sigma_{yz} \\ \sigma_{xz} & \sigma_{yz} & \sigma_{zz} \end{bmatrix}. \tag{5.12}$$

If stress is a tensor, we should be able to rotate it to a new axis system using rotation matrices the way we rotated the moment of inertia I or the effective mass tensor m^*. Consider an object with only shear forces applied in the x–y plane as above. Thus,

$$\sigma = \begin{bmatrix} 0 & F_{xy}/A & 0 \\ F_{xy}/A & 0 & 0 \\ 0 & 0 & 0 \end{bmatrix}. \tag{5.13}$$

Let's rotate the stress tensor by $45°$ around the z-axis to a new coordinate system x'–y' as shown. The rotation matrix is

$$R_z = \begin{bmatrix} \cos 45° & \sin 45° & 0 \\ -\sin 45° & \cos 45° & 0 \\ 0 & 0 & 1 \end{bmatrix} = \begin{bmatrix} \sqrt{2}/2 & \sqrt{2}/2 & 0 \\ -\sqrt{2}/2 & \sqrt{2}/2 & 0 \\ 0 & 0 & 1 \end{bmatrix} \tag{5.14}$$

which, in the new coordinate system, gives a stress tensor

$$\sigma' = R_z \begin{bmatrix} 0 & F_{xy/A} & 0 \\ F_{xy/A} & 0 & 0 \\ 0 & 0 & 0 \end{bmatrix} R_z{}^T = \begin{bmatrix} F_{xy/A} & 0 & 0 \\ 0 & -F_{xy/A} & 0 \\ 0 & 0 & 0 \end{bmatrix}. \tag{5.15}$$

Thus in the new x'–y' coordinate system we have purely normal stresses, tensional on the y' face and compressional on the x' face. The shear forces resolve into compressional and tensional forces as shown in Figure 5.6. Shear forces that lead to brittle fracture will cause a fracture plane oriented at $45°$ to the surfaces to which the shear forces are applied!

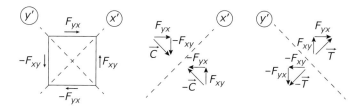

Figure 5.6 Symmetric shear forces on a cube can be resolved into purely compressive forces on one body diagonal and purely tensional forces on the perpendicular body diagonal.

5.4.2   Strain tensor

Strain is the deformation per unit length of a body under stress. The strain in the x–y plane can be defined in terms of the deformation in length and orientation of two perpendicular lines of length Δx and Δy attached to atomic sites inside the material. *Tensile strain* is the change in length in the x-direction of the Δx segment or the change in length in the y-direction of the Δy segment. *Shear strain* is the change in angle of the Δx segment or of the Δy segment. We can indicate by $\vec{u}\,(x, y)$ the *deformation field*, a vector field indicating the deformation of each atomic site in the x- and y-directions as a function of x and y positions. For example, as shown in Figure 5.7, we can define the *tensile strain in the x-direction* as

$$\epsilon_{xx} = \lim_{\Delta x \to 0} \frac{u_x(x + \Delta x, y) - u_x(x, y)}{\Delta x}. \tag{5.16}$$

We assume here that the change in the orientation of Δx ($= \alpha$) is small so that $\cos \alpha \cong 1$. The *shear strain* is defined similarly

$$\frac{\partial u_x}{\partial y} = \lim_{\Delta y \to 0} \frac{u_x(x, y + \Delta y) - u_x(x, y)}{\Delta y} \tag{5.17a}$$

and

$$\frac{\partial u_y}{\partial x} = \lim_{\Delta x \to 0} \frac{u_y(x + \Delta x, y) - u_y(x, y)}{\Delta x}. \tag{5.17b}$$

In the limit where the deformation angles are small, so that $\tan \beta \cong \beta$ and $\tan \alpha \cong \alpha$, we have

$$\beta = \frac{\frac{\partial u_x}{\partial y} \Delta y}{\Delta y} \longrightarrow \frac{\partial u_x}{\partial y} \tag{5.18a}$$

$$\alpha = \frac{\frac{\partial u_y}{\partial x} \Delta x}{\Delta x} \longrightarrow \frac{\partial u_y}{\partial x}. \tag{5.18b}$$

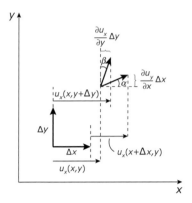

Figure 5.7 Deformation in solid under stress is a combination of tensile and shear strain.

Question 5.3

Draw a diagram like Figure 5.7, indicating the displacements in the y-direction: $u_y (x, y)$, $u_y (x, y + \Delta y)$, and $u_y (x + \Delta x, y)$. Demonstrate by trigonometry that Eqs. (5.18a) and (5.18b) give the right expressions for the angles α and β in radians using the small angle approximation.

The *engineering shear strain,* which is the total change in the angle under deformation $\gamma = \sigma_{shear}/G$ from Eq. (5.6), is the sum of these two angles

$$\gamma_{xy} = \alpha + \beta. \tag{5.19}$$

In general, $\alpha \neq \beta$ and therefore the deformation is not symmetric with an interchange of x and y. To make the strain tensor symmetric, we define the elements of the strain tensor ϵ to be

$$\epsilon_{ij} = \frac{1}{2}\left(\frac{\partial u_i}{\partial x_j} + \frac{\partial u_j}{\partial x_i}\right). \tag{5.20}$$

Thus, $\epsilon_{xy} = \frac{1}{2}\left(\frac{\partial u_x}{\partial y} + \frac{\partial u_y}{\partial x}\right) = \frac{1}{2}\gamma_{xy}$ while $\epsilon_{xx} = \frac{1}{2}\left(\frac{\partial u_x}{\partial x} + \frac{\partial u_x}{\partial x}\right) = \frac{\partial u_x}{\partial x}$.
The three-dimensional strain tensor is thus

$$\epsilon = \begin{bmatrix} \epsilon_{xx} & \epsilon_{xy} & \epsilon_{xz} \\ \epsilon_{xy} & \epsilon_{yy} & \epsilon_{yz} \\ \epsilon_{xz} & \epsilon_{yz} & \epsilon_{zz} \end{bmatrix} \tag{5.21}$$

which is symmetric.

5.4.3 Elasticity constitutive relation

The elastic properties of any material are given by the elasticity function that relates the stress tensor to the strain tensor:

$$\boldsymbol{\sigma} = \boldsymbol{E}\boldsymbol{\epsilon}. \tag{5.22}$$

Note that since $\boldsymbol{\sigma}$ and $\boldsymbol{\epsilon}$ are second-order tensors with elements σ_{ij} and ϵ_{kl}, the elasticity function $\boldsymbol{E}$ is a *fourth-order tensor* with $3^4 = 81$ elements E_{ijkl}. We can find the ijth element of $\boldsymbol{\sigma}$ from

$$\sigma_{ij} = \sum_{k=1}^{3} \sum_{l=1}^{3} E_{ijkl}\, \epsilon_{kl}. \tag{5.23}$$

Question 5.4

Show that the summation notation for tensor multiplication gives the same result as our "row times column" technique by comparing

$$L_i = \sum_{j=1}^{3} I_{ij}\, \omega_j$$

with the result we had in Example 4.2 for the angular momentum vector $\vec{L}$ equaling the inertial tensor $\boldsymbol{I}$ times the angular velocity vector $\vec{\omega}$:

$$\vec{L} = \begin{bmatrix} 29569 & 0 & -10576 \\ 0 & 35675 & 0 \\ -10576 & 0 & 17356 \end{bmatrix} \begin{bmatrix} 0 \\ 0 \\ 0.5 \end{bmatrix} = \begin{bmatrix} -5288 \\ 0 \\ 8678 \end{bmatrix}.$$

From the summation notation in Eq. (5.23), write an expression for σ_{12} in terms of elements of ϵ and elements of the elasticity tensor.

Since both $\boldsymbol{\sigma}$ and $\boldsymbol{\epsilon}$ are symmetric tensors, we can reduce the number of independent elements of the elasticity tensor to a symmetric 6×6 tensor with 21 independent elements as below:

$$\begin{bmatrix} \sigma_{xx} \\ \sigma_{yy} \\ \sigma_{zz} \\ \sigma_{xy} \\ \sigma_{yz} \\ \sigma_{xz} \end{bmatrix} = \begin{bmatrix} E_{xxxx} & E_{xxyy} & E_{xxzz} & 2E_{xxxy} & 2E_{xxyz} & 2E_{xxzx} \\ & E_{yyyy} & E_{yyzz} & 2E_{yyxy} & 2E_{yyyx} & 2E_{yyzx} \\ & & E_{zzzz} & 2E_{zzxy} & 2E_{zzyx} & 2E_{zzzx} \\ & & & 2E_{xyxy} & 2E_{xyyz} & 2E_{xyzx} \\ & \cdots & & & 2E_{yzyz} & 2E_{yzzx} \\ & & & & & 2E_{zxzx} \end{bmatrix} \begin{bmatrix} \epsilon_{xx} \\ \epsilon_{yy} \\ \epsilon_{zz} \\ \epsilon_{xy} \\ \epsilon_{yz} \\ \epsilon_{zx} \end{bmatrix}. \tag{5.24}$$

The *stiffness matrix* $\boldsymbol{E}$ is symmetric; we have only displayed the elements in the upper diagonal. The lower left of the matrix is just the same as the upper right transposed.

5.4.4   Elastic matrix for isotropic materials

For cubic materials the number of independent elements reduces even further to only three independent constants. We can get these constants in terms of things we have seen before if we consider the *compliance* matrix $C = E^{-1}$

$$
\begin{bmatrix} \epsilon_{xx} \\ \epsilon_{yy} \\ \epsilon_{zz} \\ \epsilon_{xy} \\ \epsilon_{yz} \\ \epsilon_{xz} \end{bmatrix} = \begin{bmatrix} \frac{1}{E} & -\frac{v}{E} & -\frac{v}{E} & 0 & 0 & 0 \\ -\frac{v}{E} & \frac{1}{E} & -\frac{v}{E} & 0 & 0 & 0 \\ -\frac{v}{E} & \frac{v}{E} & \frac{1}{E} & 0 & 0 & 0 \\ 0 & 0 & 0 & \frac{1}{2G} & 0 & 0 \\ 0 & 0 & 0 & 0 & \frac{1}{2G} & 0 \\ 0 & 0 & 0 & 0 & 0 & \frac{1}{2G} \end{bmatrix} \begin{bmatrix} \sigma_{xx} \\ \sigma_{yy} \\ \sigma_{zz} \\ \sigma_{xy} \\ \sigma_{yz} \\ \sigma_{zx} \end{bmatrix}. \tag{5.25}
$$

Here E is the Young's modulus, G is the shear modulus, and v is the *Poisson's ratio*, which measures how much a material stretched in one direction deforms in a transverse direction

$$
\epsilon_{lateral} = -v\epsilon_{logitudinal}. \tag{5.26}
$$

Of these three elastic parameters, in an isotropic material, only two are independent. For example, in terms of G and v, the Young's modulus can be found from

$$
E = 2G(1 + v). \tag{5.27}
$$

The second element of the strain column vector can be written, for example, as

$$
\epsilon_{yy} = \frac{\sigma_{yy}}{E} - \frac{v}{E}(\sigma_{xx} + \sigma_{zz}). \tag{5.28}
$$

In the case where there is no stress applied in the x- or z-direction, this gives us back our familiar expression that $\sigma_{yy} = E\epsilon_{yy}$. The fourth element of the strain vector leads to

$$
\epsilon_{xy} = \frac{1}{2G}\sigma_{xy}. \tag{5.29}
$$

Since we defined the strain tensor so that $\epsilon_{xy} = \frac{1}{2}\gamma_{xy}$ where γ_{xy} is the engineering strain, we find

$$
\frac{1}{2}\gamma_{xy} = \frac{1}{2G}\sigma_{xy} \Rightarrow \sigma_{xy} = G\gamma_{xy}, \tag{5.30}
$$

as we found before.

We can find the bulk modulus $B = -\frac{\Delta p}{\Delta V/V}$ in terms of the compliance matrix elements by noting that in terms of the hydrostatic pressure $\sigma_{xx} = \sigma_{yy} = \sigma_{zz} = -\Delta p$ while, since a fluid does not support transverse strains, $\sigma_{xy} = \sigma_{xz} = \sigma_{yz} = 0$. For a test cube with an edge length of L, the compressibility $\frac{\Delta V}{V} = 3\frac{\Delta L}{L} = \epsilon_{xx} + \epsilon_{yy} + \epsilon_{zz}$, giving

$$B = \frac{E}{3(1-2v)}. \tag{5.31}$$

Question 5.5

For a test cube with dimensions (L, L, L) we have $\Delta V = (L + \Delta L)(L + \Delta L)(L + \Delta L) - L^3$. Show that if ΔL is small so you can neglect terms of order $(\Delta L)^2$ and $(\Delta L)^3$, then $\frac{\Delta V}{V} = 3\frac{\Delta L}{L}$. Using this show how Eq. (5.31) follows from the definition of the compliance matrix in Eq. (5.25).

Chapter 5 problems

1. Describe in your own words the difference between the Young's modulus, E, and the shear modulus, G. Look up values of E and G for different materials. Give two examples of applications of materials chosen for their values of E and/or G.

2. An aluminum rod 5 cm × 5 cm projects horizontally 7 cm from a wall. A 2000 kg object is suspended from the end of the rod. Find the vertical deflection of the rod if the shear modulus of aluminum is 3×10^{10} N/m^2.

3. The Young's modulus of structural steel is 200×10^9 N/m^2 and its yield strength is 250×10^6 N/m^2. How much work is done by a machine that stretches a 1 meter long rod of structural steel 3 cm in diameter from zero elongation until it enters the plastic deformation region?

4. A beam 10 feet long with a mass of 2500 kg is suspended horizontally by two 2 m long steel wires of diameter 1 cm, one 2 feet from one end and the other 3.5 feet from the other end. Using the values of Problem 3 for the elastic constants of steel, would either wire have exceeded its linear elastic limit? How long would each wire be if the beam were removed? If a typical engineering safety margin is a factor of three, what diameter of cables would you recommend be used?

5. At the end of the chapter, we showed that the bulk modulus B, defined in terms of the compressibility $\Delta V/V$ by $\Delta p = -B\frac{\Delta V}{V}$, can be expressed as

$$B = \frac{E}{3(1-2v)},$$

where v is the *Poisson's ratio*. For metals, $v \approx 0.3$, so $B \approx \frac{E}{1.2}$. There are some polymers, however, that can be made with the Poisson's ratio $v \cong 0.5$. What does this give for a bulk modulus B? What would the volume change ΔV be for such a material as a function of pressure? How would you specify the compressibility of such a polymer? Can you think of any useful device that could be made with such a material? In your own words, what is the difference between materials with different Poisson's ratio? Can $v > 0.5$ or $v < 0$? If so, what mechanical properties would materials like that have?

6. Consider an ideal gas with an equation of state given by $pV = NRT$, where N is the number of moles of the gas, T is the temperature in kelvin, and $R = 8.314$ J/mole·K is the universal gas constant. Find the bulk modulus B of an ideal gas at room temperature, $T = 300$ K, and atmospheric pressure, $p = 1.013 \times 10^5$ N/m². You may assume the modulus is measured *isothermally*, with no change in temperature. How much would the volume of the gas change if the pressure were isothermally increased by 20%?

7. The non-zero elements of the elastic tensor for three different non-cubic materials are shown in the table below.

Material	Stiffness matrix elements (GPa)								
	E_{xxxx}	E_{yyyy}	$2E_{xyxy}$	$2E_{yzyz}$	$2E_{yzyz}$	$2E_{zxzx}$	E_{xxyy}	E_{xxzz}	E_{yyzz}
Zinc (hexagonal)	143.0	143.0	50.0	40.0	40.0	63.0	17.0	33.0	33.0
Aligned Graphite/ Epoxy (hexagonal)	14.5	14.5	160.0	7.07	7.07	3.62	7.26	5.53	5.43
Calcium Formate (orthorhombic)	49.2	24.4	35.4	10.5	12.2	28.3	24.8	24.5	14.5

The stiffness matrix is symmetric, so the upper-right off-diagonal elements in the last three columns have identical elements in the lower-left part of the matrix, $E_{xxyy} = E_{yyxx}$, for example. Since these materials all have symmetry axes which are orthogonal, the other elements of E are zero.

Consider a cubic centimeter of each of these materials with normal faces aligned with the x, y, and z symmetry axes. Calculate the displacement of the z-normal face if it has a uniform force applied of 10^5 N in a direction that is inclined 65° from the z-axis toward the y-axis. What difference would it make if it were inclined toward the x-axis instead? (Hint: You can use MATLAB to calculate the compliance matrix which is the inverse of E. Show how you would use E^{-1} to find the elements of the strain tensor given the stress tensor elements.)

8. A helical coil spring changes a linear force into a torsional twisting shear stress and a smaller transverse shear stress. Focusing only on the torsional strain component, for a spring made of wire of diameter d formed into N turns of radius R the torque

due to an axial force P is $\tau = RP$ and an axial displacement of δ is related to the total angular twist by $\delta = R\theta$.

a. Show from Equation (5.7) that the axial displacement is given by

$$\delta = \frac{64 \, NR^3 P}{d^4 G}.$$

b. The extension of a spring ΔL under the influence of a force is described by $P = k\Delta L$ where k is the spring constant. Calculate the spring constant for a helical spring which has $N = 20$ circular turns of wire of radius $R = 1$ cm and is made of $d = 0.5$ mm diameter spring steel with a shear modulus $G = 7.93 \times 10^{10}$ N/m^2.

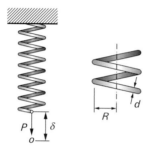

9. Euler–Bernoulli beam-bending. The bending deflection of a fixed-end cantilevered beam with a shear end load $\vec{F}$ is given by the Euler–Bernoulli equation

$$\frac{d^2}{dy^2}\left(EI \frac{d^2 w}{dy^2} \right) = F,$$

where $w(y)$ is the beam deflection as a function of y as shown in the figure, I is the second moment of the cross-section area, and E is the Young's modulus. The beam material along the dashed neutral axis is neither stretched nor compressed. The material outside of the neutral axis is stretched and the material inside the neutral axis is compressed as shown, thus the Young's modulus and not the shear modulus is the appropriate parameter for the beam bending. If the beam is uniform with square cross-section, $I = \frac{a^4}{12}$, where a is the length of each side of the beam cross-section, and the solution of the Euler–Bernoulli equation with the appropriate boundary conditions is

$$w(y) = \frac{F}{EI} \frac{y^2(3L - y)}{6}.$$

a. Apply this equation to the problem in Example 5.1 to find how much total x-displacement would result from beam-bending and compare this displacement with the displacement calculated from the shear modulus displacement.

b. Compare the bending displacement and the shear strain displacement in the case where the rod length is 10 cm and all the other parameters are the same.

c. Compare the bending and shear strain displacements if the rod length is 5 cm.

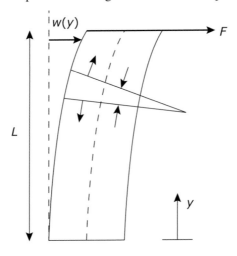

6 Harmonic oscillation

A mass–spring system as shown in Figure 6.1 is a model for a harmonic oscillator. By Hooke's Law, force on the spring should equal the *spring constant k* times the elongation of the spring:

$$F = -k(d - d_o). \tag{6.1}$$

In Equation (6.1), the force is positive in the downward direction, so if $d > d_o$ as in the figure, the force is up. By creating a new variable $x = (d - d_o)$ and applying Newton's Second Law, we get the following differential equation:

$$m\frac{d^2x}{dt^2} = -kx. \tag{6.2}$$

Since $\frac{d}{dx} \cos ax = -a \sin ax$ and $\frac{d}{dx}(-a \sin ax) = -a^2 \cos ax$, one solution of the differential equation (Eq. (6.2)) is $x(t) = A \cos \sqrt{\frac{k}{m}}t$ telling us that the mass will oscillate back and forth with time. Similarly, $x(t) = B \sin \sqrt{\frac{k}{m}}t$ is another solution. We can also find solutions in terms of complex exponentials $x(t) = Ae^{i\sqrt{\frac{k}{m}}t}$ since $\frac{d^2}{dx^2}\left(Ae^{i\sqrt{\frac{k}{m}}t}\right) = \left(i\sqrt{\frac{k}{m}}\right)^2 Ae^{i\sqrt{\frac{k}{m}}t} = -\frac{k}{m}Ae^{i\sqrt{\frac{k}{m}}t}$. This is not surprising considering the Euler relation between complex exponentials and sinusoidal functions $e^{\pm i\theta} = \cos \theta \pm i \sin \theta$. While a complex distance x is clearly non-physical, we can use the complex exponential

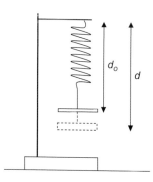

Figure 6.1 Elastic deformation of spring showing increase in length under added weight.

function for a solution with the provision that we will take the real part of the solution at the end to get a physical displacement.

The most general solution of a second-order differential equation like Equation (6.2) has two constants of integration that need to be determined by the initial conditions, for example the position and velocity at $t = 0$, $x(0)$ and $v(0)$. These two constants can be expressed in terms of the amplitudes A and B of the sine and cosine terms of the solution, or in terms of the *amplitude D* and *phase δ* of the oscillation.

$$x = A \cos(\omega_o t) + B \sin(\omega_o t)$$
$$x = D \cos(\omega_o t + \delta) \tag{6.3}$$
$$x = \mathcal{Re}\{De^{i(\omega_o t + \delta)}\}$$

where $\omega_o = \sqrt{\frac{k}{m}}$.

Question 6.1

Find an expression for D and δ in the last two relations in Eq. (6.3) in terms of A and B in the first relation.

These solutions all represent *simple harmonic motion*, lossless oscillation at a frequency ω_o. The mass will continue to move back and forth forever at the same frequency with constant oscillation amplitude, the maximum displacement from $x = 0$. The phase, often expressed as an angle fraction of the 360° full cycle, represents the offset if the amplitude at $t = 0$ is not the maximum amplitude. Since the model has no energy loss, mechanical energy is transferred back and forth between 100% kinetic energy at $x = 0$ and 100% elastic potential energy when the mass comes to rest momentarily when x equals the maximum amplitude.

Question 6.2

Find the expressions for the *velocity* and *acceleration* of a simple harmonic oscillator in each of the forms of Eq. (6.3). Show that the velocity v is zero when the oscillator position x is maximum.

6.1   Damped harmonic oscillator

We can extend the model to include losses by introducing a frictional force that opposes the motion. If the friction is due to air resistance, the force would be proportional to v^2 as we have seen before. If the friction is due to viscous forces, as when the oscillator has a damper in a fluid dashpot, or if the damping is due to internal dissipative forces inside the spring material, the frictional force is more closely approximated by a force which is proportional to the velocity: $F_f = -m\gamma v = -m\gamma \frac{dx}{dt}$. (The mass m is introduced for dimensional purposes.) Using this form for the frictional force (instead of the v^2 dependence from air resistance) gives us the following linear differential equation which we can solve analytically:

$$m\frac{d^2x}{dt^2} = -kx - m\gamma\frac{dx}{dt}. \tag{6.4}$$

Using the complex exponential form $x(t) = De^{i(\omega_o t + \delta)}$, we find that the exponentials cancel from each term and the equation reduces to an algebraic equation:

$$\omega_o{}^2 = \frac{k}{m} + i\gamma\omega_o \tag{6.5}$$

with a positive frequency solution $\omega_o = \sqrt{\frac{k}{m} - \frac{\gamma^2}{4}} + i\frac{\gamma}{2}$.

Question 6.3

Show that substituting $x(t) = De^{i(\omega_o t + \delta)}$ in Eq. (6.4) gives the quadratic equation for ω_o in Eq. (6.5), and show that the solution of Eq. (6.5) is given by $\omega_o = \sqrt{\frac{k}{m} - \frac{\gamma^2}{4}} + i\frac{\gamma}{2}$.

The damping force has two effects. First the resonant frequency (the real part of ω_o) is modified, usually by a small amount if γ is small. Second, the imaginary part of the frequency results in an exponentially decreasing oscillation amplitude:

$$x(t) = \mathcal{Re}\left\{De^{-\frac{\gamma}{2}t}e^{i\left(\sqrt{\frac{k}{m}-\frac{\gamma^2}{4}}t+\delta\right)}\right\} = De^{-\frac{\gamma}{2}t}\cos\left(\sqrt{\frac{k}{m} - \frac{\gamma^2}{4}}t + \delta\right). \tag{6.6}$$

The effect of the damping on the oscillator motion is illustrated in Figure 6.2. The envelope of the displacement of the damped oscillator is given by the two curves $x = De^{-\frac{\gamma}{2}t}$ and $x = -De^{-\frac{\gamma}{2}t}$ and the oscillation damps out exponentially in time.

Example 6.1 A mass $m = 40$ kg is attached to a spring with a spring constant $k = 5$ N/m. (a) Find the motion of the mass if it is displaced from its equilibrium position by 1 cm and released from rest at $t = 0$. (b) Compare the motion to the case where the mass is displaced by 0.5 cm and released at $t = 0$ with a velocity of 0.3 cm/s upward. (c) Finally, find the motion if the mass is displaced from rest at $x(0) = 1.0$ cm as in (a), but the motion is damped with a damping constant $\gamma = 0.05$ s^{-1}.

Solution: Starting from the complex exponential form $x(t) = De^{i(\omega_o t + \delta)}$, we observe that the velocity is given by $v(t) = \frac{dx}{dt} = (i\omega_o)De^{i(\omega_o t + \delta)}$. When we compare with real measurements, we need to take the real parts:

$$x_o = x(0) = \mathcal{Re}\{De^{i\delta}\} = \mathcal{Re}\{D(\cos\delta + i\sin\delta)\} = D\cos\delta$$

$$v_o = v(0) = \mathcal{Re}\{(i\omega_o)De^{i\delta}\} = \mathcal{Re}\{i\omega_o D(\cos\delta + i\sin\delta)\} = -\omega_o D\sin\delta.$$

Continued

(a)

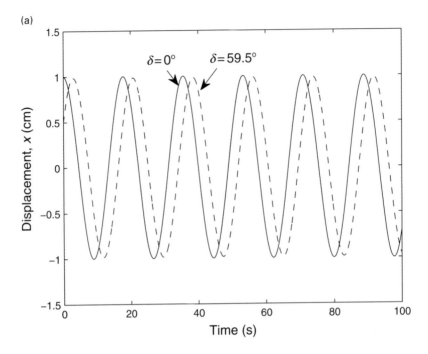

(b)

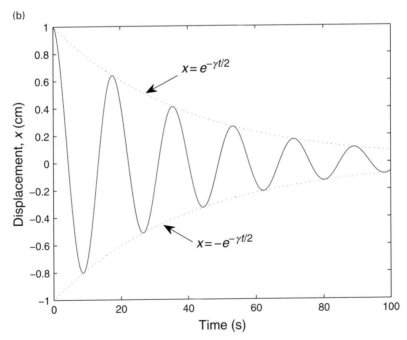

Figure 6.2 (a) Position as a function of time for a harmonic oscillator with two different phases δ imposed by initial conditions as in Example 6.1. (b) Time dependence of harmonic oscillator with a damping parameter $\gamma = 0.05 \text{ s}^{-1}$ from a velocity-dependent term in the differential equation.

Example 6.1 (*cont.*)

Dividing, we find $\frac{v_o}{x_o} = -\omega_o \tan \delta$. With no damping, $\omega_o = \sqrt{\frac{k}{m}} = 0.354 \frac{\text{rad}}{\text{s}}$ and for (a), $v_o = 0 \Rightarrow \delta = 0$ and $D = \frac{x_o}{\cos \delta} = 1$ cm. For (b), $\delta = \tan^{-1}\left(-\frac{v_o}{\omega_o x_o}\right) = \tan^{-1}\left(-\frac{0.3}{(0.354)(0.4)}\right) = -59.5°$ and $D = \frac{x_o}{\cos \delta} = \frac{0.5}{\cos(-59.5°)} = 0.985$ cm. For (c), $v_o = 0 = -\omega_o D \sin \delta \Rightarrow \delta = 0$ and $D = \frac{x_o}{\cos \delta} = \frac{1.0}{\cos 0} = 1.0$ cm and $\omega_o = \sqrt{\frac{k}{m} - \frac{\gamma^2}{4} + i\frac{\gamma}{2}} = \sqrt{\frac{5}{40} - \frac{0.05^2}{4}} + i\frac{0.05}{2} = 0.352 + 0.025i \, \text{s}^{-1}$. These can be plotted with the following MATLAB code. The resulting plots are in Figure 6.2.

```
k=5;m=40;gamma=0.05;
x_oa=1.0; v_oa=0;w_oa=sqrt(k/m);
x_ob=0.5; v_ob=0.3;w_ob=sqrt(k/m);
x_oc=1.0; v_oc=0;w_oc=sqrt(k/m-gamma^2/2)+i*gamma/2;
t=0:.01:100;

delta_a=0;Da=1.0;
x_a=real( Da*exp(i*(w_oa*t+delta_a)) );
delta_b=atan( -v_ob/(x_ob*w_ob) );
Db=x_ob/cos(delta_b);
x_b=real( Db*exp(i*(w_ob*t+delta_b)) );
delta_c=0;Dc=1.0;

figure(1); plot(t,x_a,'-',t,x_b, '--');
xlabel('Time (s)'); ylabel('Displacement x (cm)')

figure(2); plot(t, x_c,'-',t,exp(-t*gamma/2),':',t,
-exp(-t*gamma/2),':')
xlabel('Time (s)'); ylabel('Displacement x (cm)')
```

Question 6.4

Given the form for the position as a function of time for a harmonic oscillator $x(t) = De^{i(\omega_o t + \delta)}$ describe in your own words the effect of the parameters D, ω_o, and δ. How would a change in each parameter affect the motion of the object?

6.2   Driven harmonic oscillator

If we put an external sinusoidal driving force on the spring–mass system above, as in Figure 6.3, we have a model for a driven harmonic oscillator. The displacement of the mass will then be a function of the drive strength and frequency as well as the spring

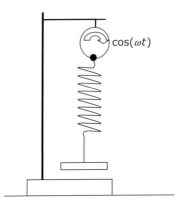

cos(ωt)

Figure 6.3 A harmonic oscillator with a time-harmonic applied driving force.

constant, mass, and damping parameter of the spring system. The response is divided into two regimes. There is the transient response when the driving force is first applied. This is, in general, complicated and depends on the initial state of the oscillator.

However, if we let the transient response die out and look only at the steady-state response, we can assume that the spring will respond at the frequency of the driving force. If the driving force is $F(t) = F_o \cos \omega t$, we expect that the steady-state response of the oscillator will be of the form $x(t) = D \cos(\omega t + \delta)$, i.e., the oscillator will oscillate at the frequency of the driving force ω instead of at its resonant frequency ω_o. The oscillation amplitude D and the phase difference δ between the oscillator and the driving force will depend on the driving amplitude and frequency as well as the resonant frequency and damping of the oscillator.

We will use the exponential form for the driving force $F(t) = F_o e^{-i\omega t}$ and for the displacement $x(t) = \tilde{D}e^{-i\omega t}$. This form for $x(t)$ gives the oscillatory part of the solution at the frequency ω of the driving force. (The negative sign in the exponent of $e^{-i\omega t}$ will correspond to the notation we shall be using when we study the wave properties of materials.) The parameter $\tilde{D}$ will be a complex number expressed in polar form as $|\tilde{D}|e^{i\delta}$ which will represent both the amplitude $D = |\tilde{D}|$ and relative phase δ of the oscillator with respect to the driving force. The equation of motion is then:

$$m\frac{d^2}{dt^2}\tilde{D}e^{-i\omega t} = -k\tilde{D}e^{-i\omega t} - m\gamma\frac{d}{dt}\tilde{D}e^{-i\omega t} + F_o e^{-i\omega t}. \tag{6.7}$$

The advantage of the complex exponential form is that the derivatives all have the same $e^{-i\omega t}$ factor, which can be canceled out to give an algebraic equation for $\tilde{D}$ in terms of F_o:

$$\tilde{D}(\omega) = \frac{F_o/m}{\omega_o^2 - \omega^2 - i\gamma\omega} \tag{6.8}$$

where $\omega_o = \sqrt{k/m}$. The real and imaginary parts of $\tilde{D}$ for $F_0 = 1$ N and the same oscillator parameters as in Figure 6.2(b) are shown in Figure 6.4.

This function form is known as a *Lorentzian oscillator* and is found to describe phenomena as varied as lattice vibrations (*phonons*) in polar insulators or semiconductors

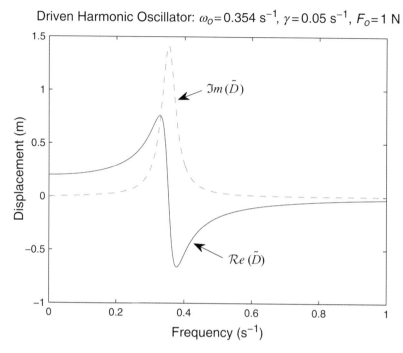

Figure 6.4 The real and imaginary parts of the complex amplitude of the harmonic motion of a driven harmonic oscillator. The real amplitude reverses phase and the imaginary part (the oscillator damping) reaches a peak at a frequency equal to the resonant frequency ω_o.

and optical properties of electronic transitions in gases. A Lorentzian oscillator is characterized by three parameters: a resonant frequency ω_o (in units of s^{-1}), a damping constant γ (in units of s^{-1}), and an oscillator strength $S = \frac{F_o/m}{\omega_o^2}$ (in units corresponding to the observed quantity – meters in this case where we are modeling displacement). The oscillator strength gives the measured response at $\omega \rightarrow 0 : \tilde{D}(\omega = 0) = S$. At the resonant frequency of the system, the real part is nearly singular with an abrupt drop (and change in sign if the oscillator strength is large enough). The imaginary part is always positive for the $e^{-i\omega t}$ dependence we used (if we used $e^{+i\omega t}$ it would be always negative) with a peak at the resonant frequency. The full width at half maximum of the peak in the imaginary part is the damping parameter γ. If $\gamma \ll \omega_o$, the area under the imaginary part of the Lorentzian oscillator is independent of γ and directly related to the oscillator strength S:

$$\int_0^\infty \mathcal{I}m\left(\frac{S\omega_o^2}{\omega_o^2 - \omega^2 - i\gamma\omega}\right)d\omega = \frac{\pi}{2}S\omega_o. \tag{6.9}$$

In Figure 6.5, we plot the amplitude, $|\tilde{D}| = \sqrt{\mathcal{R}e(\tilde{D})^2 + \mathcal{I}m(\tilde{D})^2}$, and phase angle, $\delta = \tan^{-1}(\mathcal{I}m(\tilde{D})/\mathcal{R}e(\tilde{D}))$, of $\tilde{D}$ using the MATLAB `abs()` and `angle()` functions. We see that the amplitude peaks at ω_o – the system responds most strongly

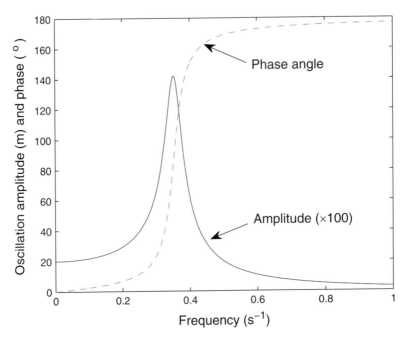

Figure 6.5 The amplitude and phase angle of the driven harmonic oscillator as a function of frequency for the driven harmonic oscillator.

at its resonant frequency – and the phase changes from in-phase for $\omega < \omega_o$ to $180°$ out-of-phase for $\omega > \omega_o$.

6.3   Lorentzian oscillator: an LCR resonant circuit

We have derived the Lorentzian oscillator form for a simple driven spring–mass system. But, in fact, the same Lorentzian form can be found in a wide variety of phenomena that are characterized by a resonant frequency and a damping mechanism. One of the most important examples is an LCR circuit.

An electrical circuit with a capacitance C, an inductance L, and resistance R is well-known to have a resonant frequency $\omega_o = 1/\sqrt{LC}$. In fact, the response of the LCR circuit is described by a Lorentzian oscillator. If we apply Kirchhoff's Voltage Law (KVL) to the circuit in Figure 6.6, remembering that the voltage drop across a resistor is $V_R = IR$, the voltage drop across a capacitor is $V_C = Q/C$, and the voltage drop across an inductor is $V_L = L\,dI/dt$, we find

$$V_o \cos \omega t = L\,dI/dt + Q/C + IR \tag{6.10}$$

with the current I taken as positive in the clockwise direction. Noting that $I = dQ/dt$ and expressing the applied voltage as $V_o\, e^{-i\omega t}$ (with our usual understanding that we will take the real part of our solution to find physical quantities), we arrive at the following equation:

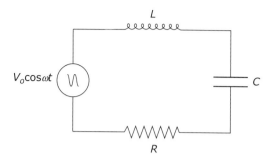

Figure 6.6 Electric circuit with inductor (L), capacitor (C), and resistor (R) in series driven by sinusoidal power supply.

$$V_o e^{-i\omega t} = L \, d^2Q/dt^2 + Q/C + R \, dQ/dt. \tag{6.11}$$

Ignoring the transient response when the voltage source is turned on, we make our usual assumption that the steady-state response of the system will have the same time dependence as the voltage source: $Q(t) = \tilde{Q}_o e^{-i\omega t}$ where we have labeled the amplitude of the charge as $\tilde{Q}_o$ to remind ourselves that the amplitude can be complex to allow for a phase shift between the voltage and the capacitor charge (or the current $\tilde{I} = dQ/dt = (-i\omega)\tilde{Q}_o e^{-i\omega t}$). Taking the time derivatives and canceling $e^{-i\omega t}$ out of each term we find

$$V_o = -\omega^2 L\tilde{Q}_o + \tilde{Q}_o/C - i\omega R\tilde{Q}_o. \tag{6.12}$$

Question 6.5

Using the standard electrical engineering method for finding the voltage drop across a component $V = \tilde{I}_o Z$ where the *impedance Z* is given by $Z_R = R$ for a resistor, $Z_L = j\omega L$ for an inductor, and $Z_C = 1/j\omega C$ for a capacitor and with $j = i = \sqrt{-1}$, show that you get an equation like Eq. (6.12) by applying KVL to the circuit above. Is there anything different about the equation you get? (Hint: In electrical engineering phasor notation $i = \tilde{I}_o e^{j\omega t}$ so $i = \frac{dq}{dt} \Rightarrow \tilde{I}_o = j\omega\tilde{Q}_o$.)

Solving for $\tilde{Q}_o$ and substituting $\omega_o = 1/\sqrt{LC}$, we find that $\tilde{Q}_o(\omega)$ has the Lorentzian

form

$$\tilde{Q}_o = \frac{V_o/L}{\omega_o^2 - \omega^2 - i(R/L)\omega} \tag{6.13}$$

with real and imaginary parts as shown in Figure 6.7 for $V_o = 10 \, \text{V}$, $R = 50 \, \Omega$, $L = 100 \, \text{mH}$, and $C = 7 \, \mu\text{F}$.

A more familiar plot would show the amplitude $|\tilde{Q}_o| = \sqrt{\mathcal{R}e(\tilde{Q}_o)^2 + \mathcal{I}m(\tilde{Q}_o)^2}$ and the phase $\angle\tilde{Q}_o = \tan^{-1}\left(\mathcal{I}m(\tilde{Q}_o)/\mathcal{R}e(\tilde{Q}_o)\right)$ as in Figure 6.8.

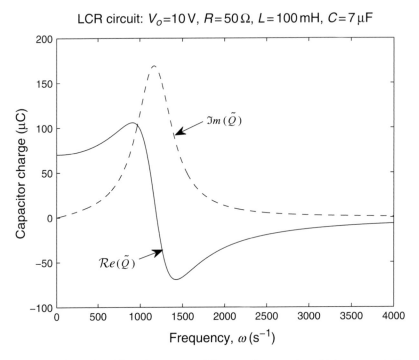

Figure 6.7 The real and imaginary parts of the complex quantity which represents the amplitude of the change on the capacitor in the series LCR circuit of Figure 6.6 as a function of frequency in radians/s.

If we want to find the charge at any particular time we simply multiply $\tilde{Q}_o(\omega)$ by $e^{-i\omega t}$ and take the real part, but in actual practice the value of the AC voltage, current, or charge at any particular time is rarely of interest. What is desired is usually the peak-to-peak value (or the amplitude) and the phase with respect to the driving voltage as in Figure 6.8. All these quantities are easy to find from the Lorentzian form, particularly with modern computing tools such as MATLAB that do the complex calculations for you.

A note on sign conventions

Electrical engineering conventions differ from what we have done in two ways. First, because the lowercase i symbol is commonly used for current, electrical engineers use the lowercase j symbol to represent $\sqrt{-1}$. Second, due to the Laplace transform use of $e^{st} = e^{+j\omega t}$, electrical engineers take all their steady-state time dependences to be of the form $e^{+j\omega t}$ instead of the $e^{-i\omega t}$ convention that we have been using, which is the common usage in physics and other engineering disciplines. This makes the imaginary parts of all frequency domain response functions negative in the electrical engineering literature. Thus, if you want to convert equations from the electrical engineering literature to the physics/mathematics form, or vice versa, you can make both changes at the same time by replacing every j in the formulas by $-i$. There is no physical or mathematic significance to this: both j and i represent $\sqrt{-1}$. However, if you use

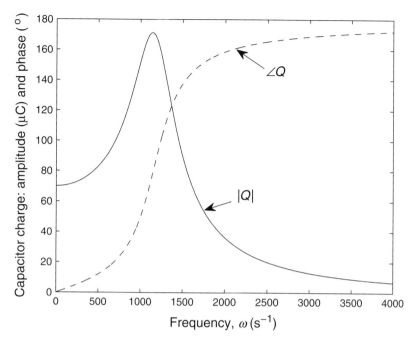

Figure 6.8 The amplitude and phase (with respect to the driving voltage) of the charge on the capacitor in the LCR circuit of Figure 6.6 as a function of frequency.

response functions with positive imaginary parts (such as we have been using) in traveling waves using the electrical engineering $e^{j(-kx+\omega t)}$ notation, you will find waves that increase exponentially rather than being exponentially damped as they propagate through absorbing media!

Chapter 6 problems

1. Show that each of the three forms for the displacement $x(t)$ in Equation (6.3) is a solution of Equation (6.2) and find the values for the constants A and B or D and δ in the case in which the displacement and velocity at $t = 0$ are $x(0) = 5$ cm and $v(0) = 3$ cm/s.

2. Investigate the effect of the different variables on the driven harmonic oscillator. Plot the relevant quantities for different values of the driving force, spring stiffness, mass, and the damping parameter, α.

3. For the LCR circuit Lorentzian oscillator discussed in this chapter, plot the real and imaginary parts and the amplitude and phase of the current $I = dQ/dt$ and compare your plots with the plots for the charge in Figures 6.7 and 6.8. Do your results make sense? Also, calculate the response for the overdamped oscillator by increasing the value of the resistance to 2000 Ω and discuss the similarities and differences with the results of Figures 6.7 and 6.8. How do you interpret your results?

4. Explore the LCR circuit Lorentzian oscillator for other sets of values for L, C and R. Write down some of your conclusions.

5. What is the effect on the resonant frequency
   a. if two springs of the same spring constant are connected in parallel?

   b. if two springs of the same spring constant are connected in series?

   c. if the mass is doubled?

6. Consider a damped harmonic oscillator. The oscillating mass, m, is 4 kg, the spring constant, k, 16 N/m, and the damping force, F_d, is proportional to the velocity ($F_d = -m\alpha v$). If the initial amplitude is 20 cm and falls to half after 6 complete oscillations, calculate
   a. the damping coefficient, α and

   b. the energy "lost" during the first 6 oscillations.

7. Consider a damped harmonic oscillator with a mass of 2 kg, a spring constant of 10 N/m, and a damping force proportional to the velocity ($F_d = -m\alpha v$) with a coefficient, α, equal to 0.15.
   a. What is the time it will take for the amplitude of the oscillation to decrease to 1/4 of the original value?

   b. How many oscillations did it take to reach this amplitude?

8. For a damped mechanical harmonic oscillator, use your own values and
   a. plot the velocity, $v(t)$, vs. time curve. What does $v_{max}(t)$ look like?

   b. plot the acceleration, $a(t)$, vs. time curve. What does $a_{max}(t)$ look like?

9. For a driven oscillator, use your own values and
   a. plot the velocity, $v(t)$, vs. time curve. What does $v_{max}(t)$ look like?

   b. plot the acceleration, $a(t)$, vs. time curve. What does $a_{max}(t)$ look like?

   c. plot the max $x(t)$ and max $v(t)$ as a function of the driving frequency, ω.

10. Using MATLAB, create a plot like Figure 6.8 for the amplitude and phase of the current $i(t)$ against frequency for a series LCR circuit as in Figure 6.6 with $V_o = 10$ V, $R = 50$ Ω, $L = 100$ mH, and $C = 7$ μF. Repeat your plot for $R = 10$ Ω and $R = 250$ Ω. What is the maximum amplitude of the current in amps for each case?

7 Waves

When oscillatory motion in time at one position is coupled to surrounding points, the oscillation can propagate as a *wave*, carrying energy from one point to another without actual transfer of material. Waves are one of the most pervasive phenomena in the universe. Understanding the fundamentals of wave propagation and interactions with other waves and with matter is critical to understanding virtually all modern technology. This topic of waves is therefore one of the most crucial in science and engineering and will provide a bridge to our discussion of quantum physics in the following chapters.

7.1 One-dimensional waves: waves in strings

We will begin our consideration of waves by looking at waves in strings. Everyone is familiar with the movement of a deformation through a string; these waves are the basis of stringed musical instruments from violins to banjos. If we look at a deformation moving to the right through a string under tension T as in Figure 7.1, we note that although the wave deformation is propagating in the $+\hat{x}$-direction, each particle of the string is moving in the $\pm\hat{y}$-direction – the wave is *transverse*. The equation of motion of a small section dx of the string can be found by applying Newton's Laws. The mass of the section of string is given by $dm = \rho\,dx$ where ρ is the linear mass density of the string given in terms of the mass of the string M and length L by $\rho = M/L$.

For deformations that are not too large, we can assume that the tension in the string T is constant through the string and is always exerted tangentially to the string. Any force in the $\hat{y}$-direction must be due to a change in the slope of the string from the left side of the mass element dm to the right side of the mass element as shown in Figure 7.2. In terms of the slope on the left $S_1 = \frac{dy}{dx}\big]_x$ and the slope on the right $S_2 = \frac{dy}{dx}\big]_{x+dx}$ we find the force in the $\hat{y}$-direction is given by

$$F_y = TS_2 - TS_1 = T\left(\frac{dy}{dx}\bigg]_{x+dx} - \frac{dy}{dx}\bigg]_x\right).$$

Setting this equal to ma, the mass times the acceleration of the segment of the string, we find

$$\rho dx \frac{d^2y}{dt^2} = T\left(\frac{dy}{dx}\bigg]_{x+dx} - \frac{dy}{dx}\bigg]_x\right). \tag{7.1}$$

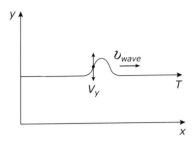

Figure 7.1 The transverse displacement as a function of distance of a string with a propagating wave disturbance. The string displacement is in the vertical direction, while the wave is propagating in the horizontal direction.

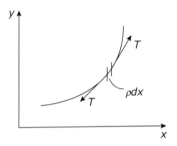

Figure 7.2 The forces acting on a small segment of the string in Figure 7.1 above. The tension in the string is assumed to be constant but the direction of the tension varies as the wave passes.

Dividing through both sides by dx, and recognizing that the change in slope divided by dx is just the mathematical definition of the second derivative of y with respect to x we get

$$\frac{d^2 y}{dt^2} = \frac{T}{\rho} \frac{\left(\frac{dy}{dx} \big]_{x+dx} - \frac{dy}{dx} \big]_x \right)}{dx} = \frac{T \, d^2 y}{\rho \, dx^2}. \tag{7.2}$$

This equation is identical to the *wave equation*

$$\frac{d^2 y}{dt^2} = c^2 \frac{d^2 y}{dx^2} \tag{7.3}$$

with $c = \sqrt{T/\rho}$. The wave equation expresses the coupling between the motion in the y-direction and the string behavior in the x-direction as a relation between the second derivative of y with respect to t in terms of the second derivative of y with respect to x.

The solutions of the wave equation can be any function of x with x replaced by $(x - ct)$ or $(x + ct)$. For example, one could take a Gaussian line shape

$$y(x) = De^{-x^2} \tag{7.4}$$

and change it to a function of x and t

$$y(x, t) = De^{-(x \pm ct)^2}. \tag{7.5}$$

If we plug this into the wave equation above and apply the chain rule for differentiation, we find

$$\frac{d^2}{dt^2} De^{-(x\pm ct)^2} = 2D(\pm c)^2 e^{-(x\pm ct)^2} [2(x\pm ct)^2 - 1],$$

while

$$\frac{d^2}{dx^2} De^{-(x\pm ct)^2} = 2De^{-(x\pm ct)^2} [2(x\pm ct)^2 - 1],$$

demonstrating that

$$\frac{d^2y}{dt^2} = c^2 \frac{d^2y}{dx^2}.$$

The wave form $y(x,t) = De^{-(x\pm ct)^2}$ allows us to identify the significance of the constant c. At $t = 0$, $y(x,t)$ has a maximum at $x = 0$. For $t > 0$, $y(x,t) = De^{-(x-ct)^2}$ has a maximum at $(x - ct) = 0$, i.e. at positive x. As t increases the maximum of $y(x,t)$ moves to larger values of x so that $x = ct$. Thus,

$$c = dx/dt \tag{7.6}$$

and c is identified as the speed at which the peak of $y(x,t)$ moves to the right – the *wave velocity*. If $y(x,t) = De^{-(x+ct)^2}$ then as t increases x must decrease to keep $(x + ct) = 0$. So the form $y(x,t) = De^{-(x-ct)^2}$ represents a wave moving to the right (to higher values of x) at speed c as t increases, while $y(x,t) = De^{-(x+ct)^2}$ represents a wave moving to the left at speed c.

By comparing the wave equation we found for the string with this general form, we find that the wave velocity in a string is given by

$$c = \sqrt{T/\rho}. \tag{7.7}$$

The velocity increases as we increase the tension in the string and decreases as the mass density of the string increases.

This analysis is not limited to a Gaussian shape; any function $f(x \pm ct)$ inserted in the wave equation will give

$$f''(x \pm ct) = (\pm c)^2 f''(x \pm ct) \tag{7.8}$$

where f'' represents the second derivative of the function with respect to its argument.

Question 7.1

Use MATLAB to plot the function $y(x,t) = 5\, e^{-(x-0.25t)^2}$ as a function of x from $x = -2$ to $x = 6$ at $t = 0$, $t = 4$, and $t = 8$ s. What is the velocity of the wave? Take a derivative to find an expression for $\frac{dy}{dx}$ and plot $\frac{dy}{dx}$ for $t = 4$ s on the same graph. What does $\frac{dy}{dx}$ represent?

7.2   Harmonic waves

In particular, harmonic waves of the form

$$y(x,t) = A \cos\left(k(x \pm ct)\right) \tag{7.9}$$

are the most commonly considered wave solutions. Since sines and cosines (or complex exponentials) are *complete functions*, any function $f(x \pm ct)$ can be represented by a sum of sine and cosines (or a sum of complex exponentials). Expressing a function in terms of its harmonic components is called *Fourier analysis*.

If we examine the cosine form above, we can interpret the elements of this wave expression. Setting $t = 0$ and plotting $y(x,0) = A \cos(kx)$ vs. x, we find Figure 7.3. At $x = 0$, the wave has the value of $y(0,0) = A \cos(0) = A$, where A, the maximum value of $y(x,t)$, is the *wave amplitude*. When $kx = 2\pi$, the wave has gone through a complete cycle and returned to the value $y(\lambda,0) = A \cos(2\pi) = A$. The value of x corresponding to one complete cycle in x is called the *wavelength* λ. In terms of λ, the constant k (the *wave vector*) is given by

$$k = \frac{2\pi}{\lambda} \left(\frac{\text{radians}}{\text{m}}\right). \tag{7.10}$$

When $x = \lambda$ (or 2λ, 3λ, etc.), $kx = 2\pi$ radians (or 4π, 6π, etc.), and the cosine function has completed a complete cycle.

Similarly, if we look at the wave as a function of time at $x = 0$, we end up with the familiar harmonic oscillator expression: $y(0, t) = A \cos(kct) = A \cos \omega t$, where kc, the wave vector times the wave velocity, equals the angular frequency ω in radians/s: $kc = \omega$. The *period* T of the wave – the time it takes for the wave to complete one complete cycle (2π radians) at a fixed value of x is given by $T = 2\pi/\omega$. The frequency f in cycles/s (hertz) is given by $f = 1/T = \omega/2\pi$.

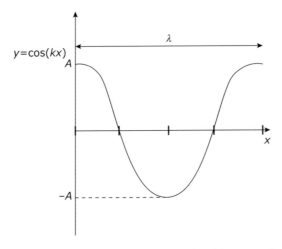

Figure 7.3 Wave position as function of position at $t = 0$.

In terms of frequency in hertz and wavelength in meters, we find the wave velocity is given by

$$c = \frac{\omega}{k} = \frac{2\pi f}{2\pi/\lambda} = \lambda f. \qquad (7.11)$$

Since $f = 1/T$ this gives that the wave speed $c = \lambda/T$; the wave travels a distance of one wavelength in a time equal to one period of the wave.

Note on units: The wave speed equation $c = \lambda/T$, where the wavelength λ is in units of length (meters) and the period T is in units of time (seconds), yields the correct units of velocity, meters/second. Since a complete cycle of a sine or cosine function is 2π radians, the wave vector $k = 2\pi/\lambda$ is in units of radians/meter (often abbreviated as m^{-1}) and the circular frequency $\omega = 2\pi f$ is in units radians/second (abbreviated as s^{-1}). Thus we find dimensionally that the wave velocity $c = \omega\left(\frac{radians}{second}\right)/k\left(\frac{radians}{meter}\right) = \frac{\omega}{k}\left(\frac{meters}{second}\right)$.

We have expressed the wave in the form of a cosine curve

$$y(x,t) = A\,\cos\,(kx \pm \omega t),$$

but to allow for the possibility that the wave may not start at its maximum amplitude at $x = 0$ and $t = 0$, we should express $y(x,t)$ as a sum of cosine and sine terms

$$y(x,t) = A\,\cos\,(kx \pm \omega t) + B\,\sin\,(kx \pm \omega t). \qquad (7.12)$$

Alternatively, we could express the relative phase by adding a phase shift δ to the argument of the cosine

$$y(x,t) = D\,\cos\,(kx \pm \omega t + \delta) = D\,\cos\delta\cos\,(kx \pm \omega t) - D\,\sin\delta\sin\,(kx \pm \omega t), \qquad (7.13)$$

where we get the last equality by applying the trigonometric identity

$$\cos\,(\theta \pm \varphi) = \cos\theta\cos\varphi \mp \sin\theta\sin\varphi$$

From this we can identify the amplitudes A and B of the cosine and sine terms above in terms of the wave amplitude D and phase shift δ as $A = D\cos\delta$ and $B = -D\sin\delta$ with $D^2 = A^2 + B^2$. We can also express the wave as a complex exponential as we did with the harmonic oscillators in the last chapter

$$y(x,t) = Re\{\tilde{D}e^{i(kx-\omega t)}\} \qquad (7.14)$$

where the complex amplitude $\tilde{D} = De^{i\delta}$. We can represent waves traveling in both positive and negative x-directions by allowing the wave vector k to be positive for waves traveling to the right and negative for waves traveling to the left, noting that: $Re\{e^{i(-kx-\omega t)}\} = Re\{e^{i(kx+\omega t)}\} = \cos\,(kx + \omega t)$.

Question 7.2

Use Euler's relation to verify $Re\{e^{i(-kx-\omega t)}\} = Re\{e^{i(kx+\omega t)}\} = \cos\,(kx + \omega t)$.

In general, this last form for the wave in terms of the complex exponential is the most useful form. In most cases, what we are interested in is the amplitude of the wave $|\tilde{D}|$ and the relative phase δ. If we do want to know the real amplitude of the wave at a particular value of x and t, we can calculate the wave amplitude $\tilde{D}$ using the complex exponential notation, and then multiply by $e^{i(kx-\omega t)}$ and take the real part to find the physically measurable value of $y(x, t)$.

7.3   Addition of waves: interference, reflection, standing waves, and beats

If we have two waves traveling in a medium at the same time with

$$\tilde{y}_1(x, t) = \tilde{D}_1 e^{i(\pm k_1 x - \omega_1 t)} \tag{7.15a}$$

and

$$\tilde{y}_2(x, t) = \tilde{D}_2 e^{i(\pm k_2 x - \omega_2 t)} \tag{7.15b}$$

the total displacement is the sum of the two wave displacements

$$\tilde{y}(x, t) = \tilde{y}_1(x, t) + \tilde{y}_2(x, t). \tag{7.16}$$

The phenomenon of *wave interference* is when waves arrive at a point either in-phase and add together (constructive interference) or out-of-phase and cancel out either partially or completely (destructive interference).

For waves in a string, the most common way to end up with two waves in the same string is by reflection from one end of the string. A wave stream traveling in the string hits one end of the string and reflects back down the string. The most common reflection condition is where the end of the string is rigidly fixed so the string cannot move at its end. In this case the wave energy that is carried to the end of the string is reflected without loss of energy into a wave traveling at the same frequency and velocity in the opposite direction. The total wave displacement of the incident plus reflected wave is

$$y(x, t) = y_{in}(x, t) + y_{ref}(x, t) = \mathcal{Re}\{\tilde{D}_{in} e^{i(+kx-\omega t)} + \tilde{D}_{ref} e^{i(-kx-\omega t)}\}. \tag{7.17}$$

Since the string is fixed at the end, we have $y(x, t) = 0$ for all times t at the value of x corresponding to the fixed end of the string. Without loss of generality, we can assume that the fixed end of the string is at $x = 0$ which gives us

$$y(x = 0, t) = 0 = \mathcal{Re}\{\tilde{D}_{in} e^{-i\omega t} + \tilde{D}_{ref} e^{-i\omega t}\}. \tag{7.18}$$

This must hold at all values of time t, which can only be satisfied if $\tilde{D}_{ref} = -\tilde{D}_{in}$, equal in both phase shift and amplitude. We can assume that the phase shifts are both zero, so $\tilde{D}_{in}$ and $\tilde{D}_{ref}$ are both real. So the total displacement of the string from the incident plus the reflected wave is

$$\tilde{y}(x, t) = D_{in} e^{i(kx-\omega t)} - D_{in} e^{i(-kx-\omega t)} = D_{in}(e^{ikx} - e^{-ikx})e^{-i\omega t} = D_{in}(2i \sin kx)e^{-i\omega t}. \tag{7.19}$$

The final step comes from expanding the complex exponentials as $e^{\pm ikx} = \cos kx \pm i \sin kx$ and subtracting out the cosine terms. The disconcerting factor of i can be handled by substituting $i = e^{i\pi/2}$ which becomes a constant phase factor in the $e^{-i\omega t}$ exponential

$$y(x,t) = \mathcal{Re}\left\{2D_{in} \sin kx\, e^{-i(\omega t - \pi/2)}\right\} = 2D_{in} \sin kx \cos\left(\omega t - \pi/2\right). \tag{7.20}$$

This describes a displacement that oscillates in time between a maximum value $+2D_{in} \sin kx$ and a minimum value $-2D_{in} \sin kx$. Of particular note is that when $kx = 0, \pm\pi$, $\pm 2\pi, \pm 3\pi, \ldots$ (when $x = \frac{n\lambda}{2}$ where n is any integer) the displacement is always zero! Instead of progressing to the right or the left, the string oscillates in a fixed x-position. This wave pattern is called a *standing wave* and the x-positions where the displacement is always zero are called the *nodes* of the standing wave.

If the other end of the string at $x = -d$ is also clamped, we have the additional condition:

$$y(-d, t) = 2D_{in} \sin\left(-kd\right) \cos\left(\omega t - \pi/2\right) = 0 \tag{7.21}$$

which can only be satisfied if $kd = -\pi, -2\pi, -3\pi, \ldots$ or, equivalently, $d = -\lambda/2, -\lambda$, $-3\lambda/2, \ldots$ If the length of the string is fixed at d, only certain wavelengths, or equivalently only certain frequencies, can be supported as standing waves in the string. In terms of the tension and mass density of the string which determine the velocity, we find the allowed frequencies that are supported by a string fixed at both ends are

$$f_n = \sqrt{T/\rho}\,\frac{n}{2d}. \tag{7.22}$$

Question 7.3

Use $\lambda = c/f$ and the expression for the wave velocity c in Eq. (7.7) to show that $kd = -n\pi$ with $n = 1$, 2, 3... leads to Eq. (7.22).

The dependencies here are familiar to anyone who plays a violin, guitar, or any other stringed instrument. A heavier string has a lower frequency (*pitch*) and a lighter string has a higher pitch. The string can be tuned by tightening the tension of the strings with the tuning pins, raising the pitch, or loosening the tension to drop the pitch. The instrument is played by changing the length of the string with your fingers on the fingerboard, a shorter string has a higher frequency/pitch. The integer n indicates the *order* of the resonance. Most often the instrument is played using the first-order ($n = 1$) resonance, but a skillful player can get a higher frequency *harmonic* out of the string by resting his finger lightly on the middle of the string, forcing the string to vibrate in the $n = 2$ harmonic.

If two slightly different frequency waves are propagating in the same medium, an observer at one position will observe a total wave which oscillates between constructive

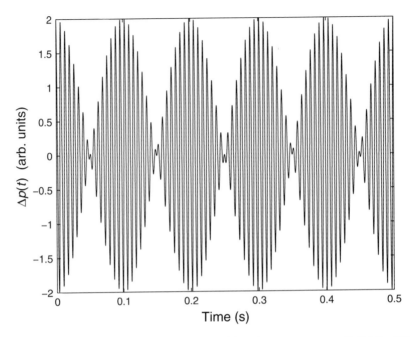

Figure 7.4 Sound wave intensity as function of time from two sources of slightly different frequencies showing the beat frequency in the sound intensity which results from the alternating constructive and destructive interference of the two component waves.

and destructive interference at a rate that depends on the difference between the frequencies. For example, consider two sources of sound with stable phase and slightly different frequency: $\omega_1 = \overline{\omega} + \frac{\Delta\omega}{2}$ and $\omega_2 = \overline{\omega} - \frac{\Delta\omega}{2}$ observed at a position $x = 0$. The total displacement is

$$y(t) = y_1(t) + y_2(t) = Ae^{-i\left(\overline{\omega} + \frac{\Delta\omega}{2}\right)t} + Ae^{-i\left(\overline{\omega} - \frac{\Delta\omega}{2}\right)t} = Ae^{-i\overline{\omega}t}\left(e^{-i\frac{\Delta\omega}{2}t} + e^{+i\frac{\Delta\omega}{2}t}\right) = 2Ae^{-i\overline{\omega}t}\cos\frac{\Delta\omega}{2}t.$$

(7.23)

The effect is to hear a tone of frequency $\overline{\omega}$ with the intensity modulated at a frequency $\frac{\Delta\omega}{2}$ – referred to as the *beat frequency*. Figure 7.4 plots the sound wave which results from two sources at 156 and 146 Hz.

Question 7.4

From Figure 7.4, find the frequency in cycles/s (Hz) of the beat frequency. Show that this is consistent with Eq. (7.23).

7.4   Sound waves in fluids

A sound wave in a fluid is both a longitudinal fluid displacement wave and a coupled longitudinal fluid pressure wave which creates the force that causes the fluid displacement. (Since a fluid is free to move in a direction perpendicular to the applied pressure, it does not support transverse stress or transverse waves.) Consider the equation of motion of a small volume of the fluid with density $\rho = \rho_o + \Delta\rho$ accelerated by the force due to a pressure $P = P_o + \Delta P$. Before the wave arrives, the fluid element to be accelerated at $x = x_o$ is of cross-section area A and width dx and the base density is given by $\rho_o = \frac{M}{A dx}$ where M is the mass of the molecules in the volume. When the wave arrives, it displaces the molecules by an amount $u(x)$. The molecules in our volume element are displaced to a position $x = x_o + u(x)$ and the width of the volume element is changed from dx to $dx + du$ as shown in Figure 7.5. The new density of the fluid in the volume element is

$$\rho = \frac{M}{A(dx + du)} = \rho_o \frac{dx}{dx + du} = \rho_o \left(1 + \frac{du}{dx}\right)^{-1} \cong \rho_o \left(1 - \frac{du}{dx}\right),\qquad(7.24)$$

where the last step comes from the binomial approximation assuming that $\frac{du}{dx} \ll 1$.

Hydrostatic pressure in a fluid is assumed to be the same in all directions, which means that the net force on the volume element is given by the cross-section area A multiplied by the pressure at S minus the pressure at T,

$$F_{net} = A(P_S - P_T) = -A\frac{\partial P}{\partial x}dx.\qquad(7.25)$$

Applying Newton's Second Law, $F = ma$, we find

$$-A\frac{\partial P}{\partial x}dx = M\frac{d^2 u}{dt^2} = \rho_o A\, dx \frac{d^2 u}{dt^2}\qquad(7.26)$$

which, on canceling the Adx on both sides yields

$$-\frac{\partial P}{\partial x} = \rho_o \frac{d^2 u}{dt^2}.\qquad(7.27)$$

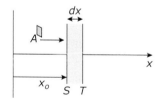

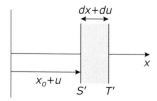

Figure 7.5 Propagation of sound wave in a fluid. A test volume of the fluid $A\Delta x$ changes both its position and its width.

Expanding the partial derivative on the left-hand side, and using the above expansion for the density $\rho = \rho_o(1 - \frac{du}{dx})$, we get

$$-\frac{\partial P}{\partial x} = -\frac{\partial P}{\partial \rho}\frac{\partial \rho}{\partial x} = -\frac{\partial P}{\partial \rho}\frac{\partial}{\partial x}\left\{\rho_o\left(1 - \frac{du}{dx}\right)\right\} = \frac{\partial P}{\partial \rho}\rho_o\frac{\partial^2 u}{\partial x^2}. \tag{7.28}$$

Canceling out the background density ρ_o we find from Eq. (7.27)

$$\frac{\partial^2 u}{\partial t^2} = \frac{1}{\partial \rho/\partial P}\frac{\partial^2 u}{\partial x^2} = \frac{\partial P}{\partial \rho}\frac{\partial^2 u}{\partial x^2}, \tag{7.29}$$

which we recognize as the wave equation for the displacement with a wave velocity $v = \sqrt{\frac{\partial P}{\partial \rho}}$.

The bulk modulus of the fluid $B = -\frac{\Delta P}{\Delta V/V} = -\frac{dP}{dV/V} = \rho_o\frac{dP}{d\rho}$, where the last equality comes from

$$\rho = \frac{M}{V} \Rightarrow d\rho = -\frac{M}{V^2}dV = -\rho_o\frac{dV}{V}. \tag{7.30}$$

Expressing the derivative of pressure with respect to density in terms of the bulk modulus B and density ρ_o, $\frac{\partial P}{\partial \rho} = B/\rho_o$, we get the wave equation

$$\frac{\partial^2 u}{\partial t^2} = \frac{B}{\rho_o}\frac{\partial^2 u}{\partial x^2}, \tag{7.31}$$

from which we determine that the wave velocity in a fluid is $v = \sqrt{B/\rho_o}$. Comparing with the wave velocity in a string $c = \sqrt{T/\rho}$ we see that the bulk modulus B plays the same role in sound waves as the tension plays in waves on a string, and the velocity is inversely proportional to the square root of the linear density (kg/m) in a string and to the square root of the fluid volume density (kg/m^3) in a fluid.

Sound in a fluid moves as a wave of displacements of the particles of the fluid. The wave equation above has solutions – function of the variable $(x - vt)$ – that we are familiar with from waves on a string. In particular, the harmonic wave solutions can be expressed in the complex exponential form

$$u(x, t) = u_m e^{i(kx - \omega t)}. \tag{7.32}$$

These displacements are in the same direction as the wave is moving, which is the definition of a *longitudinal wave*. Accompanying this motion of displacement, there is a variation of pressure as the fluid is alternately compressed and expanded. Using the defining equation of the bulk modulus $B = -\frac{\Delta P}{\Delta V/V}$ and recognizing that the change in volume can be expressed in terms of the change in the longitudinal displacement and the cross-section area of the beam as $\Delta V = A \Delta u$ while the volume $V = A \Delta x$, we find the relation between the pressure wave and the displacement wave

$$\Delta P = -B\frac{\Delta V}{V} = -B\frac{A\Delta u}{A\Delta x} \to -B\frac{\partial u}{\partial x}. \tag{7.33}$$

Applying this relation to the harmonic displacement wave above, we get

$$\Delta P = -B\frac{\partial u}{\partial x} = -B(ik)u_m e^{i(kx - \omega t)} = -iB\left(\frac{\omega}{v}\right)u_m e^{i(kx - \omega t)} \tag{7.34}$$

where for the last equality we have used the relation $v = \omega/k$. Note that, since it follows from the wave equation that the sound wave velocity $v = \sqrt{B/\rho_o}$, we can eliminate the bulk modulus B from the equation in favor of another material parameter, the speed of sound v, and get the following equations for the coupled longitudinal displacement and pressure waves:

$$u(x,t) = u_m\, e^{i(kx-\omega t)} \tag{7.35a}$$

$$\Delta P(x,t) = -i\rho_o v\omega u_m\, e^{i(kx-\omega t)}. \tag{7.35b}$$

We recognize the factor of $-i$ in the expression for ΔP in Eq. (7.35b) as a phase factor $e^{-i\frac{\pi}{2}}$ which represents that the pressure wave is $-90°$ out of phase from the displacement wave. The quantity $\rho_o v$, the product of the medium density and the wave velocity, is the *acoustic characteristic impedance* of the medium and plays a key role in the reflection and transmission of sound at the interface between two materials.

7.5 Three-dimensional plane waves

So far, we have restricted our sound waves to move only in the x-direction, as might represent the one-dimensional propagation of sound in a tube. The solutions above for the displacement and pressure waves represent the propagation of the wave in the positive and negative x-direction, depending on the sign of k. (Positive k represents waves propagating in the $+x$-direction, and negative k values represent waves propagating in the $-x$-direction.)

If we remove the restriction that the wave propagate only in the $\pm x$-direction, we need to extend the wave equation to depend on x, y, and z. This involves replacing the second derivative with respect to x with the *scalar Laplacian* ∇^2. In rectangular coordinate systems, the wave equation becomes:

$$\nabla^2 u = \frac{\partial^2 u}{\partial x^2} + \frac{\partial^2 u}{\partial y^2} + \frac{\partial^2 u}{\partial z^2} = \frac{1}{v^2}\frac{\partial^2 u}{\partial t^2}. \tag{7.36}$$

The solutions to this three-dimensional wave equation in rectangular coordinates are *plane waves*:

$$u(x,y,z,t) = u_m\, e^{i(\vec{k}\cdot\vec{r}-\omega t)} \tag{7.37a}$$

$$\Delta P(x,y,z,t) = -i\rho_o v\omega u_m\, e^{i(\vec{k}\cdot\vec{r}-\omega t)} \tag{7.37b}$$

where $\vec{r} = x\hat{x} + y\hat{y} + z\hat{z}$ is a position vector and $\vec{k} = k_x\hat{x} + k_y\hat{y} + k_z\hat{z}$ is *the three-dimensional wave vector*. The magnitude of $\vec{k}$ is given in terms of the wave velocity and the frequency by

$$|\vec{k}| = \frac{\omega}{v} = \frac{2\pi}{\lambda} \tag{7.38}$$

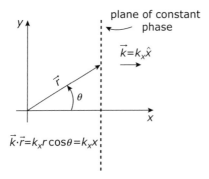

Figure 7.6 Planes of constant phase for three-dimensional wave with $\vec{k} = k_x \hat{x}$ are identified by $\vec{k} \cdot \vec{r} = k_x x$.

and the direction of $\vec{k}$ is in the direction of the wave velocity and perpendicular to planes of constant phase. For example, if $\vec{k}$ is in the $\hat{x}$-direction, $\vec{k} = k_x \hat{x}$, then, as shown in Figure 7.6, the phase factor $\vec{k} \cdot \vec{r} = k_x \hat{x} \cdot \vec{r} = k_x \hat{x} \cdot (x\hat{x} + y\hat{y} + z\hat{z}) = k_x x$, and all points in three-dimensional space with the same value of x have the same phase.

The energy flow of an acoustic wave involves both the kinetic energy of the fluid particles related to the displacement $u(\vec{r}, t)$ and the elastic potential energy related to the pressure deviation $\Delta P(\vec{r}, t)$. The average total energy flow per cross-sectional area of a harmonic wave such as the one we have been considering is:

$$S = \frac{1}{2}\rho_o v \omega^2 u_m{}^2 = \frac{1}{2}\frac{(\Delta P_m)^2}{\rho_o v} \frac{\text{W}}{\text{m}^2}, \tag{7.39}$$

where ΔP_m is the amplitude of the pressure wave, given by Eq. (7.37b) in terms of the maximum displacement by

$$\Delta P_m = \rho_o v \omega u_m. \tag{7.40}$$

7.6   Propagation in absorbing media

If a wave propagates through a medium that absorbs energy from the wave, the wave amplitude must decrease with distance as the wave loses energy to the medium. The energy can either go into heating up the medium (*absorption*) or in generating waves that are directed in other directions (*scattering*). Both absorption and scattering can be described by adding an imaginary part to the wave vector $\vec{k} = \vec{k}_r + i\vec{k}_i$. Applying this to a sound pressure wave, for example, gives

$$\Delta P(\vec{r}, t) = \Delta P_m \, e^{i(\vec{k}_r + i\vec{k}_i)\cdot\vec{r} - i\omega t} = \Delta P_m e^{-\vec{k}_i \cdot \vec{r}} e^{i(\vec{k}_r \cdot \vec{r} - \omega t)}. \tag{7.41}$$

This is easier to see if we apply our usual convention that the physical wave is found by taking the real part of the complex wave form to find

$$\Delta P(\vec{r}, t) = \Delta P_m e^{-\vec{k}_i \cdot \vec{r}} \cos\left(\vec{k}_r \cdot \vec{r} - \omega t\right). \tag{7.42}$$

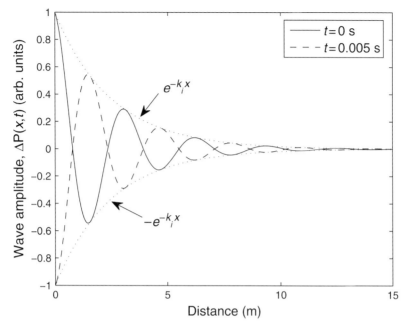

Figure 7.7 Pressure variation as a function of distance for acoustic wave propagating in an absorbing medium plotted for two different times.

We recognize this as a wave propagating in the direction of $\vec{k}_r$, the real part of the wave vector, with an angular frequency ω and a wavelength $\lambda = \frac{2\pi}{|\vec{k}_r|}$. However, the amplitude of this wave is not constant in space, but the imaginary part of the wave vector causes the wave amplitude to decrease exponentially as the wave propagates through the factor $e^{-\vec{k}_i \cdot \vec{r}}$. The simplest case to consider is the case where the wave propagates in the $\hat{x}$-direction, $\vec{k}_r = k_r \hat{x}$ and the imaginary part of the wave vector is also in the $\hat{x}$-direction, $\vec{k}_i = k_i \hat{x}$, so $\vec{k} \cdot \vec{r} = (k_r + ik_i)\hat{x} \cdot (x\hat{x} + y\hat{y} + z\hat{z}) = (k_r + ik_i)x$. The wave is then represented by

$$\Delta P(x, t) = \Delta P_m \, e^{-k_i x} e^{i(k_i x - \omega t)}. \tag{7.43}$$

A snap-shot of this wave at two different times, $t = 0$ and $t = \frac{T}{2} = \frac{\pi}{\omega}$ is shown in Figure 7.7: the wave propagates inside an envelope which decreases in distance by $\pm e^{-k_i x}$.

In a wave like this with $\vec{k}_i$ parallel to $\vec{k}_r$ (known as a *homogeneous wave*) the amplitude decreases exponentially in the direction of propagation. If $\vec{k}_i$ is not parallel to $\vec{k}_r$, the wave decays exponentially in a different direction than the direction of propagation. We will see an example of such an *inhomogeneous wave* when we consider propagation across a boundary into an absorbing medium.

Example 7.1 A plane wave of 100 Hz sound propagates in the positive x-direction in an absorbing medium. The wave vector of the pressure wave can be expressed as $k = 20 + 0.2i$ (m^{-1}). (a) In terms of the maximum pressure variation ΔP_m at $x = 0$, write an equation for $\Delta P(x,t)$. (b) What is the wavelength of the sound in the material? (c) What is the sound velocity in the material? (d) How far does the sound propagate before the pressure intensity is decreased by a factor of 4?

Solution: (a) In terms of the maximum amplitude, a plane wave is given by $\Delta P(\vec{r}, t) = \Delta P_m\, e^{i(\vec{k}\cdot\vec{r} - \omega t)}$. In this case, $\vec{k} = (20 + 0.2i)\hat{x}$ which gives the wave as a function of position and time as $\Delta P(x, t) = \Delta P_m\, e^{i((20+0.2i)\hat{x}\cdot\vec{r} - \omega t)} = \Delta P_m\, e^{i((20+0.2i)x - \omega t)} = \Delta P_m\, e^{-0.2x + i(20x - \omega t)} = \Delta P_m\, e^{-0.2x} e^{i(20x - \omega t)}$.

This is the form for a damped propagating wave with an amplitude $\Delta P_m\, e^{-0.2x}$ and an oscillatory part $e^{i(20x - \omega t)}$.

(b) If we take the real part of the wave above we find $\Delta P(x, t) = \Delta P_m\, e^{-0.2x} \cos(20x - \omega t)$. Clearly the wavelength, the distance that the wave travels between successive maxima is given by $\lambda = \frac{2\pi}{Re(k)} = \frac{2\pi}{20} = 0.314$ m.

(c) The sound velocity is given by the wavelength λ (in m/cycle) times the frequency f (in cycles/s), or equivalently: $v = \omega/Re(k) = 200\pi/20 = 31.4$ m/s.

(d) The pressure intensity is given by the *amplitude* of the wave $\Delta P_m\, e^{-0.2x}$, where ΔP_m is the amplitude at $x = 0$. The question then is at what distance x does $\Delta P_m\, e^{-0.2x} = \Delta P_m/4$, or $x = \frac{\ln(0.25)}{-0.2} = 6.93$ m.

If we were asked for the instantaneous $\Delta P(x,t)$ at a given value of x and t, we would use the equation $\Delta P(x, t) = \Delta P_m\, e^{-0.2x} \cos(20x - \omega t)$ or, equivalently, put the values of x and t into $\Delta P(x, t) = \Delta P_m\, e^{i(kx - \omega t)}$ (with the full complex form for k), and then take the real part of the expression.

7.7   Reflection and refraction of plane waves at an interface between two media

Plane waves are a good approximation for waves that are produced by an extended coherent-phase source and can also approximate a small section of an expanding spherical wave if the wave is far enough from the source that the curvature over a small area can be neglected. If we consider a wave incident on a plane interface between two different media a long distance from the source, we can approximate the problem by considering the behavior of the plane wave incident on the interface (with wave vector $\vec{k}_i$), the plane wave reflected from the interface (with wave vector $\vec{k}_r$), and the plane wave which propagates through the interface into the new medium (with wave vector $\vec{k}_t$). We show these three wave vectors in the plane that contains the normal to the interface and the incident wave vector $\vec{k}_i$ in Figure 7.8.

Note that all three plane waves – the incident wave, the reflected wave, and the transmitted wave – exist for all values of y and also all values of z perpendicular to

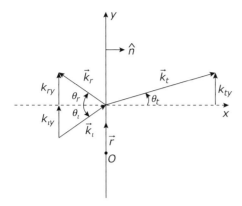

Figure 7.8 Wave vectors $\vec{k}_i$, $\vec{k}_r$, and $\vec{k}_t$ for plane waves incident on, reflected from, and transmitted through an interface between two media.

the plane of the figure. If we assume that the boundary is located at $x = 0$, only the transmitted wave exists for $x > 0$ in Medium 2, while the incident and reflected wave are defined in Medium 1 for $x < 0$. On the interface between the two media, we can apply *boundary conditions* that relate the waves in Medium 1 to the waves in Medium 2. The boundary conditions depend on what kind of wave we are discussing and the physics of the wave interaction along the interface.

For sound waves, the boundary conditions are usually taken as (1) on the interface the wave pressure in Medium 1 ΔP_1 is equal to the pressure in Medium 2 ΔP_2, and (2) on the interface the normal component of the particle velocity is equal in Medium 1 and Medium 2. The first condition follows directly from Newton's Third Law: the force that Medium 1 exerts on Medium 2 ($Area \cdot \Delta P_1$) is equal to the force that Medium 2 exerts on Medium 1 ($Area \cdot \Delta P_2$). Canceling out the area gives $\Delta P_1 = \Delta P_2$ on the interface. The second condition follows from the continuity of the media – if the molecules of the two media are moving at different velocities normal to the interface it will create a separation between the two media. This phenomenon of *cavitation* can happen in exceptional circumstances – for example at the interface between water and a powerful boat propeller, but if we restrict our attention to sound waves that are not too intense, we can write our second boundary condition as $\frac{du_n}{dt}]_1 = \frac{du_n}{dt}]_2$ on the interface, where u_n is the particle displacement normal to the interface and the subscripts 1 and 2 refer to Medium 1 and Medium 2.

Applying the first condition to the incident, reflected, and transmitted waves, we find:

$$\Delta P_i e^{i(\vec{k}_i \cdot \vec{r} - \omega t)} + \Delta P_r e^{i(\vec{k}_r \cdot \vec{r} - \omega t)} = \Delta P_t e^{i(\vec{k}_t \cdot \vec{r} - \omega t)} \tag{7.44}$$

where ΔP_i, ΔP_r, and ΔP_t are the complex amplitudes of the incident, reflected, and transmitted waves respectively. The second condition gives:

$$\frac{d}{dt}\vec{u}_i \cdot \hat{n} e^{i(\vec{k}_i \cdot \vec{r} - \omega t)} + \frac{d}{dt}\vec{u}_r \cdot \hat{n} e^{i(\vec{k}_r \cdot \vec{r} - \omega t)} = \frac{d}{dt}\vec{u}_t \cdot \hat{n} e^{i(\vec{k}_t \cdot \vec{r} - \omega t)} \tag{7.45}$$

where $\vec{u}_i$, $\vec{u}_r$, and $\vec{u}_t$ are the longitudinal molecular displacements of the waves and $\hat{n}$ is a unit vector normal to the interface.

These equations must hold at all times and at all locations on the interface. The time and spatial dependence of the plane waves is entirely contained in the complex exponentials $e^{i(\vec{k}\cdot\vec{r}-\omega t)}$, so for the two conditions to hold at all times and at every point on the interface we must have

$$e^{i(\vec{k}_i\cdot\vec{r}-\omega t)} = e^{i(\vec{k}_r\cdot\vec{r}-\omega t)} = e^{i(\vec{k}_t\cdot\vec{r}-\omega t)} \qquad (7.46)$$

where $\vec{r}$ is the position vector for any point on the interface. This relation immediately gives us the condition that the frequency ω of the incident, reflected, and transmitted waves has to be the same – a result we have anticipated by giving them all the same symbol ω. The frequency of a wave is not changed by reflection from or transmission through an interface. This allows us to cancel the $e^{-i\omega t}$ time dependence out of each term and consider spatial and amplitude relationships independent of the time dependence.

Considering next the spatial dependence, after canceling out the $e^{-i\omega t}$ time dependence, we have

$$e^{i(\vec{k}_i\cdot\vec{r})} = e^{i(\vec{k}_r\cdot\vec{r})} = e^{i(\vec{k}_t\cdot\vec{r})} \Rightarrow \vec{k}_i\cdot\vec{r} = \vec{k}_r\cdot\vec{r} = \vec{k}_t\cdot\vec{r} . \qquad (7.47)$$

If we set the origin of the coordinate system at point O on the interface in the plane of the three wave vectors as shown in Figure 7.8, then the vector $\vec{r}$ for any point in the interface is parallel to the interface in the $\hat{y}$-direction as shown. Taking the dot product of $\vec{r}$ with the wave vectors $\vec{k}_i$, $\vec{k}_r$, and $\vec{k}_t$ and canceling out the magnitude of $\vec{r}$ at the point under consideration gives us then:

$$\vec{k}_i\cdot\hat{y}|\vec{r}| = \vec{k}_r\cdot\hat{y}|\vec{r}| = \vec{k}_t\cdot\hat{y}|\vec{r}| \Rightarrow k_{iy} = |\vec{k}_i|\sin\theta_i = k_{ry} = |\vec{k}_r|\sin\theta_r = k_{ty} = |\vec{k}_t|\sin\theta_t \qquad (7.48)$$

with the angles of incidence θ_i, reflection θ_r, and transmission θ_t measured from the normal to the interface as shown in Figure 7.8. So the y-components of the wave vectors must be equal to satisfy either boundary condition at all points on the interface. Recalling that $|\vec{k}| = 2\pi/\lambda = \omega/v$ where ω is the (angular) frequency of waves and v is the wave velocity, we see that the magnitudes of the wave vectors are equal only in media with the same wave velocity. Since the incident and reflected waves are in the same medium, we have $|\vec{k}_i| = |\vec{k}_r|$ and $\sin\theta_i = \sin\theta_r$ leading to the relation $\theta_i = \theta_r$: *for plane waves incident on a plane interface between two media the angle of incidence is equal to the angle of reflection.*

In general, the wave velocity – and thus the magnitude of the wave vectors – in two different media will be unequal. If the magnitudes of the wave vector are unequal but their y-components are the same, this implies that the direction of the wave vectors – the propagation direction of the plane wave – will change on crossing an interface between two media with different wave velocities. The relation between the angle of incidence and the angle of transmission is given by the following relation:

$$|\vec{k}_i|\sin\theta_i = \frac{\omega}{v_1}\sin\theta_i = |\vec{k}_t|\sin\theta_t = \frac{\omega}{v_2}\sin\theta_t \Rightarrow \frac{\sin\theta_i}{v_1} = \frac{\sin\theta_t}{v_2} . \qquad (7.49)$$

This relationship between the angle of incidence θ_i and the angle of transmission θ_t must hold for all plane waves incident on an interface. In optics this equation, expressed in terms of the refractive indices $n = c/v$ (where $c = 2.9979 \times 10^8$ m/s is the speed of light in vacuum), is referred to as *Snell's Law*.

Example 7.2 Consider a plane sound wave propagating in water with a velocity $v_1 = 1480$ m/s incident at an angle of $40°$ from the normal on the boundary with air with $v_2 = 343$ m/s. Find the angle of propagation of the transmitted plane wave in air. Now reverse the problem with the wave in air incident at an angle of $40°$ from the normal on the boundary with air. What is the angle of the propagated wave?

   Solution: From Eq. (7.48) with $\theta_i = 40°$ and $v_1 = 1480$ m/s we find $\sin\theta_t = \frac{v_2}{v_1}\sin 40° = \frac{343}{1480}\sin 40° = 0.149$ from which we find $\theta_t = \sin^{-1}(0.149) = 8.57°$. For the case with sound propagating from air into water we have $\sin\theta_t = \frac{1480}{343}\sin 40° = 2.77$. We know, of course, that there is no angle which has a sine of greater than 1.0! In fact, there is no transmitted wave from air into water when the angle of incidence is $40°$. In fact, there is **no** transmitted wave if $\sin\theta_t > 1$, or if the angle of incidence is greater than a critical angle θ_c where $\theta_c = \sin^{-1}\left(\frac{343}{1480}\right) = 13.4°$. For angles of incidence $\theta_i > \theta_c$ the wave is totally reflected (with $\theta_r = \theta_i$), a phenomenon known in optics – where the velocity of light inside a material is almost always less than the velocity of light in air – as *total internal reflection*. Does this result help explain why it is so difficult to shout directions from the edge of a pool to swimmers underwater?

   All that remains is to find the relationship between the wave amplitudes of the incident, reflected, and transmitted waves. We will consider the case of normal incidence where $\theta_i = 0°$ and, from the condition of the continuity of phase $\vec{k}_i \cdot \vec{r} = \vec{k}_r \cdot \vec{r} = \vec{k}_t \cdot \vec{r}$, we have $\theta_r = \theta_t = 0°$ as well. The generalization to an arbitrary angle of incidence θ_i will be left as a problem.

   At normal incidence, $\vec{u}_i \cdot \hat{n} = |\vec{u}_i|\cos\theta_i = |\vec{u}_i|$ with similar relations for the reflected and transmitted waves, with the exception that the reflected wave is propagating anti-parallel (rather than parallel) to $\hat{n}$, so $\vec{u}_r \cdot \hat{n} = -|\vec{u}_i|$. Recalling the relationship between the pressure change and particle displacement in Eq. (7.40), $\Delta P_m = \rho_0 v\,\omega\,u_m$, our conditions on the continuity of ΔP and the normal component of particle velocity when we cancel out the exponential factors, lead to the two equations

$$\Delta P_i + \Delta P_r = \Delta P_t \tag{7.50}$$

$$\frac{\Delta P_i}{\rho_1 v_1} - \frac{\Delta P_r}{\rho_1 v_1} = \frac{\Delta P_t}{\rho_2 v_2}. \tag{7.51}$$

Substituting the first equation for ΔP_t in the second equation and rearranging terms, we find the amplitude reflection coefficient

$$r = \frac{\Delta P_r}{\Delta P_i} = \frac{\rho_2 v_2 - \rho_1 v_1}{\rho_2 v_2 + \rho_1 v_1}. \tag{7.52}$$

Substituting instead for ΔP_r we find the amplitude transmission coefficient

$$t = \frac{\Delta P_t}{\Delta P_i} = \frac{2\rho_2 v_2}{\rho_2 v_2 + \rho_1 v_1}. \tag{7.53}$$

In these equations, we note the central role of the *acoustic impedance*, ρv, the product of the equilibrium medium density and the sound velocity in the medium, which arises in the relation between the particle displacement (or velocity) and the wave pressure change and also in the equation for wave energy flow.

Example 7.3 Calculate the fraction of sound energy normally incident from air onto water that is reflected from the water surface. At sea level and 15 °C the density of air is $\rho = 1.225 \text{ kg/m}^3$ and the sound velocity is 343 m/s. Water at 15 °C has $\rho = 1000 \text{ kg/m}^3$ and the sound velocity is 1480 m/s.

Solution: The acoustic impedence ρv of air is $\rho_1 v_1 = 1.225 \times 343 = 420 \text{ kg m}^{-2} \text{ s}^{-1}$ while for water $\rho_2 v_2 = 1000 \times 1480 = 1480 \times 10^3 \text{ kg m}^{-2} \text{ s}^{-1}$. The intensity ratio of the reflected pressure wave is given by

$$r = \frac{\Delta P_r}{\Delta P_i} = \frac{\rho_2 v_2 - \rho_1 v_1}{\rho_2 v_2 + \rho_1 v_1} = \frac{1480 - 0.420}{1480 + 0.420} = 0.9994.$$

To consider energy reflection we recall the relation between acoustic sound energy and the pressure variation in Eq. (7.39): $S = \frac{1}{2}\frac{(\Delta P_m)^2}{\rho_o v}\frac{W}{m^2}$. Since the incident and reflected wave are both in air with the same acoustic impedance $\rho_1 v_1$, the energy reflection R is given by

$$R = \frac{S_r}{S_i} = \frac{2\rho_1 v_1 (\Delta P_r)^2}{2\rho_1 v_1 (\Delta P_i)^2} = \frac{(\Delta P_r)^2}{(\Delta P_i)^2} = r^2 = 0.9989.$$

The transmitted energy T is most easily found by considering that, since the interface is of vanishingly small volume and cannot absorb energy, $T + R = 1 \Rightarrow T = 1 - R = 1 - 0.9989 = 0.0011$.

This result can also be found by calculating the transmission ratio of the pressure wave $t = \frac{\Delta P_t}{\Delta P_i} = \frac{2\rho_2 v_2}{\rho_2 v_2 + \rho_1 v_1} = \frac{2 \cdot 1480}{1480 + 0.420} = 1.9994$, but in this case the incident and transmitted waves are not in the same medium so the energy transmission is $T = \frac{S_t}{S_i} = \frac{2\rho_1 v_1 (\Delta P_t)^2}{2\rho_2 v_2 (\Delta P_i)^2} = \frac{\rho_1 v_1}{\rho_2 v_2} t^2 = \frac{0.420}{1480}(1.9994)^2 = 0.0011$, as before.

7.8   Interference on reflection from a layer

Figure 7.9 indicates the multiple reflections when a plane wave is incident from Medium 1 on a layer of Medium 2 of thickness D over a substrate of Medium 3 (assumed to be semi-infinite). The total reflected waves can be found by adding the amplitude of (1) the wave reflected from the first interface (ΔP_1) to (2) the wave that

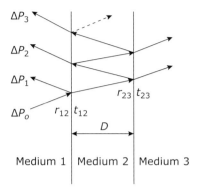

Figure 7.9 Multiply-transmitted and reflected rays resulting from a plane wave incident on a plane interface.

transmits through the first interface, transmits across the layer, reflects off the second interface, transmits across the film again and then transmits through the first interface (ΔP_2), to (3) the wave that transmits through the first interface, bounces back and forth twice in the layer, and then transmits through the first interface (ΔP_3), to all the other multiply-reflected waves. If we consider the initial wave to be normally incident on the layer, the phase differences between the multiply-reflected waves must be kept track of by including the factor of e^{ik_2D} for each pass through the film as below:

$$
\begin{aligned}
\Delta P_r &= \Delta P_1 + \Delta P_2 + \Delta P_3 + \dots \\
&= \Delta P_o r_{12} + \Delta P_o t_{12} e^{ik_2D} r_{23} e^{ik_2D} t_{21} + \Delta P_o t_{12} e^{ik_2D} r_{23} e^{ik_2D} r_{21} e^{ik_2D} r_{23} e^{ik_2D} t_{21} + \dots \\
&= \Delta P_o r_{12} + \Delta P_o t_{12} e^{ik_2D} r_{23} e^{ik_2D} t_{21} \left(1 + r_{21} e^{ik_2D} r_{23} e^{ik_2D} + (r_{21} e^{ik_2D} r_{23} e^{ik_2D})^2 + \dots \right) \\
&= \Delta P_o r_{12} + \frac{\Delta P_o t_{12} e^{ik_2D} r_{23} e^{ik_2D} t_{21}}{(1 - r_{21} e^{ik_2D} r_{23} e^{ik_2D})}
\end{aligned}
$$

(7.54)

where the last step follows from the binomial theorem $\frac{1}{1-n} = 1 + n + n^2 + n^3 + \dots$

Combining the exponentials together and dividing out by the initial pressure wave, we find

$$
r_{13} = \frac{\Delta P_r}{\Delta P_i} = r_{12} + \frac{t_{12} e^{i2k_2D} r_{23} t_{21}}{(1 - r_{21} e^{i2k_2D} r_{23})}
$$

(7.55)

where r_{12}, r_{21}, t_{12}, and t_{21} are found from the results of the previous section and $k_2 = \omega / v_2$. This is a complex quantity – the magnitude gives the amplitude of the reflected signal with respect to the incident wave and the phase gives the relative phase of the reflected and incident waves. When $e^{i2k_2D} = \pm 1$ the amplitude of the reflected wave will oscillate from a maximum (+1) to a minimum (−1). These *film interference fringes* are commonly observed with light waves in, for example, the colored patterns of reflected light from soap films or bubbles.

Example 7.4 A Pyrex glass plate of thickness $d = 3$ mm is immersed in a large container of water. An ultrasound wave is normally incident from the water onto the plate. Pyrex glass has a density of 2230 kg/m^3 and a longitudinal ultrasonic velocity of 5640 m/s, while water has a density of 1000 kg/m^3 and an ultrasound velocity of 1480 m/s. Plot the acoustic power reflection vs. the frequency of the ultrasound wave as the frequency of the ultrasound is varied from 0.2–20 MHz.

Solution: The acoustic impedance of the Pyrex glass layer is $\rho_2 v_2 = 2230 \times 5640 = 12.58 \times 10^6$ kg m^{-2} s^{-1} while the acoustic impedance of water, which is the incident and backing material, is $\rho_1 v_1 = \rho_3 v_3 = 1.48 \times 10^6$ kg m^{-2} s^{-1}. This gives us

$$r_{12} = \frac{\rho_2 v_2 - \rho_1 v_1}{\rho_2 v_2 + \rho_1 v_1} = \frac{12.58 - 1.48}{12.58 + 1.48} = 0.7894,$$

$$r_{23} = r_{21} = \frac{\rho_2 v_2 - \rho_3 v_3}{\rho_2 v_2 + \rho_3 v_3} = \frac{1.48 - 12.58}{12.58 + 1.48} = -0.7894,$$

$$t_{12} = \frac{2\rho_2 v_2}{\rho_2 v_2 + \rho_1 v_1} = \frac{2 \cdot 12.58}{12.58 + 1.48} = 1.7894,$$

$$\text{and} \quad t_{21} = \frac{2\rho_1 v_1}{\rho_2 v_2 + \rho_1 v_1} = \frac{2 \cdot 1.48}{12.58 + 1.48} = 0.2106.$$

The wave vector in the glass layer is $k^2 = 2\pi/\lambda_2 = 2\pi f/v_2$ where f is the ultrasound frequency and v_2 is the velocity of sound in the glass. Note that since the sound is *normally incident*, it only excites the longitudinal compressive sound wave in the glass. (There are also shear modes of acoustic propagation in solids which have different velocities than the longitudinal mode.)

The pressure reflection amplitude found from Eq. (7.55) $r_{13} = \frac{\Delta P_r}{\Delta P_i} = r_{12} + \frac{t_{12} e^{i2k_2 D} r_{23} t_{21}}{(1 - r_{21} e^{i2k_2 D} r_{23})}$ is a complex quantity because the reflected wave is different in *amplitude* as well as *phase* from the incident wave, as represented by the amplitude and phase of r_{13}. We can extract either by our usual methods: $|r_{13}| = \sqrt{r_{13} r_{13}{}^*}$ where $r_{13}{}^*$ is the complex conjugate of r_{13} and $\delta = \tan^{-1}\left[\frac{Im(r_{13})}{Re(r_{13})}\right]$, or by using the MATLAB abs() and angle() functions. The energy reflection coefficient is found from $S = \frac{1}{2}\frac{(\Delta P_m)^2}{\rho_o v}$ in terms of the *amplitude* of the pressure wave ΔP_m as $R = \frac{S_r}{S_i} = \frac{2\rho_1 v_1 (\Delta P_{mr})^2}{2\rho_1 v_1 (\Delta P_{mi})^2} = \left(\frac{\Delta P_{mr}}{\Delta P_{mi}}\right)^2 = |r_{13}|^2$. The MATLAB code to calculate this is below, and a plot of R vs. ultrasound frequency f is in Figure 7.10.

```
v1=1480; rho1=1000; rhov1=rho1*v1; rhov3=rhov1;
v2=5640; rho2=2230; rhov2=rho2*v2;

r12=(rhov2-rhov1)/(rhov2+rhov1);
r23=(rhov3-rhov2)/(rhov2+rhov3);
r21=r23;
t12=2*rhov2/(rhov2+rhov1);
t21=2*rhov1/(rhov2+rhov1);
```

Continued

Example 7.4 (*cont.*)

```
D=1e-3;
f=[ 0.5:.001:20] *1e6;
k2=2*pi*f./ v2;
r13=r12+(t12 *exp(i*2*k2*D)*r23*t21)./ (1-r21*exp(i*2*k2*D)
  *r23);

plot(f/1e6, abs(r13).^2)
xlabel('Ultrasound Frequency (MHz)'); ylabel('Power Reflection
  Coefficient')
```

7.9   Spherical waves

For sound waves, the effects of interference are often obscured by the wave bending
(*diffraction*) effects that are more pronounced at the longer wavelengths of sound. To
understand diffraction effects, we must begin by looking at spherical wave fronts.

   Since real sound sources are not infinite planes, it is necessary to consider the wave
intensity from a finite source of arbitrary shape. We will begin by considering a *point
source* which produces spherical waves expanding out equally in all directions from the

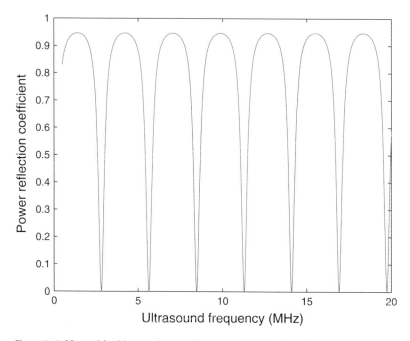

Figure 7.10 Normal incidence ultrasound power reflection from 3 mm Pyrex glass plate immersed
in water as a function of ultrasound frequency from Example 7.4.

origin. To solve this problem, we need to consider the Laplacian operator in spherical coordinates:

$$\nabla^2 u = \frac{\partial^2 u}{\partial r^2} + \frac{2}{r}\frac{\partial u}{\partial r} + \frac{1}{r^2 \sin\theta}\frac{\partial}{\partial\theta}\left(\sin\theta\frac{\partial u}{\partial\theta}\right) + \frac{1}{r^2 \sin^2\theta}\frac{\partial^2 u}{\partial\varphi^2} \qquad (7.56)$$

where (r, θ, φ) are the spherical coordinates, the radial distance from the origin r, the polar angle θ, and the azimuthal angle φ.

If we assume a spherically symmetrical point source at the origin (no θ or φ dependence), our wave equation reduces to

$$\frac{\partial^2 u}{\partial r^2} + \frac{2}{r}\frac{\partial u}{\partial r} = \frac{1}{v^2}\frac{\partial^2 u}{\partial t^2} \qquad (7.57)$$

and the solutions are *spherical waves* centered on the origin with a displacement and pressure waves given by

$$u(r,t) = a_m \frac{e^{i(kr\pm\omega t)}}{r} \qquad (7.58a)$$

$$\Delta P(r,t) = \pm i\omega(\rho_o v)u(r,t) = \pm i\omega(\rho_o v)a_m \frac{e^{i(kr\pm\omega t)}}{r} = b_m \frac{e^{i(kr\pm\omega t)}}{r} \qquad (7.58b)$$

where the $\pm$ sign refers to, respectively, converging (+) or expanding (−) spherical waves. The wave amplitude a_m has dimensions of m² and b_m has dimension N/m to keep the dimensions of the displacement $u(r, t)$ as meters and $\Delta P(r, t)$ as N/m². We note the appearance of the characteristic impedance $(\rho_o v)$ as a parameter relating the displacement to the pressure.

The $1/r$ factor causes the net energy flow per area to decrease as $1/r^2$ as the wave expands, as necessary to keep the total power constant as the wave expands spherically across the spherical area $A = 4\pi r^2$

$$S = \frac{1}{2}(\rho_o v)\omega^2\frac{a_m^2}{r^2} = \frac{1}{2}\frac{(\Delta P_m^2)}{(\rho_o v)} = \frac{1}{2}\frac{(b_m^2)}{(\rho_o v)r^2}\,\text{W/m}^2. \qquad (7.59)$$

7.10   Huygens's Principle and wave diffraction

One of the most fundamental wave properties that we are familiar with from our experience every day is the phenomenon of *wave diffraction*, the bending of waves around corners which makes us the unwilling recipient of conversations and noises in the halls or adjacent rooms. The first attempt to describe the phenomenon of wave diffraction was by Christiaan Huygens, a contemporary of Newton, who reasoned that waves propagate as if every point on the wave front (a plane of constant phase) could be considered to be the source of an expanding spherical wave, as shown in Figure 7.11. If we consider the plane of tangency to the expanding spheres to be the plane of constant phase of the propagating wave, we see that it represents the propagation of a plane wave

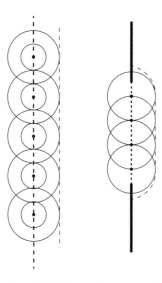

Figure 7.11 Huygens's construction for free-space propagation of plane wave (left) and propagation of plane wave through an aperture (right).

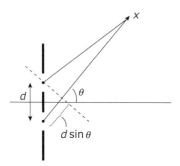

Figure 7.12 Relative phase between waves propagating through two slits collected at point X at an angle θ from the midpoint between the slits.

as well as the bending of the wave propagating through a slit, *single-slit diffraction*. There is the small matter of the wave traveling in the opposite direction which was not resolved until the mathematical theory of diffraction was developed much later by Augustin-Jean Fresnel and Gustav Kirchhoff, among others.

Huygens's Principle does give us a good qualitative understanding of diffraction through slits. Consider, for example, a plane wave incident on a screen with two slits in it as shown in Figure 7.12. The wave that passes through the lower slit must travel a distance $\Delta r = d \sin \theta$ farther than the wave that passes through the upper slit to arrive at the point indicated. If the difference in path length is an even number of wavelengths, the waves will constructively interfere and add, while if the path difference is a half-wavelength (or one-and-a-half, etc.) the waves will experience destructive interference and cancel out. The resulting pattern of bright and dim bands on a screen behind the slits is given by the *grating equation* (where n is an integer $= 0, 1, 2, 3 \ldots$):

$$d \sin \theta = n\lambda \quad \text{(for maximum)} \tag{7.60a}$$

$$d \sin \theta = \left(n + \frac{1}{2}\right)\lambda \quad \text{(for minimum).} \tag{7.60b}$$

7.11   Fresnel–Kirchhoff diffraction

According to the Huygens Principle, the propagation of a wave front can be modeled by assuming that each point on the wave front is the source of a spherical wave. The field transmitted through the aperture can be calculated by adding spherical waves with an origin at each point in the aperture. This principle is put in mathematical form by the *Fresnel–Kirchhoff diffraction formula*

$$I(P) = \frac{-i}{2\lambda} \iint\limits_{A} \frac{Ce^{ik(r+s)}}{rs}[\cos(n, s) - \cos(n, r)]dS. \tag{7.61}$$

In this equation, r is the distance from the source to a point in the aperture and s is the distance from that point to the detector. The constant C is the amplitude of the incident spherical wave, and the integral is over the area A of the aperture. The term $\cos(n, r)$ is the cosine of the angle between a forward pointing normal to the aperture $\boldsymbol{n}$ and the $\boldsymbol{r}$ vector and $\cos(n, s)$ is the cosine of the angle between the same normal and the s vector as shown in Figure 7.13 below.

If the transmitter is a long distance away from the aperture $C\frac{e^{ikr}}{r}$ is essentially a constant over the aperture and can be taken out of the integral. This implies that the phase and amplitude of the incident wave are constant across the aperture. In addition $\cos(n, r) \cong \cos(\pi) = -1$ and Equation (7.61) becomes

$$I(P) = \frac{-iCe^{ikr}}{2\lambda r} \iint\limits_{A} \frac{e^{iks}}{s}[\cos(n, s) + 1]dS. \tag{7.62}$$

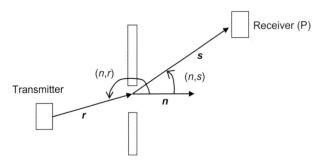

Figure 7.13 Geometry for Fresnel–Kirchhoff diffraction.

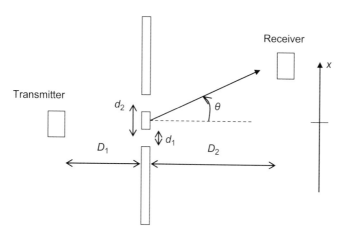

Figure 7.14 Two-slit experiment (top view).

The $[\cos(n, s) + 1]$ factor in Equation (7.62) resolves the problem of the back-propagating wave in the Huygens model. The back-propagating wave has the (n, s) angle $= 180°$ so $[\cos(n, s) + 1] = 0$.

Note that in the Fresnel–Kirchhoff–Huygens formulation in Equation (7.61) or Equation (7.62) there is no essential difference between the situation in Figure 7.13 (which is traditionally referred to as *single-slit diffraction*) and the situation in Figure 7.14 (which is traditionally called *double-slit interference*) except the shape of the aperture integrated over. The same equations cover propagation through apertures of any shape!

Example 7.5 Ultrasound with frequency 40 kHz propagates in air from a source a long distance away to a screen with apertures in it. The speed of ultrasound in air is $v = 343$ m/s. Use MATLAB to calculate the intensity of the pressure wave on a line 15 cm beyond the screen as a function of the horizontal distance x for the cases (a) when the screen has one long vertical slit of width $d_1 = 7$ mm as in Figure 7.13, and (b) when the screen has two long vertical slits of width $d_1 = 7$ mm separated by a distance of $d_2 = 2.5$ cm as measured from the center of the slits as in Figure 7.14.

Solution: We assume that the source is far enough away from the screen that the source wave arrives at every point on the screen with a constant phase and amplitude – essentially a plane wave. We can then use Eq. (7.62) with the prefactor to the integral $\frac{-iCe^{ikr}}{2\lambda r}$ considered a constant over the screen. We will consider first the contribution just from the Huygens sources in the slits in a horizontal plane in the middle of the slits at the level of the detectors, and then we will extend it to the entire length of the vertical slits (assumed to be 10 cm high). The MATLAB code for the Huygens sources only in a horizontal plane is below, and the results for both cases are shown in Figure 7.15.

Continued

Example 7.5 (*cont.*)

```
%Example 7.5: Single and double slit ultrasound diffraction

v=343.;  f=40e3;
yl=v/f;                  % wavelength
k=2*pi/yl;

D2=0.15;                 % distance to detector plane
w=0.007;                 % width of slit
d=.025;                  % separation of two slits
YA=-0.2:.001:0.2;        % sampling positions on detector plane
N=max(size(YA));

for n=1:N
     Y=YA(n);            % select one detector position
                 % single slit
        yp=-w/2: 0.01*w: w/2;           % location of point
                                          sources in aperture
        s=sqrt(D2^2+(Y-yp).^2);         % distance from point
                                          sources to detector
        cosns=D2./ s;
        Ip=exp(i*k*s).* (cosns +1)./ s; % Fresnel-Kirchhoff
                                          diffraction
        Amp1(n)=sum(Ip)/max(size(yp));  % normalized to number
                                          of sources
                 % double slit
        yp1=[ -d/2-w/2:0.01*w:-d/2+w/2] ;
        yp2=[ d/2-w/2:0.01*w:d/2+w/2] ;
        yp=[ yp1,yp2] ;
        s=sqrt(D2^2+(Y-yp).^2);         % distance from point
                                          sources to detector
        cosns=D2./ s;
        Ip=exp(i*k*s).* (cosns +1)./ s; % Fresnel-Kirchhoff
                                          diffraction
        Amp2(n)=sum(Ip)/max(size(yp));  % normalized to number
                                          of sources
end

figure(2)
plot(YA,abs(Amp1), YA, abs(Amp2))
xlabel('Position on Screen (m)'); ylabel('Acoustic Intensity')
```

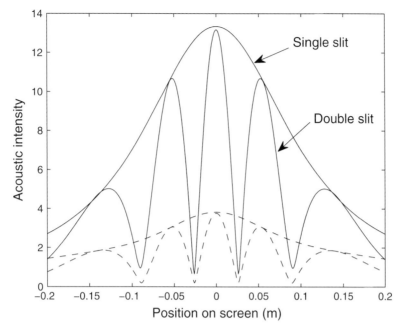

Figure 7.15 Diffraction from a single 7 mm wide slit and from two 7 mm slits separated by 2.5 cm as in Example 7.5. The solid lines are the acoustic intensity calculated assuming the Huygens sources are confined to a horizontal plane. The dashed lines are with the sources distributed across the 10 cm slit height with the detector height at the midpoint of the slit. The intensity is normalized to the number of (evenly distributed) Huygens sources considered.

Chapter 7 problems

1. A sinusoidal transverse wave is moving to the right and is described by the general equation $y(x, t) = A\sin(kx \pm \omega t + \varphi)$. If its wavelength is 30 cm, its period 15 s, its amplitude 5 cm, and its position at $x = 0$ and $t = 0$ is 1 cm, determine the following:

   a. the correct sign in front of ω

   b. the wave number, k

   c. the angular frequency, ω

   d. the phase, φ

   e. the speed of the wave

   f. the transverse velocity of the particle at $x = 0$ and $t = 3$ s

   g. the transverse acceleration of the particle at $x = 0$ and $t = 3$ s.

2. How would the wave in Problem 7.1 be written in the form $y(x, t) = \mathcal{R}e\{\tilde{A}e^{i(kx-\omega t)}\}$? What are the values and signs of $\tilde{A}, k,$ and ω? Could the wave also be written in the form $\mathcal{R}e\{\tilde{A}e^{i(kx+\omega t)}\}$? What would the values and signs of of $\tilde{A}, k,$ and ω be in this case?

3. Calculate how much faster (or slower) sound travels in water than in air? (Look up all relevant constants needed for the calculation.)

4. What do you have to do to a guitar string to
   a. increase the frequency of its fundamental by a factor of 1.5?

   b. increase the wavelength of its fundamental by a factor of 1.5?
   You should identify at least two factors that can be changed and give a *quantitative* answer for the changes necessary.

5. The human ear can detect frequencies from 20 Hz to 20 kHz. How can you reconcile the fact that we can hear beats of frequencies much below 20 Hz?

6. In terms of the wave amplitude, the total wave energy/power, the wave velocity, the wave frequency, and the wavelength (or wave number), discuss what changes and what remains the same as a mechanical wave propagates in a medium. Does your answer change on whether you are thinking about a plane wave or a spherical wave? What if the medium is absorbing? Finally, what about if the wave passes across an interface from one medium to another? You may find it convenient to express your answer as a chart where you consider the five wave properties while propagating as a plane wave, spherical wave, a plane wave in an absorbing medium, a spherical wave in an absorbing medium, and a wave that moves from one medium to another medium.

7. Plot the wave that results from the interference of two identical waves traveling in the same direction with a phase difference of $3\pi/2$.

8. If you were to plot, as a function of tension of the string T, the wavelength λ, of the fundamental standing wave formed in a given length of string L, what will the graph look like?

9. If $y(x, t) = 0.2\sin(\pi x/2)\sin(18\pi t)$ represents the third harmonic of a standing wave on a string under 250 N tension, what is
   a. the speed of the wave on the rope?

   b. the mass of the rope?

10. Two identical waves moving in opposite directions on a string interfere and give a standing wave whose amplitude is 1 cm, the distance between consecutive nodes is 8 cm, and the time it takes a particle on the string to move from one extreme of the anti-node to the other is 5 ms. If one of the waves is given by $y(x, t) = A\sin(kx+\omega t)$, calculate
    a. the wave vector, k

    b. the amplitude, A

    c. the angular frequency, ω

11. If a 1 m long string being driven by a 120 Hz oscillator produces the fourth harmonic of a standing wave, when the tension on the string is 4 N and its linear mass density is 2 g/m, calculate the tension needed to produce the third harmonic.

12. Consider a small sound transducer that produces a spherically symmetrical sound wave. Use energy considerations to explain how you would expect the sound wave power (watts per square meter) would decrease with the distance r from the transducer. From the equations in this chapter, how would you expect the amplitude of the pressure wave would decrease with r? How do you reconcile your two answers? Would your answers change if the transducer produced a wave that propagated only into the forward-facing hemisphere?

13. MATLAB exercise: Plot the normal incidence reflection of 5 kHz sound waves in air (density $\rho_1 = 1.3$ kg/m^3 and sound velocity $v_1 = 340$ m/s) as a function of layer thickness D from 0 to 50 cm of a layer of material with density $\rho_2 = 3$ kg/m^3 and sound velocity $v_2 = 300$ m/s on a substrate of material with $\rho_3 = 25$ kg/m^3 and sound velocity $v_3 = 1000$ m/s.

14. MATLAB exercise: Consider 40 kHz ultrasound propagating in air ($v_1 = 343$ m/s). By breaking up the slit into point sources of Huygens's wavelets separated by 0.1 mm, calculate the intensity of the Fresnel–Kirchhoff diffraction on a plane 30 cm from the slit as a function of lateral distance -50 cm $< x < 50$ cm for the following slit configurations: (a) a single slit of width 0.2 cm, (b) a single slit with a width of 0.5 cm, (c) a single slit with a width of 2 cm, (d) two equal slits of width 1 cm with their centers separated by 4 cm, and (e) two unequal slits of width 0.5 cm and 1 cm with their centers separated by 4 cm. For this problem, you can consider all the Huygens's wavelets to be at a single height equal to the height that the diffracted waves are detected (i.e., you can ignore contributions to the diffraction pattern from sources above or below the detection level – for extra credit drop this restriction).

15. Consider an acoustic plane wave incident at an angle θ_i onto a plane interface between Medium 1 with acoustic impedance $\rho_1 v_1$ and Medium 2 with acoustic impedance $\rho_2 v_2$, so that the dot product of the wave displacement with a unit vector normal to the interface is $\vec{u}_i \cdot \hat{n} = |\vec{u}_i| \cos \theta_i$.

    a. Apply the boundary conditions in Eq. (7.43) and Eq. (7.44) to the incident, reflected, and transmitted waves to derive the equations for the wave reflection and transmission parameters in terms of the media properties and the incident angle θ_i.

    $$r = \frac{\Delta P_r}{\Delta P_i} = \frac{\frac{\cos \theta_i}{\rho_1 v_1} - \frac{\sqrt{1 - \frac{v_2^2}{v_1^2} \sin^2 \theta_i}}{\rho_2 v_2}}{\frac{\cos \theta_i}{\rho_1 v_1} + \frac{\sqrt{1 - \frac{v_2^2}{v_1^2} \sin^2 \theta_i}}{\rho_2 v_2}}$$

    $$t = \frac{\Delta P_t}{\Delta P_i} = \frac{\frac{2 \cos \theta_i}{\rho_1 v_1}}{\frac{\cos \theta_i}{\rho_1 v_1} + \frac{\sqrt{1 - \frac{v_2^2}{v_1^2} \sin^2 \theta_i}}{\rho_2 v_2}}.$$

    b. Use MATLAB to plot the energy reflection and transmission coefficients R and T as a function of incident angle θ_i when a sound wave in diesel oil with $\rho_1 = 800$ kg/m^3 and $v_1 = 1250$ m/s is incident on an interface with water with

$\rho_2 = 1000$ kg/m^3 and $v_2 = 1480$ m/s. Repeat the plot when the incident Medium 1 is water and the transmitted Medium 2 is diesel oil.

16. Doppler effect. Sound emitted by a moving source has its frequency shifted because the source moves a finite distance in the period T of the sound wave. If the source has component of velocity in the direction of the observer, the wavelength is changed by the distance the sources moves closer or farther from the receiver by $\lambda' = T(c - v_s \cos \theta_s)$, where c is the speed of sound in air, v_s in the speed of the source, and θ_s is the angle between the velocity vector of the source and the direction to the observer. The frequency that is heard by the observer f_o is changed from the frequency emitted by the source f_s by

$$f_o = \frac{f_s}{1 - \frac{v_s}{c} \cos \theta_s} \, .$$

If the source is moving toward the observer, $\theta_s < 90°, f_o > f_s$, and the observer hears a higher pitch than the source. If the source is moving away from the observer, $\theta_s > 90°, f_o > f_s$, and the observer hears a lower pitch. If the source is moving transverse to the direction to the observer $\theta_s = 90°$ and there is no *Doppler shift*. [Because of relativistic effects, electromagnetic radiation does have a *transverse Doppler shift* that needs to be considered in the design of GPS systems.]

On the other hand, if the source is stationary and observer is moving with a velocity component in the direction of the source the receiver will hear a sound frequency $f_o = f_s(1 + \frac{v_o \cos \theta_o}{c})$ where v_o is the observer's velocity and θ_o is the angle between the *observer's velocity* and the direction from the observer to the source. If both source and observer movement is considered, the shift is the product of the two shifts and

$$f_o = f_s \left(\frac{c + v_o \cos \theta_o}{c - v_s \cos \theta_s} \right) \, .$$

In the diagram below, the source is a 1200 Hz siren on a police car traveling at 50 mph. The observer is a person running toward the path of the police car at 10 mph. Use MATLAB to calculate and plot the frequency heard by the observer (a) if the observer stops running at a distance $y = 10$ ft from the path of the police car as the police car moves from $x = 300$ ft to $x = -300$ ft (beyond the intersection), (b) if the police car stops at $x = 20$ ft and the observer runs from $y = 100$ ft to $y = 0$, and (c) if both the police car and the observer continue moving, starting from $x = 300$ ft and $y = 100$ ft.

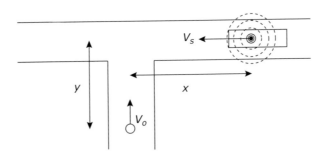

8   The quantum puzzle

By the end of the nineteenth century all the important elements of classical physics were fully mature. Newton's Laws had been applied with success to everything from industrial machines to planetary orbits. Classical thermodynamics had extended energy conservation arguments into thermal processes, and scientists including J. Willard Gibbs and Ludwig Boltzmann had developed the statistical interpretation of entropy. James Clerk Maxwell had explained all electromagnetic phenomena, from AC generators/motors to light, with the four Maxwell's equations and three material-dependent constitutive relations. There was serious discussion that physics was on the verge of being "finished" – becoming a closed system of knowledge like classical geometry with only essentially routine tasks left of determining how the known laws could be applied to engineering problems.

Yet by 1930 the foundations of mechanics, electromagnetics, and atomic-scale systems had been shaken to the core by the new revolutions in the Theory of Relativity, the photon picture of light, and quantum physics. We are still reaping the benefits of the application of these new scientific paradigms in our technology today, including semiconductor electronic devices, lasers, and GPS systems. But the description of reality that these theories provide is still deeply anti-intuitive and disturbing.

Nevertheless, quantum physics is one of the most extensively experimentally verified theories that we know. Even when quantum theory and common sense seem to diverge, quantum theory gives the right answers when tested by experiment. In the next chapter we will develop the framework of quantum theory that has been so successful when people "shut up and calculate," as one scientist suggested as the way through the philosophical thicket of quantum physics. But in this chapter, we will spend a brief time considering the historical development and continuing paradoxical implications of quantum theory.

8.1   The photoelectric effect and photons

In 1905, Albert Einstein, who was a key figure in the revolution of modern physics, was working as a clerk in the patent office in Zurich, Switzerland. Einstein had finished his studies in physics, but had not obtained a coveted position as a university professor. He was intensely interested in the latest developments in classical physics and was aware of a few of the remaining loose ends that were resisting explanation in the classical model.

His roommate was a student at the University of Zurich, and Einstein reviewed with interest his roommate's notes about the lectures that he was attending. Since European scientists were the best in the world in the early twentieth century, the University of Zurich was awash with the latest successes and remaining puzzles of the new statistical and electromagnetic theory.

Einstein's relatively routine day job left him a lot of time to consider the implications of classical physics and to explore some of his own intuitions. For example, years before, he had recognized that there was an inconsistency with Maxwell's picture of electromagnetic waves: if an observer were traveling at the speed of light next to a light beam he would observe static electric and magnetic fields that were not supported by charges or magnets. The time dependence that gave rise to the fields due to Faraday and Maxwell's Laws would disappear in the reference frame of that observer. Working out this paradox would lead to the development of the Special Theory of Relativity that was published in 1905. But we are going to look in detail at the topic of another paper that Einstein published in 1905: the explanation of the photoelectric effect for which Einstein won the Nobel Prize in Physics in 1921.

The photoelectric effect was one of the "loose ends" that didn't seem consistent with Maxwell's theory of electromagnetism. The photoelectric effect is illustrated in Figure 8.1(a). When light at a single wavelength λ is directed onto the surface of a metal (a "photocathode") in a vacuum chamber, electrons are ejected from the surface of the metal and a current can be measured between the photocathode and a nearby anode electrode. If the voltage V between the photocathode and the anode is made negative enough, equal to $-V_{th}$, where V_{th} is the *threshold voltage*, the current is cut off completely, as shown in the graph in Figure 8.1(b).

If the intensity of the light is increased from I_1 to I_2 as shown in Figure 8.1(b), the current at the same voltage is increased, but the threshold voltage is not changed. Note that this is not what one might expect from classical electromagnetic theory. Classical theory would say that the electric field is given by $\varepsilon(x,t) = \varepsilon_o \cos(kx - \omega t)$, assuming that the incident electromagnetic (EM) wave is a plane wave. If the intensity (W/m^2) of the light is increased, the wave amplitude E_o is increased, and the electron should be subjected to a larger force and have a larger change in energy $\Delta E = W_{ab} = \int_a^b q\,\varepsilon_o \cos(kx - \omega t)\,dx$. This is not what is observed. Instead, when the intensity is changed at a constant wavelength the threshold voltage remains unchanged.

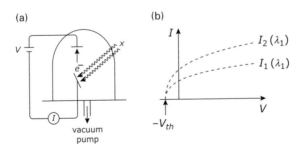

Figure 8.1 (a) Photoelectric effect experimental set up (left) and (b) data for the photo-emitted current as a function of accelerating/retarding voltage for two intensities of incident light (right).

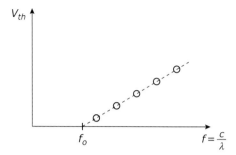

Figure 8.2 The cut-off voltage for the photoelectric effect plotted as a function of the frequency of the incident light.

To extend the puzzle, the threshold voltage depends, not on the intensity, but linearly on the frequency ($f = c/\lambda$) of the light, as shown in Figure 8.2. The minimum frequency that will eject an electron from the surface, f_o, depends on the metal of the photocathode (f_o is smaller for a clean sodium surface than for gold, for example) but **not** on the intensity of the light. This behavior is simply not consistent with the Maxwellian electromagnetic wave model and Newtonian force–energy relations.

In his 1905 paper, Einstein adapted an innovation of Max Planck that had used to address a problem in the statistical mechanics of *blackbody radiation*. Blackbody radiation is the electromagnetic radiation emitted by a hot glowing body – a heated fireplace poker glowing red-hot, for example, or the Sun glowing yellow-/white-hot. Planck could get a good fit to experimental data for blackbody radiation, but only if he assumed that the energy of the oscillators in the heated body depended on their frequency by $E = hf$, where f is the frequency of the oscillator in hertz and h is a constant that Planck calculated from experimental results with a value $h = 6.63 \times 10^{-34}$ J s. Planck had no theory for why this should be, but with this assumption, he was able to fit the blackbody spectra.

Einstein adapted and generalized Planck's model. He postulated that the electromagnetic field is composed of *photons*, particle-like excitations with an energy given by the Planck formula:

$$E_p = hf = h\frac{c}{\lambda} \tag{8.1}$$

where we have replaced the frequency f by the speed of light divided by the wavelength.

How does this model explain the photoelectric effect? We assume that there is some energy of confinement that keeps electrons in the metal from streaming out into the vacuum. This energy is called the *work function* Φ and is a characteristic of the metal. When a photon strikes the surface, it transfers energy $E_p = hc/\lambda$ to a *single* electron. If E_p is less than the work function Φ, the electron does not have enough energy to overcome the energy that confines the electrons inside the metal and there is no photoemission. On the other hand, if E_p is greater than Φ the electron can escape into the vacuum with a kinetic energy $KE = hc/\lambda - \Phi$. Photo-emitted electrons will be able to reach the anode until the anode voltage becomes negative enough that the potential energy $U = qV$

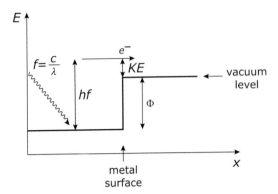

Figure 8.3 Einstein's explanation of the photoelectric effect. If the photon energy $E_p = hf$ exceeds the metal work function Φ, photoelectrons will be emitted with kinetic energy $KE = hf - \Phi$.

exceeds the kinetic energy of the photo-emitted electron, at which point all the electrons will be driven back to the cathode. The threshold voltage is given by the condition:

$$qV_{th} = hf - \Phi. \tag{8.2}$$

This elegant picture explains everything that was known (and is known) about the photoelectric effect. If the light intensity is increased the number of photons – and consequently the number of photo-emitted electrons – increases, but the energy of each electron is still fixed at $hf - \Phi$, so the threshold voltage V_{th} isn't changed. As the optical frequency is decreased to less than f_o, the photon energy is less than the work function and no electrons are emitted. At frequencies above f_o the threshold voltage varies linearly with photon frequency as is indicated by the equation above.

Question 8.1

A *photomultiplier tube* detects light by measuring the current of electrons photo-emitted from a metal surface by the light. What is the longest wavelength of light that could be detected by a photomultiplier tube with a photocathode made of sodium metal with a work function $\Phi = 2.28$ eV? What work function would be necessary to detect infrared light with $\lambda = 5$ μm?

The only problem with Einstein's explanation was that it contradicted over 150 years of evidence that light is a wave. Newton had suggested that light was composed of "corpuscles" (particles), but the documentation of optical wave interference effects by Thomas Young and others ultimately carried the argument. When Maxwell showed how electromagnetic theory led to a wave equation with the velocity of propagation equal to the observed speed of light, the question was considered settled. Electromagnetic wave theory had explained countless phenomena and is still used extensively and successfully to model optical effects and systems. Now Einstein had opened the question again. While one can describe photons as "wave packets" as illustrated in Figure 8.4, in actuality the wave picture of light and the photon picture are apparently

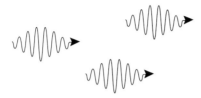

Figure 8.4 The photon picture of light as a collection of wave packets.

equally valid and which one you use depends on what phenomena you are studying. One physicist is reputed to have said, "I use the wave theory on Monday, Wednesday, and Friday, and the photon picture on Tuesday, Thursday, and Saturday. On Sunday, I'm agnostic."

The elements of the photon theory are listed below:

(1) **Light consists of a collection of particle-like photons.** The photon flux (rate at which a source radiates photons) can be found in terms of the energy flux found from the Poynting vector $\vec{S} = \frac{1}{2} \vec{E} \times \vec{H}$ (W/m^2) by $N_{ph} = |\vec{S}|/hf$ (photons/s).

> **Question 8.2**
>
> Consider an electromagnetic wave of visible light with wavelength $\lambda = 500$ nm and a flux of 1 μW/m^2. How many photons per second are arriving in an area of 1 square centimeter?

(2) **Each photon carries energy** $E_p = hf = \hbar\omega$, where we have introduced the angular frequency ω (radians/s) in place of the frequency f in hertz. The constant $\hbar = \frac{h}{2\pi} = 1.0546 \times 10^{-34}$ J·s (pronounced "h-bar") is so much more widely used than h that it is not at all unusual to see equations that contain factors of $2\pi\hbar$. "Planck's constant" may refer to $\hbar$ as often as to h. The photon energy for different electromagnetic regimes is listed in the table below.

Range	Typical wavelength	Photon energy, $E_p = hc/\lambda$
Microwave	$\lambda = 3$ cm	$E_p = 6.62 \times 10^{-24}$ J $= 41 \times 10^{-6}$ eV
Infrared	$\lambda = 3$ μm	$E_p = 6.62 \times 10^{-20}$ J $= 0.413$ eV
Visible	400 nm $< \lambda <$ 650 nm	3.1 eV $> E_p >$ 1.9 eV
Ultraviolet	$\lambda = 200$ nm	$E_p = 6.2$ eV
X-ray	$\lambda = 1$ nm	$E_p = 1240$ eV
Gamma rays	$\lambda < 1$ pm	$E_p > 124$ keV

(3) **Photons are localized.** Localized to how small a volume? Well, our explanation of the photoelectric effect is that each photon gives its energy to one electron, which is smaller than an atom. (Atoms have a radius on the order of 0.1 nm.) The photon is essentially a *point excitation*.

   This extreme localization is what makes the wave packet picture – which seems to resolve the problem of wave–particle duality – not fully satisfactory.

A Fourier construction from a sum of waves should have a spatial extent on the order of the wavelength, 400–600 nm for a visible photon and much larger for infrared quantum detectors which use interband electron excitations in narrow band gap semiconductors to detect infrared radiation with wavelength as long as 12 microns. If we assume that the energy of an optical photon is spread out over an area on the order of λ^2, we can calculate how long it would take to collect enough energy at an atomic site to promote an electron across the work function Φ. If an optical beam at a wavelength of 500 nm has an energy density of 1 μW/m^2, we find that the photon flux is 2.5×10^{12} photons/s-m^2 which corresponds to about 0.6 photons/s in an area on the order of the wavelength squared (500 nm)^2. If we calculate the amount of this energy that is collected at an atomic site (0.1 nm)^2, it should take on the order of 10^8 seconds to collect enough energy at the atomic site to excite an electron above a 2.4 eV work function. Instead, electrons are photo-emitted from the metal without any measurable delay after the light is turned on. If there is even one photon in the beam, it can transfer all of its energy instantaneously to a single electron.

Question 8.3

What is the energy in eV of a photon of radiation with wavelength $\lambda = 500$ nm? Calculate the time required for an optical field of energy density of 1 μW/m^2 to accumulate 2.4 eV of energy in an area of an atomic site (0.1 nm)^2, knowing that 1 electron volt (eV) $= 1.6 \times 10^{-19}$ J.

---

(4)   **Photons carry momentum $p = \hbar k$.** In addition to carrying energy, photons also carry momentum $\vec{p} = \hbar \vec{k}$ where $\vec{k}$ is the photon *wave vector* with a magnitude $|\vec{k}| = 2\pi/\lambda$ and a direction that is the same as the direction of propagation of the photon. This was demonstrated by Arthur Compton in 1923 who was studying the scattering of X-rays by electrons in a carbon target. He found that the X-ray wavelength was shifted between the incident and scattered beam, and the wavelength shift depended on the scattering angle θ as shown in Figure 8.5. The angular dispersion and wavelength shift of the scattered X-rays could be predicted by assuming a two-particle collision between an X-ray photon and an electron initially at rest with a conservation of energy given by

$$\hbar\omega_i = \hbar\omega_s + \frac{1}{2}m_e v^2 \tag{8.3}$$

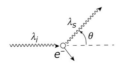

Figure 8.5 Compton scattering of an X-ray photon by an electron, illustrating the photon scattering angle θ.

and a conservation of momentum given by

$$\hbar \vec{k_i} = \hbar \vec{k_s} + m_e \vec{v} \tag{8.4}$$

where the subscripts on ω and $\vec{k}$ refer to the incident (i) and scattered (s) photons, while m_e and v refer to the mass and the velocity of the electron after the scattering process. Expressing the momentum equation in their x- and y-coordinates and solving these equations for the wavelength shift yielded the equation

$$\Delta \lambda = \lambda_i - \lambda_s = \frac{h}{mc} (1 - \cos \theta) \tag{8.5}$$

where θ is the scattering angle of the photon, which accurately described Compton's experimental results.

Question 8.4

What angle of scattering gives the largest shift of wavelength for an X-ray Compton scattering off an electron. What is the largest shift of wavelength $\Delta \lambda$ that will be observed if an X-ray with $\lambda = 1$ nm scatters off an electron with $m = 9.11 \times 10^{-31}$ kg?

The momentum of a photon flux can also be demonstrated in a *Crookes radiometer*, the familiar device, illustrated in Figure 8.6(a), with four vanes in an evacuated glass globe – each with one side white and one side black –which rotates when light shines on it. In fact, the mass-market radiometers that one can buy in a novelty store do not actually work because of photon momentum. Inexpensive radiometers are not evacuated to a high vacuum, and their motion is dominated by the effect of collision with the residual gas atoms. Light is preferentially absorbed by the black side – which is why that side looks black! The black side therefore heats up above the temperature of the white side. Gas atoms hitting the warmer black sides are more energetically reflected than the atoms hitting the cooler white side. So the inexpensive radiometer rotates with "the black side pushing." However, if a similar device is mounted in ultra-high vacuum where the gas pressure is negligible, it rotates the other way, with "the white side pushing." This can be understood by looking at the momentum processes as illustrated in Figure 8.6(b).

When the incident light is reflected on the white side, the photon momentum change is $-2\hbar k$ since the final momentum is equal but in an opposite direction to the initial momentum and $\Delta p = p_f - p_i = -\hbar k - \hbar k = -2\hbar k$ while on the black side where there is no reflected photon the change in momentum is $-\hbar k$. Thus the force on the white side F_1 is twice the force on the black side F_2 and the radiometer rotates with the white side pushing.

You may well ask how we know that the photon energy depends on the frequency ω while the momentum depends on the wave vector k since the two are related by the speed of light: $\omega = ck$. Why not just express the energy and momentum in

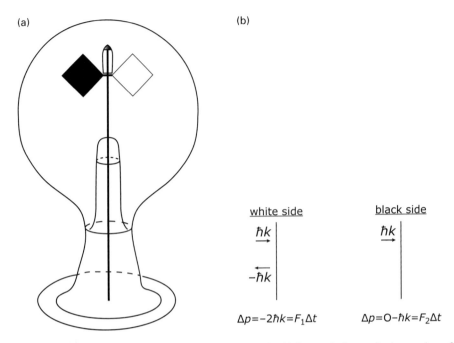

Figure 8.6 (a) A Crookes radiometer spins when light is shining on it due to the interaction of photons with the white and dark sides of the rotatable vanes. (b) A model for the movement of radiometer vanes due to photon momentum.

terms of the frequency? There would be no way to distinguish between the two for light in a vacuum where the velocity is always $c = 2.9979 \times 10^8$ m/s. But if light travels in a medium where its velocity is not c, it is found that the momentum, in fact, depends inversely on the wavelength, which decreases when the velocity decreases, and not on the frequency, which doesn't change. Photons that are traveling in a material have more momentum than photons of the same frequency traveling in vacuum!

8.2   Louis de Broglie and matter waves

In 1924, Louis de Broglie in his PhD Thesis speculated as follows: if a photon has a momentum $p = \hbar k = 2\, h/\lambda$, why not define a *particle wavelength* in terms of the particle momentum mv as follows: $\lambda = {}^h/_p = \frac{h}{mv}$. de Broglie explored a few ramifications of his theory and where it might be observed, but didn't have any evidence that the effect was real. Since the wavelength for any macroscopic particle was far too small to observe, it was clear that any wave effects would only be observed for atomic-scale particles. The best chance would be with electrons, the lightest particle known at the time.

For example, consider a 5 eV beam of electrons created by accelerating electrons across a 5 V potential in an electron gun. The energy of such electrons propagating in a vacuum is all kinetic energy

$$KE = \frac{1}{2} m v^2 = (5 \text{ eV}) \left(1.6 \times 10^{-19} \frac{\text{J}}{\text{eV}} \right) = 8 \times 10^{-19} \text{ J} \tag{8.6}$$

and their momentum is

$$p = mv = \sqrt{2 m (KE)} = \sqrt{2 (9.11 \times 10^{-31} \text{ kg})(8 \times 10^{-19} \text{ J})} = 1.2 \times 10^{-24} \text{ kg} \frac{\text{m}}{\text{s}}. \tag{8.7}$$

The wavelength of these 5 eV electrons is thus:

$$\lambda = {}^h\!/_p = {}^{(6.63 \times 10^{-34})}\!/_{(1.2 \times 10^{-24})} = 5.3 \times 10^{-10} \text{ m} = 0.55 \text{ nm}. \tag{8.8}$$

Question 8.5

The mass of a proton is 1833 times greater than the mass of an electron. What would be the wavelength of a proton with kinetic energy of 5 eV? What would be the kinetic energy (in eV) of a proton with a de Broglie wavelength of 0.55 nm?

This distance, 0.55 nm, is on the order of the distance between atoms in a solid. So it is not impossible that some wave-like effects might be observed if an electron beam was impinged on a solid. Consider the experiment shown in Figure 8.7(a). If an electron beam is directed on a single crystal material, electron waves that reflect off the first layer of atoms could interfere with electrons that penetrate the first layer and reflect off the second layer. This is a well-known wave property of thin-film interference as seen, for example, in the reflection of light from oil slicks.

The analysis is indicated in Figure 8.7(b). If we look at two points on the wave front, the wave that goes through the first layer and is reflected off the second layer goes a distance of 2ℓ farther than the wave that reflects off the first layer, where $\ell = d \sin \theta$, d is the distance between the layers, and θ is the angle between the beam and the normal to the layers. (We have assumed that the angle of reflection is the same as the angle of incidence, a result which can be demonstrated for waves by examining the boundary conditions at the interface.)

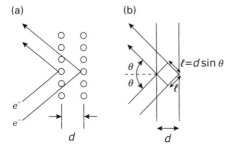

Figure 8.7 (a) Electrons impinging on a crystal lattice and (b) the wave picture of interference of the two "electron waves" reflected from successive layers separated by the crystal lattice spacing d.

The total difference in the path traveled by the beam reflecting off the front layer and the beam reflecting off the second layer is $\Delta x = 2\ell = 2d \sin \theta$. When $\Delta x = n\lambda$ where n is an integer (1, 2, 3 ...) the two waves return in-phase and reinforce each other. On the other hand, when $\Delta x = (n + \frac{1}{2})\lambda$ the waves return 180° out-of-phase and cancel out. This results in a series of bright and dim spots depending on the angle of incidence of the beam and on the wavelength of the light – what you see when you see light reflected off a thin oil film on water. This was also a well-understood effect when X-rays were reflected from crystals – X-ray diffraction from crystals was pioneered by Max von Laue who received the Nobel Prize in 1914 for his discovery.

Following the publication of de Broglie's theory, European physicists looked for some experiments that might confirm these effects when electrons – as opposed to X-rays – were reflected from crystalline materials. They found data that appeared to show such effects in the work of two American Bell Labs physicists, Clinton Davisson and Lester Germer, who had published results on the reflection of low energy electron beams from the surface of nickel single crystals. Davisson and Germer, who were simply interested in characterizing nickel surfaces, were surprised to find that their work was being touted as evidence of a new wave theory of particles. After a trip to a European conference, they returned to Bell Labs to refine their experiments and became early supporters of the new theory.

The equation that governs the behavior and time evolution of the matter waves was developed by Erwin Schrödinger in 1926, but how to interpret the matter waves was a puzzle. To make this clear, it is useful to look at the electron double-slit experiments that were done in the 1930s. The electron double-slit experiment is shown in Figure 8.8.

Electrons are emitted by a hot filament and accelerated by a voltage V in an *electron gun*. The electrons, with a wavelength that can be controlled by the accelerating potential, then propagate through a vacuum to a double slit. Behind the slit is a photographic plate that records where the electrons hit. The intensity of electrons hitting the screen as a function of their position on the photographic plate perpendicular to the

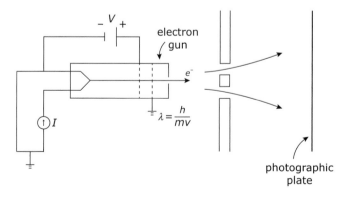

Figure 8.8 Electron double-slit experiment. A mono-energetic electron beam impinges on a plate with two slits separated by a small distance d.

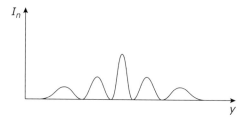

Figure 8.9 Interference pattern created by electrons propagating through a two-slit apparatus as in Figure 8.8.

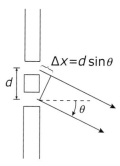

Figure 8.10 Explanation for variation of electron current intensity with angle for propagation of electrons through a double-slit apparatus.

slits is shown in Figure 8.9. The oscillating intensity is characteristic of a wave interference pattern. The electrons going through the two slits travel different distances with the path difference $\Delta x = d \sin \theta$ as shown in Figure 8.10.

The angles where the waves coming from the two slits arrive in-phase ($\Delta x = n\lambda$) correspond to the peaks in the intensity, while the dips are where the waves arrive out-of-phase ($\Delta x = \left(n + \frac{1}{2}\right)\lambda$). The appearance of the interference effects is an unambiguous indication of wave effects – the electron waves from one slit interfere with the electron waves from the other slit.

However the story actually gets much stranger. By reducing the temperature of the filament, it was possible to reduce the number of electrons emitted until it could be calculated that only one electron was in the chamber at a time. One might think that the electron must go through one or the other slit and the interference should disappear. When the experiments were done, however, and the photographic plate was removed from the chamber after many days of collecting data from the single electrons traveling through the slits, in fact, the interference pattern was still there! One has to imagine the electron wave traveling simultaneously through both slits and interfering with itself before hitting the photographic plate.

It was also possible to put small coils around the slits, so that the slit that the electron passed through could be deduced by the small voltage spike induced in the coil when

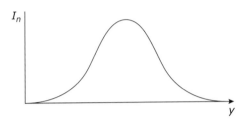

Figure 8.11 Electron current intensity as a function of position for electrons propagating through a double-slit apparatus when a detector coil measures which slit each electron passes through.

the electron passed through it. This was done, and again the experimenters waited for many days while the electrons built up the intensity pattern on the photographic plate. This time, though, the experimenters could observe a pulse on one or the other coil – confirming that the electron was only passing through one slit. In this case, however, when the photographic plate was developed, the pattern was a featureless curve characteristic of diffraction through a single slit as illustrated in Figure 8.11– there were no interference fringes!

If this sounds like the electrons knew when they were being observed and only acted as a wave when no one was watching, this is not far from the quantum interpretation. In measuring the voltage induced in the coil when the electron passed through the investigators were disturbing the system – in the language of quantum physics they destroyed the coherence of the electron wave through their measurement by changing the electron's momentum – and thus eliminated the interference effects. This is an example of the Heisenberg Uncertainty Principle, formulated by Werner Heisenberg in 1927. The Uncertainty Principle states that a particle's position and momentum cannot be simultaneously measured to arbitrary accuracy. In fact, the product of the uncertainty in position and the uncertainty in momentum is greater than or on the order of Planck's constant

$$\Delta x \, \Delta p \gtrsim \hbar. \tag{8.9}$$

Heisenberg's reasoning came from consideration of diffraction of the electron wave through a single slit. As the slit is made smaller, the diffraction pattern becomes wider. As the position of the electron is determined more exactly (by making the slit smaller), the motion of the electron (its momentum in the x-direction) is disturbed in an unpredictable way, leading to a broad uncertainty in its position at the screen (the single-slit diffraction pattern). Another way to express the Uncertainty Principle is that in the process of observing a quantum system, we inevitably disturb the system. To determine where an electron is to within 1 nm, for example, we would need to reflect a photon with a wavelength on the order of 1 nm off of it. But, as we have seen, a 1 nm photon is an X-ray with a large momentum $p = \hbar k$, and in this (Compton) scattering process the electron momentum will be changed unpredictably. If we know the electron's position exactly, the momentum is indeterminate.

Question 8.6

Heisenberg's Uncertainty Principle can also be expressed in terms of the uncertainty in energy ΔE and the uncertainty in time Δt as

$$\Delta E \, \Delta t \gtrsim \hbar. \tag{8.10}$$

Thus only a state with an infinite lifetime can have a precisely defined energy. What is the uncertainty in energy (in eV) of a state with an energy of 1.5 eV that has a lifetime of $\Delta t = 1 \times 10^{-15}$ s?

Experiments like the double-slit experiment posed the question of what, exactly, the electron wave function represented. Schrödinger and others tended to believe that the electron was the wave function. But the wave function of the electron in the double-slit experiment can be quite large – hundreds of nanometers or even microns wide, while the actual spot on the photographic plate was comparable to the size of an atom of the emulsion material – on the order of few nanometers. So what is the connection between the wave function and the actual position of the observed electron?

The interpretation worked out in the 1920s and 1930s by Niels Bohr, Werner Heisenberg, Max Born, and others is known as the Copenhagen interpretation of quantum mechanics. The elements of the Copenhagen interpretation are as follows:

1. Particle dynamics are controlled by the wave function $\Psi(x, t)$ as expressed by Schrödinger's equation.
2. The modulus squared of the wave function gives the *probability* of finding the particle at any given point at time t: $P(x, t) = |\Psi(x, t)|^2$.
3. When a large number of particles are observed, their spatial distribution will be given by $|\Psi(x, t)|^2$, but when any single particle is observed, the wave function "collapses" and the particle will be observed at only one point with a probability controlled by the wave function.

In brief, the Copenhagen interpretation implies that we can't know for certain where an electron or other quantum particle will be in the future. All we can know is the probability that it will be at any given location. For large numbers of particles, this doesn't have any practical importance – for example we don't worry much that all the air particles in the room might, by their random processes, end up in only one half of the room and suffocate half the class – but philosophically it's strange. Einstein, among others, refused to accept the Copenhagen interpretation of quantum physics; he famously said, "God does not play dice with the universe." On the other side, quantum physics has been extensively tested experimentally and the predictions of the probabilities of the various quantum outcomes are always verified by experiment.

In fact, there is a spooky quality to quantum physics related to the special role of observation. There appear to be two ways that the universe evolves under quantum conditions. If a particle is not observed, its motion is described by the wave function as predicted by Schrödinger's equation. But as soon as the particle is observed, the wave function "collapses" and the particle is found to be at only one point. So the unobserved universe evolves in one way, but the act of observation causes the range of probability to be resolved into only one result.

This problem is captured nicely in the "thought experiment" postulated by Erwin Schrödinger: a cat is sealed in a chamber with a capsule of poison gas that will be released when a radioactive atom decays. Since radioactive decay is a quantum process, the Copenhagen interpretation implies that as long as we don't observe the chamber, the wave function consists of a term where the radioactive atom hasn't decayed (and the cat is alive) and another (increasing) term where the atom has decayed (and the cat is dead). When the chamber is observed, the observer will see only a live cat or a dead cat, but until the chamber is observed, the wave function of the system consists of a combination of a live cat and a dead cat. (Presumably the fact that the cat observes the system doesn't count.) Schrödinger thought that this was a *reductio ad absurdum* of the probabilistic interpretation of quantum theory, but, in fact, this is what the Copenhagen interpretation implies.

In fact, not only does the act of observation change the state of the universe at the point of observation, it can also cause instantaneous change at remote points. This consequence was pointed out in another thought experiment, the Einstein–Podolsky–Rosen (EPR) paradox. The model here is an atom that emits two electrons with opposite spin traveling in opposite directions. Until the system is observed, the "A" electron can be spin up and the "B" electron can be spin down, or vice versa. Each electron is described by a wave function which is 50% spin up and 50% spin down. When electron "A" is observed and found to be spin up, however, instantaneously the electron "B" wave function collapses and B is spin down. This occurs instantaneously even if the electrons have traveled light years apart!

Einstein found this possibility to be unsupportable. Relativity theory is based on the constancy of the speed of light and holds that no signal may exceed the speed of light. Einstein felt that quantum physics and the Schrödinger equation were incomplete and there were *hidden variables* controlled by undiscovered equations that controlled the state of electron within the envelope of the wave function.

It turns out that the predictions of quantum physics in the EPR paradox can actually be tested experimentally to determine if hidden variables exist. In 1964, John Steward Bell, a physicist from Northern Ireland, derived an inequality predicted by quantum theory, but that could not be explained by any purely local theory with hidden variables. In fact, when the experiments were performed, the result was: quantum theory 1; hidden variables 0. Quantum theory was confirmed even in situations like the EPR paradox, where it appears to be completely at odds with normal reality. As David Mermin, a solid state physicist, has noted, "Anyone who isn't bothered by Bell's theorem has rocks in their head."

The implication of Bell's theorem and the EPR paradox is that quantum systems can remain "entangled" even if they are spatially separated by long distances. And, in fact, "entangled" photons can be created and used for applications including sensing and, potentially, computing. The concept of a "quantum computer" based on entangled quantum systems has been extensively explored in recent years. It has been suggested that such a system could be superior to current digital computers for tasks such as determining the prime factors of large numbers.

Another puzzle of quantum physics is the question of how particles are controlled by the wave function. How exactly does the electron "know" the amplitude of the wave function in the double-slit experiment and how does it arrange itself to conform to the wave function amplitude squared in determining the probability of striking the photographic plate? Note that the same problem is contained in the photon picture of electromagnetic waves. How does the electromagnetic field of the radiation control the probability that a photon will appear in a certain location? The Maxwellian picture was perfectly complete: the electromagnetic field propagated according to the wave equation with the appropriate boundary conditions and particles and materials responded to the local electromagnetic field according to the equation of motion of particles in a field. But if the light flux is very low, each photon propagates through the system to a spot that is probabilistically related to the electromagnetic field strength. What is the process by which the photon senses where the field is strong or weak, and how does it know that it should collapse its wave function with twice the probability where the field energy is twice as strong?

After considering this puzzle, some physicists were inspired to come up with the "many worlds" or "multiple universes" interpretation of quantum physics. In this interpretation, there are an infinite number of interpenetrated universes – in some of them the electron in the double-slit experiment will land at the peak of the first maximum. In others it will land near the second minimum. All possible results of the collapsed wave function exist, and we exist in the one universe in which the electron ends up where it is observed. There is no way to experimentally confirm or reject the "many worlds" theory, as far as we know, so in any utilitarian sense it is a waste of time to consider such an unverifiable theory, and in the next chapter we will "shut up and calculate" observable results – probabilistic predictions of the behavior of small particles and systems – from quantum theory. There are many remarkable results and quantum theory has been exceedingly successful in prediction (of probabilities) in every system in which it has been applied. But it is a sign of how troubling the philosophical foundations of quantum theory are that such a theory as the "many worlds" interpretation of quantum mechanics has attracted so much attention and work!

Chapter 8 problems

1. What does one have to do to increase the photoelectric current? Justify your answer.

2. Look up the work function of five metals and calculate the largest wavelength required to knock off electrons from each.

3. An orbiting satellite can become charged due to the photoelectric effect. Which material (metal) would you use to coat the outer surface of the satellite to minimize the charging of the satellite? What is the longest wavelength of incident light that can eject an electron for the material you chose?

4. An X-ray of wavelength $\lambda = 0.050$ nm is scattered by a free electron. Under what conditions does it transmit the maximum amount of energy to an electron? Calculate

   a. the new wavelength of the scattered X-ray and

   b. the percent of its original energy transferred to the electron during the collision.

5. Monochromatic light of wavelength 2.4 pm collides with a cloud of free electrons. Find the wavelength of light scattered at (a) 45° and (b) 90°.

6. Find the fractional energy loss, $\Delta E/E$, of a photon when it collides with a particle of mass m interacts with a photon. Express your answer in terms of the mass, the frequency of the scattered photon f', and the photon's scattering angle.

7. If the Earth's solar constant (the amount of power arriving on Earth from the Sun's incoming solar electromagnetic radiation per unit area, incident on a plane perpendicular to the rays) is 1.361 kW/m², calculate the number of photons per second incident on an area of 1 cm² on the surface of the Earth due to the Sun's sunlight. How does this compare with the number of photons incident to the same area emitted by a 100 W light bulb located 1 m away? (Assume that the light is monochromatic and its wavelength in both cases is 500 nm.)

8. Consider two 500 W light bulbs, one UV and one infrared, emitting at 400 nm and 800 nm, respectively. Which emits photons at a higher rate? Find the ratio of their emission rates.

9. The human eye is very sensitive and the sensors in the retina *can* respond to a single photon. However, neural filters only allow a signal to pass to the brain to trigger a conscious response when at least about five to nine photons arrive within less than 100 ms. If the eye has an area of 2.00×10^{-2} cm² and it absorbs 80% of the incoming photons, from how far away can it detect an isotropic 100 W bulb that emits monochromatic light of 500 nm?

10. Consider an electromagnetic wave of visible light with wavelength $\lambda = 500$ nm and a flux of 0.01 μW/m². How many photons per second are arriving in an area of 1 square centimeter? Assuming that each photon causes one electron of charge $q = 1.6 \times 10^{-19}$ coulomb to be ejected from the surface (100% *quantum efficiency*) what current in amperes (1 ampere = 1 coulomb/s) would be measured if all the photo-emitted electrons were measured? Is this a large current compared to the usual currents in electronic devices? How does a *photomultiplier tube* increase the current produced?

11. Create a table showing the wavelengths of electrons, protons, and photons with the following energies: 0.1 eV, 1.0 eV, 10 eV, 1 keV, 1 MeV, and 1 GeV.

12. Using the relation $k = \frac{2\pi}{\lambda} = \frac{\omega}{c}$ for light propagating in vacuum, solve the energy and x- and y-momentum conservation equations (8.3) and (8.4) to find the change

in wavelength of a Compton scattered electron as given in Eq. (8.5). For an X-ray of $\lambda = 1$ nm which scatters off a stationary free electron at an angle $\theta = 45°$, use momentum and energy conservation to find the velocity and angle of the electron recoil.

13. Can we see atoms using an optical microscope? Justify your answer. What should the energy of the electrons be in an electron microscope to be able to see individual atoms, if the size of an atom is of the order of 10^{-10} m? What about being able to see the nucleus, which is of the order of 10^{-15} m? Do such microscopes presently exist?

14. What is the wavelength of a baseball and what is its diffraction angle going through a door? (Make necessary reasonable assumptions.)

15. Look up the Principles of Quantum Cryptography.

9 Quantum mechanics

The behavior of classical macroscopic particles – from baseballs to planets – is governed by Newtonian mechanics as we have described in the first chapters of this book. The behavior of very small particles, on the other hand, is governed by an entirely different collection of equations and rules that go under the name of "quantum mechanics" and these particles are referred to as quantum particles. These rules have analogies to classical mechanics, but are fundamentally different and somewhat mysterious as we discussed in Chapter 8. The fundamental principles of quantum mechanics cannot be proven to be correct. They were developed based on the results of experimental observation, by analogy with classical mechanics, and through the consideration of logical implications of the formalisms used. There is no way to derive the fundamental elements of quantum mechanics, any more than one could predict from first principles that Newtonian forces should add as vectors. On the other hand, quantum mechanics is among the most rigorously tested theories in science, and the predictions of quantum mechanics have, so far, always been in agreement with experiment. In this chapter, we will present quantum mechanics as a series of postulates and examine their implications in application to simple one-dimensional systems. In the next chapter we will examine quantum effects in real systems including atoms, molecules, crystals, and artificially engineered nano-structures.

9.1 Postulates of quantum mechanics

QM Postulate 1: A quantum particle is described by a *wave function* $\Psi(x, y, z, t)$ which is a function of spatial coordinates and time. *All physically determinable quantities can be derived from the wave function $\Psi(x, y, z, t)$; quantities which cannot be found from the wave function are not physically meaningful.*

Note that some things that one could ask about a quantum particle, for example the exact position and simultaneous exact momentum of a particle, are not derivable from the wave function and are therefore considered indeterminate (or *uncertain*). These are things which cannot ever be known about a particle, and, as such, they are meaningless to contemplate. This feature makes the quantum universe less satisfying than the mathematical exactness of the Newtonian universe. We simply have to accept that on a very small scale, where quantum mechanics reigns, the universe is fuzzier – described by relatively indistinct wave functions rather than by exactly positioned point particles.

QM Postulate 2: Physical variables are associated with operators that operate on the wave function. In some cases these operators are differential operators; in other cases they are simply multiplied by the wave function. Some of these operators are as follows:

p_x: The x-momentum of a quantum particle is associated with the operator $-i\hbar\frac{\partial}{\partial x}$.

$$-i\hbar\frac{\partial}{\partial x}\Psi(x,y,z,t) \leftrightarrow p_x\Psi(x,y,z,t).$$

The constant $\hbar$ here is Planck's constant divided by 2π, and has a value $\hbar = 1.0546 \times 10^{-34}$ J s.

E: The total energy of a quantum particle is associated with the operator $i\hbar\frac{\partial}{\partial t}$.

$$i\hbar\frac{\partial}{\partial t}\Psi(x,y,z,t) \leftrightarrow E\,\Psi(x,y,z,t).$$

U: The potential energy of a particle is associated with the operation of multiplying the wave function by U: $U\,\Psi(x, y, z, t)$.

x: The position of a particle is associated with the operation of multiplying the wave function by x: $x\Psi(x, y, z, t)$.

QM Postulate 3: The space and time evolution of the wave function is given by *Schrödinger's equation,* which is developed by analogy with a classical *energy Hamiltonian*: kinetic energy $(\frac{p^2}{2m})$ plus potential energy (U) equals the total energy (E). For simplicity we will consider first the problem in one dimension where the wave function depends only on x and t. (We will extend to three dimensions in the next chapter.) Putting the operators associated with x-momentum and energy into a Hamiltonian equation, we have

$$\frac{1}{2m}\left(-i\hbar\frac{\partial}{\partial x}\right)^2\Psi(x,t) + U(x)\Psi(x,t) = i\hbar\frac{\partial}{\partial x}\Psi(x,t).$$

Expanding the square of the momentum operator, we arrive at the *one-dimensional Schrödinger's equation.*

$$-\frac{\hbar^2}{2m}\frac{\partial^2}{\partial x^2}\Psi(x,t) + U(x,t)\Psi(x,t) = i\hbar\frac{\partial}{\partial t}\Psi(x,t) \tag{9.1}$$

$\underbrace{\qquad}$ $\underbrace{\qquad}$ $\underbrace{\qquad}$
Kinetic energy $=\frac{p_x^2}{2m}$ Potential energy   Total energy $= E$

The wave function that describes the particle under the influence of the potential $U(x)$ is found from the solution of this equation. The wave function and its first derivative $\frac{d\Psi(x,t)}{dx}$ must be continuous in space.

As an example, consider the case of a *free particle* traveling in a region where $U(x, t) = 0$. The solution of the one-dimensional Schrödinger's equation for this case is

$$\Psi(x,t) = Ae^{i(\pm kx - \omega t)}. \tag{9.2}$$

We recognize this as a plane wave traveling to the right (for $+k$) or to the left (for $-k$) with frequency ω and wavelength $\lambda = \frac{2\pi}{k}$. Putting this solution into Schrödinger's equation and applying the derivatives to the wave function we find

$$\frac{\hbar^2 k^2}{2\,m} = \hbar\omega. \tag{9.3}$$

If we identify $\hbar$ times the wave number, $k = \frac{2\pi}{\lambda}$, as the momentum and $\hbar$ times the frequency ω as the energy of the particle, we have the correct result for a free particle, where the only energy is the kinetic energy $KE = \frac{p^2}{2m}$.

Question 9.1

Show by plugging $\Psi(x, t) = Ae^{i(\pm kx - \omega t)}$ into Eq. (9.1) that this wave function is a solution to Schrödinger's equation with the condition given in Eq. (9.3).

In general, in the case where the potential U is a function of position, but not of time, we can use the mathematical technique of *separation of variables* to write the wave function as the product of a part which depends only on x, $\psi(x)$, and a part which depends only on t, $e^{-i\omega t}$,

$$\Psi(x, t) = \psi(x)e^{-i\omega t}. \tag{9.4}$$

Substituting this into Schrödinger's equation we find

$$-\frac{\hbar^2}{2m}\frac{\partial^2}{\partial x^2}\,\psi(x)\,e^{-i\omega t} + U(x)\psi(x)\,e^{-i\omega t} = \hbar\omega\,\psi(x)\,e^{-i\omega t}.$$

Identifying the total energy $E = \hbar\omega$ and canceling $e^{-i\omega t}$ from each term, we find the *time-independent Schrödinger's equation* in one dimension:

$$-\frac{\hbar^2}{2m}\frac{d^2}{dx^2}\,\psi(x) + U(x)\psi(x) = E\,\psi(x). \tag{9.5}$$

This is the most widely used form of Schrödinger's equation and solving this will give us the x-dependent wave function $\psi(x)$ with the oscillation in time $e^{-i\omega t}$ suppressed. For example, a complete solution of the time-independent Schrödinger's equation for a free particle with $U(x) = 0$ is

$$\psi(x) = A\,e^{ikx} + Be^{-ikx} \tag{9.6}$$

where the wave amplitudes A and B represent the relative amplitude of the waves traveling to the right and to the left and the wave number $k = \frac{2\pi}{\lambda} = \sqrt{\frac{2\,m\,E}{\hbar^2}}$. The energy of a one-dimensional free particle with wave number k is given by

$$E = \frac{p_x^2}{2m} = \frac{\hbar^2 k^2}{2m}. \tag{9.7}$$

Question 9.2

Show that Eq. (9.6) with $B = 0$ is a solution to the time-independent Schrödinger's equation with $E = \frac{\hbar^2 k^2}{2m}$. What is the result if the wave function is given by Eq. (9.6) with $A = B$? How would you interpret this result?

Now the question arises, what does the wave function represent? We formalize the *Copenhagen interpretation* of quantum mechanics in the next postulate.

QM Postulate 4: The absolute value squared of the wave function multiplied by dx and dt represents the *probability* that the particle described by the wave function will be located between x and $x + dx$ at a time between t and $t + dt$

$$P(x,t)dx\,dt = |\Psi(x,t)|^2\,dx\,dt = \Psi^*(x,t)\Psi(x,t)dx\,dt \tag{9.8}$$

where the asterisk represents the complex conjugate of the wave function and $P(x, t)$ represents the *probability distribution function*.

When the potential does not depend on the time t, then the probability of finding a particle at any position is independent of time, a conclusion that comes directly out of our definition of the time-independent wave function

$$P(x,t)dx\,dt = |\Psi(x,t)|^2\,dx\,dt = \psi(x)e^{-i\omega t}\,\psi^*(x)e^{+i\omega t}\,dx\,dt = |\psi(x)|^2\,dxdt. \tag{9.9}$$

Since the particle must be somewhere, the sum of probability of finding the particle at all possible values of x is unity. This statement is expressed in the requirement of wave function *normalization*

$$\int_{-\infty}^{\infty} |\psi(x)|^2\,dx = \int_{-\infty}^{\infty} \psi^*(x)\psi(x)dx = 1. \tag{9.10}$$

Let's apply this concept to our previous example of the free particle, assumed to be moving to the right so $\psi(x) = Ae^{ikx}$. We will assume the particle is located somewhere in a region of length L. Normalization of the wave function requires

$$\int_{-L/2}^{L/2} |\psi(x)|^2\,dx = \int_{-L/2}^{L/2} \psi(x)\psi^*(x)dx = \int_{-L/2}^{L/2} A\,e^{ikx}\,A^*e^{-ikx}\,dx = \int_{-L/2}^{L/2} A\,A^*dx = 1$$

$$\tag{9.11}$$

from which we find, $A = A^* = \sqrt{\frac{1}{L}}$.

Question 9.3

Show that the most general solution to Eq. (9.11) is $A = \sqrt{\frac{1}{L}}e^{i\delta}$ where we have arbitrarily taken the case where $\delta = 0$. What is the quantity δ called in a complex number $z = |z|e^{i\delta}$? How do complex numbers with different values of δ differ?

The probability of finding the free particle between x and $x + dx$ is then

$$P(x)dx = \left| \sqrt{\frac{1}{L}} e^{ikx} \right|^2 dx = \frac{dx}{L}. \tag{9.12}$$

Thus the probability of finding the particle in any interval dx is independent of x and depends only on the size of the interval as a ratio to the total size of the system. So in the case of a free particle, we can solve for the wave function, but the wave function doesn't allow us to say anything about where the particle is: the position of the particle is completely indeterminate!

What else does the wave function tell us? The next postulate describes how we can determine physical variables from the wave function.

QM Postulate 5: The *expectation value* of any physical quantity of a particle is found from the wave function and the operator associated with that physical quantity by

$$\langle Physical\ quantity \rangle = \int_{-\infty}^{\infty} \Psi^* \, (Operator) \Psi \, dx.$$

Since our interpretation of the wave function is related to the probability of finding a physical result, we interpret the expectation value as the *weighted probability* of finding a value of a physically observable quantity in a measurement. This is similar to the understanding of the *average value* of a quantity, but with an emphasis on the predictive value of a single measurement, rather than the average of a number of measurements.

As an example, let us consider the expectation value of the position and momentum for our free-particle wave functions. The definition of the weighted probability of the position x is as follows:

$$\langle x \rangle = \sum x\, P(x) \rightarrow \int_{-\infty}^{\infty} x\, \psi^*(x)\psi(x)dx = \int_{-\infty}^{\infty} \psi^*(x)x\, \psi(x)dx. \tag{9.13}$$

This result comes from multiplying each value of x by the probability of finding the particle at that value of x. The last equation indicates that this is also consistent with our definition of the expectation value if the operator for x is just x times the wave function.

Applied to the free-particle wave function (for a particle moving in the $+x$ direction), we find

$$\langle x \rangle = \int_{-L/2}^{L/2} \frac{1}{\sqrt{L}} e^{-ikx}\, x\, \frac{1}{\sqrt{L}} e^{ikx}\, dx = \frac{1}{L} \int_{-L/2}^{L/2} x\, dx = \frac{1}{2L} \left(\frac{L^2}{4} - \frac{L^2}{4} \right) = 0.$$

Since we have seen that the probability is the same for all values of x from $+L/2$ to $-L/2$, the average value of x is $x = 0$ – at the center of the region. There is an equal chance of measuring the particle at a negative value of x and at a positive value of x.

The expectation value of x-momentum, p_x, is found as:

$$\langle p_x \rangle = \int_{-L/2}^{L/2} \frac{1}{\sqrt{L}} e^{-ikx} \left(-i\hbar \frac{\partial}{\partial x} \right) \frac{1}{\sqrt{L}} e^{ikx} \, dx = \frac{1}{L} \int_{-L/2}^{L/2} (\hbar k) dx = \hbar k. \qquad (9.14)$$

We note that the expectation (average) value of both x and p_x can be found, but the *variance* of these two quantities is very different. If a number of measurements of x are taken, they will vary over the entire range from $-L/2$ to $+L/2$ with equal probability. On the other hand, each measurement of p_x will give the same result, $\hbar k$. If we look at the variance, x is completely undetermined, with an equal probability of being found at any x value, while the value of x-momentum is precisely defined $p_x = \hbar k = \sqrt{\frac{2mE}{\hbar^2}}$.

This is a consequence of the *Uncertainty Principle*, which we will present as the next postulate.

QM Postulate 6 (Heisenberg Uncertainty Principle): Physical variables whose operators do not commute can never be precisely known at the same time. The product of the uncertainties in the two quantities is greater than or of the order of $\hbar$.

As an example, we note that the operators for position x and x-momentum do not commute, since

$$x \left(-i\hbar \frac{\partial}{\partial x} \right) \Psi(x,t) \neq \left(-i\hbar \frac{\partial}{\partial x} \right) x \Psi(x,t). \qquad (9.15)$$

Therefore, we cannot precisely determine simultaneously the position and x-momentum of a particle and the product of the uncertainties of x and p_x is greater than $\hbar$

$$\Delta x \, \Delta p_x \gtrsim \hbar. \qquad (9.16)$$

Thus for the free-particle wave function where $\Delta p_x = 0$, the uncertainly in x is of the order of the system size – essentially infinite: $\Delta x \gtrsim \frac{\hbar}{\Delta p_x} \to \infty$.

Similarly, since the time t and the energy operator $i\hbar \frac{\partial}{\partial t}$ don't commute, the product of the uncertainties of time and energy cannot be simultaneously zero and

$$\Delta E \, \Delta t \gtrsim \hbar. \qquad (9.17)$$

A manifestation of this inequality is that any system that exists for only a limited time cannot have a precisely defined energy. If an electron resides in an energy level only for an infinitesimal time Δt, then the energy of that state has an uncertainty $\Delta E \gtrsim \frac{\hbar}{\Delta t}$. Once again we see that the quantum world is inherently fuzzy on a microscopic scale.

Our last postulate arises from the fact that particles have a quantum property called *spin*. The electron spin gives electrons a finite magnetic moment, which was observed by Otto Stern and Walther Gerlach in 1922 in an apparatus, shown in Figure 9.1, that passed atomic beams of silver atoms through a non-uniform magnetic field. They found that the silver atoms were deflected into two spots, indicating that the magnetic moment was quantized either "up" or "down." This result was explained by postulating that the magnetic moment of the atoms was due to the spin of silver's single unpaired outer electron.

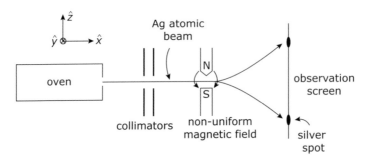

Figure 9.1 Stern–Gerlach experiment demonstrating the quantized spin of the unpaired electron on evaporated silver atoms.

Spin is a quantum property which, for an electron, has a quantum value of either $+1/2$ or $-1/2$. No matter on what axis you measure the spin, it will always be observed to be either $+1/2$ or $-1/2$ and never any value in between. In the Stern–Gerlach experiment, if one of the resultant spin-polarized beams (say in the $+\hat{z}$-direction) is passed through an identical apparatus with the same orientation, all the electrons will be deflected in the same $+\hat{z}$ direction. If the axis of the second apparatus is rotated from the z-direction toward the y-direction, two spots will appear along the axis of the second apparatus with the amplitude of each spot dependent on the angle of rotation. If the apparatus is rotated $90°$ so that is in the y-direction, two equal beams will be observed in the $\pm\hat{y}$-directions. The result of the experiment is always that the spin is either $+1/2$ or $-1/2$ regardless of the direction of measurement. This is another example of how quantum mechanical measurements give discrete results (spin $= \pm\frac{1}{2}$) that cannot be explained on the basis of unobserved variables such as a continuously variable "axis of rotation."

Quantum particles can have *half-integral* spin (electrons, proton, and neutrons each have a spin of 1/2) or *integral spin* (photons, for example, have spin of 1). Quantum particles that have a half-integral spin are referred to as *fermions*[2] and obey the Pauli Exclusion Principle as below. Quantum particles with integral spin are *bosons* and do not have an equivalent exclusion principle.

QM Postulate 7 (Pauli Exclusion Principle): Particles of half-integral spin (fermions) are restricted so that only one identical particle of each spin can occupy a given quantum state. Particles that are confined – for example, electrons in an atom – have discrete "quantum levels" that are separated in energy by "energy gaps." Each quantum energy level corresponds to a single quantum state, and thus can only be occupied by two fermions (particles with spin of 1/2 or 3/2, etc.), one spin up and one spin down.

This principle is well-known in chemistry where each quantum state $1s$, $2s$, $2p_x$, $2p_y$, $2p_z$, etc. can only accommodate two electrons. (We will look at the quantum solutions for electrons in atoms in Chapter 10.) One might ask, however, how does the Exclusion Principle apply to the free-electron states we have been using as an example so far? In one dimension, the free-electron wave functions $\psi(x) = \sqrt{\frac{1}{L}}e^{\pm ik_x x}$ are characterized by

---

[2] The origin of the names *fermion* and *boson* will become clear in Chapter 13 on Quantum Statistics.

the position x and the momentum $p_x = \hbar k_x$. A corollary of the Exclusion Principle is discussed below.

QM Corollary 7a: A single state for a freely propagating particle is defined by an area in *phase space* equal to h (Planck's constant) for each dimension of the problem. Phase space records the trajectory of a particle as the momentum vs. the position. The phase space for a one-dimensional particle is shown in Figure 9.2 with the area corresponding to a single quantum state ($= h$) shown. Only two fermions – one of each spin – can occupy an interval of phase space of size h each in one dimension. This is consistent with the Heisenberg Uncertainty Principle that the uncertainty of x times the uncertainty of p_x is *of the order* of $\hbar$. For free particles a single quantum state is defined by $\Delta x\, \Delta p_x = 6.63 \times 10^{-34}$ J·s (as opposed to $\hbar = h/2\pi = 1.054 \times 10^{-34}$ J s).

An implication of this corollary is that in a collection of N non-interacting free particles the lowest energy state of the system is not when all the particles have zero momentum. Instead, only two particles/unit length can be in the state within $\Delta p_x = h/\Delta x$ of $p_x = 0$. As illustrated in Figure 9.3, if N electrons in a system of length L are in the lowest possible energy states, all the states with momentum between $\pm p_f$ will be occupied where p_f is referred to as the *Fermi momentum*. The energy of the state at p_f is the *Fermi energy* $E_f = \frac{p_f^2}{2m}$. The lowest energy state at a temperature of absolute

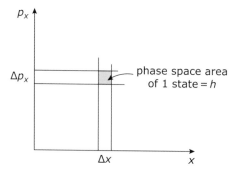

Figure 9.2 Element of one-dimensional phase space. Only one electron of each spin can occupy an area $\Delta x\, \Delta p_x = h = 6.63 \times 10^{-34}$ J·s.

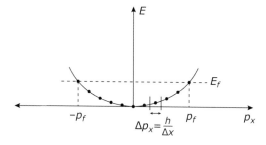

Figure 9.3 The energy of a collection of free electrons in one dimension as a function of momentum. Since only one electron of each spin per unit length can occupy each interval of momentum $\Delta p_x = h$, electrons occupy states at all energies up to the Fermi energy E_f.

zero ($T = 0$ K), has all the momentum states with $\frac{p^2}{2m} < E_f$ filled and all the states at higher energy empty. At $T = 0$, the Fermi energy is the energy between the highest filled state and the lowest empty state as shown in Figure 9.3. For N non-interacting particles with 2 spin states in a one-dimensional system of length L, we have:

$$N = \frac{2}{h}\left(\frac{L}{2} - \left(-\frac{L}{2}\right)\right)\left(p_f - (-p_f)\right) = \frac{2}{h}L\,2p_f. \tag{9.18}$$

In terms of the particle density N/L (particles/meter), we find the Fermi momentum and the Fermi energy for free particles in one dimension:

$$p_f = \frac{h}{4}\frac{N}{L}, \qquad E_f = \frac{p_f^2}{2m} = \frac{h^2}{32\,m}\left(\frac{N}{L}\right)^2. \tag{9.19}$$

In this derivation, we have neglected the particle–particle interaction. For electrons, for example, this is the Coulomb repulsive force $F = \frac{1}{4\pi\epsilon_o}\frac{e^2}{r^2}$, where $\epsilon_o = 8.85 \times 10^{-12}$ F/m is the *permittivity of free space* and e is the electron charge, which is -1.6×10^{-19} coulombs. The force is quite strong between two electrons at short range in free space, but, because the Exclusion Principle sharply reduces the number of possible scattering events, the electron–electron interactions can be effectively neglected in large assemblies of free electrons, as described by *Fermi liquid theory* discussed in Chapter 10.

In addition to the free-particle case, we will look at two examples of the application of quantum mechanics to electrons: quantum mechanical tunneling through a barrier and energy states in an infinite square well.

9.2   Quantum tunneling through a barrier

Consider the problem illustrated in Figure 9.4. A free particle in Region I ($U = 0$) with wave function $\psi(x) = Ae^{ikx}$ and energy $E = \frac{\hbar^2 k^2}{2m}$ is incident on Region II with a potential energy $U = U_o > E$. What is the quantum solution in Region II and Region III? Is there any probability that the particle will make it through the barrier to Region III?

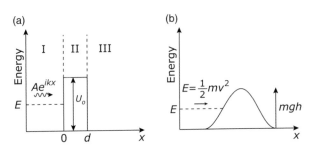

Figure 9.4 (a) A quantum particle with energy E incident on a potential barrier of strength $U_o > E$ compared with (b) a classical particle with kinetic energy $\frac{1}{2}mv^2$ incident on a hill of height h where $mgh > \frac{1}{2}mv^2$.

Note that the classical analogy, shown in the diagram on right, would be a particle with a kinetic energy $KE = \frac{1}{2}mv_o^2$ incident on a hill with height h. If $mgh > E$, there is no way that the classical particle will make it across the hill. Since the total energy $E = KE + PE = \frac{1}{2}mv_o^2$ is conserved, when the potential energy increases to a height h_{max} where $mgh_{max} = E = \frac{1}{2}mv_o^2$, the entire energy is potential energy, the kinetic energy and velocity decrease to zero, and the particle reverses directions and moves back to the left. The quantum world is a little different as shown below.

In Region I, $U = 0$ and the time-independent Schrödinger's equation is

$$-\frac{\hbar^2}{2m}\frac{d^2}{dx^2}\psi_1(x) = E\,\psi_I(x) \tag{9.20}$$

with the solution equal to a sum of plane waves traveling to the right and to the left

$$\psi_I(x) = A\,e^{ikx} + A_R\,e^{-ikx} \tag{9.21}$$

with $k = \sqrt{\frac{2mE}{\hbar^2}}$. We will ignore the reflected wave traveling to the left, setting $A_R = 0$, and just use the solution

$$\psi_I(x) = A\,e^{ikx}. \tag{9.22}$$

Similarly, in Region III the potential is zero and, if there is a wave function, the solution is also a plane wave moving to the right

$$\psi_{III}(x) = C\,e^{ikx}. \tag{9.23}$$

We are looking for the amplitude C of the wave function $\psi_{III}(x)$ in terms of the amplitude A of the incident wave. To get this we need to solve for the wave function in Region II and then apply the *boundary conditions* that hold from the requirement that the wave function and its first spatial derivative must be continuous.

First, we seek a solution to Schrödinger's equation in Region II

$$-\frac{\hbar^2}{2m}\frac{d^2}{dx^2}\psi_{II}(x) + U_o\psi_{II}(x) = E\,\psi_{II}(x). \tag{9.24}$$

Since energy is conserved, the total energy E in Region II (and Region III) is the same as in Region I, and $E - U_o$ is a negative number. Rearranging the equation in terms of the positive quantity $(U_o - E)$, we find:

$$\frac{d^2}{dx^2}\psi_{II}(x) = \frac{2m}{\hbar^2}(U_o - E)\psi_{II}(x). \tag{9.25}$$

The function that gives the same function times a positive constant when a second derivative with respect to x is taken is

$$\psi_{II}(x) = B\,e^{-\alpha x} + B_R\,e^{+\alpha x} \tag{9.26}$$

where $\alpha = \sqrt{\frac{2m(U_o - E)}{\hbar^2}}$. We arbitrarily set $B_R = 0$ and take only the first term; the wave function is a decreasing exponential as the wave function goes farther into Region II.

(If we keep both the A_R and B_R terms, we will find a more exact solution with both "incident" and "reflected" damped exponentials, but the essentials of the solution are the same.)

Now we apply the boundary conditions that the wave function must be continuous at both $x = 0$ and $x = d$

$$x = 0: \psi_I(0) = \psi_{II}(0) \Rightarrow A\,e^{ik0} = B\,e^{-a0} \Rightarrow A = B. \tag{9.27a}$$

$$x = d: \psi_{II}(d) = \psi_{III}(d) \Rightarrow B\,e^{-ad} = C\,e^{ikd} \Rightarrow C = B\frac{e^{-ad}}{e^{ikd}}. \tag{9.27b}$$

Putting these two conditions together, we find

$$\psi_{III}(x) = A\frac{e^{-ad}}{e^{ikd}}\,e^{ikx} = A\,e^{-ad}\,e^{ik\,(x-d)}. \tag{9.28}$$

(If we had included the reflected waves, we would have applied the condition that the first derivative of the wave function must also be continuous at $x = 0$ and $x = d$ to find, in addition, the amplitudes A_R and B_R in terms of A.)

In Figure 9.5, we sketch the real part of the wave function in all three regions. Note that, in contrast to the classical case, there is a finite amplitude of the wave in Region III – there is a finite probability that the particle would have *tunneled* through the barrier from Region I to Region III. Tunneling is a quantum property that has no analog in classical mechanics.

We can find the probability of tunneling by comparing the probability of the particle being at $x = d$ in Region III with the probability of it being at $x = 0$ in Region I

$$P_{tunneling} = \frac{P(x = d)}{P(x = 0)} = \frac{|\psi(x = d)|^2}{|\psi(x = 0)|^2} = \frac{A\,e^{-ad}\,e^{ik\,(x-d)}\,A^*\,e^{-ad}\,e^{-ik\,(x-d)}}{A\,e^{ikx}\,A^*\,e^{-ikx}} = e^{-2ad}. \tag{9.29}$$

If we consider an electron tunneling through a 1 nm barrier when $(U_o - E) = 1$ eV, we find the tunneling probability is:

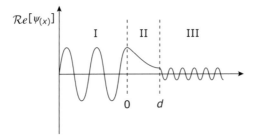

Figure 9.5 Real part of the time-independent electron wave function of a particle tunneling through a barrier.

$$P_{tunneling} = \exp\left[-2\sqrt{\frac{2m(U_o - E)}{\hbar^2}}\,d\right]$$

$$= \exp\left[-2\sqrt{\frac{2\,(9.11 \times 10^{-31}\ \text{kg})(1\ \text{eV})\left(1.6 \times 10^{-19}\,\frac{\text{J}}{\text{eV}}\right)}{(1.054 \times 10^{-34})^2}}\ 1 \times 10^{-9}\ \text{m}\right]$$

$$= 0.0000355.$$

Thus, if 100 000 electrons are incident on the barrier, about four electrons will tunnel through the barrier and appear on the other side. The high sensitivity of the tunneling probability to the barrier thickness d is used in the *scanning tunneling microscope* to examine atomic-scale details of the surface of solids, as will be discussed in Chapter 10.

9.3 One-dimensional infinite quantum well

So far, we have not encountered one of the most prominent characteristics of quantum theory: the restriction of a particle energy to certain discrete energy levels separated by gaps in the energy spectrum where no states exist. This characteristic leads to the colloquial expression "quantum leap" for an abrupt transformation (as opposed to an incremental change) of a system. *Quantization* of the energy of a system occurs when a quantum particle is spatially confined. The simplest system to study this effect is a *quantum well* where, as shown in Figure 9.6, the particle is restricted to the region between $x = 0$ and $x = d$ by a potential U_o, which is greater than the energy of the particle E.

As we have seen, the wave function in a region with a constant potential $U_o \to E$ is a decreasing exponential corresponding to the vanishing of the wave function as it extends into the region of potential U_o. The particle is therefore confined to Region II $(0 < x < d)$ with a damped extension into the Regions I and III.

To simplify the problem, we will assume that $U_o \to \infty$, a configuration known as the *infinite square well*. In Regions I and III, therefore, Schrödinger's equation gives us:

$$-\frac{\hbar^2}{2m}\frac{d^2}{dx^2}\,\psi(x) + U_o\,\psi(x) = E\,\psi(x). \tag{9.30}$$

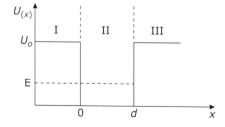

Figure 9.6 Energy potential as a function of distance for a one-dimensional square-well potential. If $U_o \to \infty$ this potential is know as an infinite square well.

When $U_o \to \infty$, the only solution for $\psi(x)$ that could allow this equation to be satisfied with a finite E is $\psi(x) = 0$ in Region I and Region III. In Region II, our solution is our standard free-particle solution:

$$\psi_{II}(x) = A\, e^{ikx} + B\, e^{-ikx} \tag{9.31}$$

with $k = \sqrt{\frac{2mE}{\hbar^2}}$. To guarantee the wave function is continuous, the boundary conditions are that $\psi_{II}(x) = 0$ at $x = 0$ and $x = d$. Except for the trivial solution of the wave function being zero everywhere, we need both the right- and left-traveling waves to have the wave function go to zero at the edges of the well.

Applying the boundary conditions at $x = 0$, we have $\psi_{II}(0) = A e^{ik0} + B e^{-ik0} = A + B = 0$, with the solution $A = -B$. This gives us

$$\psi_{II}(x) = A(e^{ikx} - e^{-ikx}) = 2iA\,\sin kx = A'\sin kx \tag{9.32}$$

where we have used Euler's relation $e^{ikx} = \cos kx + i\sin kx$ to express the difference of the complex exponentials in terms of the sine function and defined the amplitude $2iA$ as a (possibly complex) amplitude A'.

Applying the boundary conditions at $x = d$ now gives us $\psi_{II}(d) = A'\sin kd = 0$. This requires that $kd = n\pi$ where n is an integer: $n = 1, 2, 3, \ldots$ The value of k is thus *quantized*: the only allowed values of k that will yield an allowable wave function that goes to zero at the boundaries of the square well are $k_n = \frac{n\pi}{d}$. The wave number k can equal π/d, $2\pi/d$, $3\pi/d$, $\ldots$, but can never be $1.5\pi/d$ or any other value between the allowed values $k = \frac{n\pi}{d}$. Since $k = \sqrt{\frac{2mE}{\hbar^2}}$, this implies that the values of energy for a particle in an infinite square well are also quantized

$$E = \frac{\hbar^2}{2m}\left(\frac{n\pi}{d}\right)^2 \quad n = 1, 2, 3, \ldots \tag{9.33}$$

The lowest five allowed energies (*quantum levels*) for an electron in a quantum well with $d = 1$ nm are shown in Figure 9.7, where the energy is measured in electron volts (eV). One *electron volt* (eV) is the energy that an electron with a charge $q = -e = -1.602 \times 10^{-19}$ coulomb will gain when accelerated with a potential of 1 volt. Since $\Delta E = q\Delta V$, 1 eV is a measure of energy equal to 1.602×10^{-19} joules. Since energies of

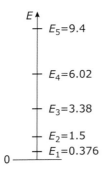

Figure 9.7 Allowed energy levels for a one-dimensional infinite square well with width $d = 1$ nm.

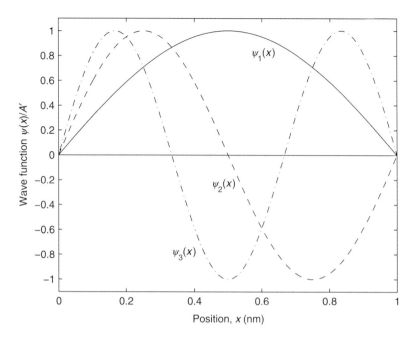

Figure 9.8 First three wave functions as a function of position for a particle in a 1 nm infinite square well.

atomic-scale systems are small, electron volts are used to characterize electron states in atoms and solids. By the Pauli Exclusion Principle, each energy level can accommodate two electrons, one spin up and one spin down.

The wave functions corresponding to the quantum energies are found by substituting the allowed values of k into $\psi_{II}(x) = A' \sin kx$. The allowed values of energy are referred to as *eigenvalues* from the German for "allowed values," and the corresponding wave functions are the *eigenfunctions*. The first three eigenfunctions of an electron in a 1 nm wide infinite square well (corresponding to the first three energies E_1, E_2, and E_3) are plotted in Figure 9.8. The wave function values are expressed as a function of the constant A', which can be calculated from the normalization criteria

$$\int\limits_{x=0}^{d} |\psi(x)|^2 dx = 1 \tag{9.34}$$

as below

$$\int\limits_{x=0}^{d} |A' \sin kx|^2 dx = |A'|^2 \left[\frac{x}{2} - \frac{1}{4k} \sin 2kx \right]_{x=0}^{d} = |A'|^2 \frac{d}{2} = 1. \tag{9.35}$$

The last step follows from the boundary condition requirement that $kd = n\pi$ and so the $\sin 2kx$ term vanishes at both the upper and lower limits. The result is $A' = \sqrt{\frac{2}{d}}$, and

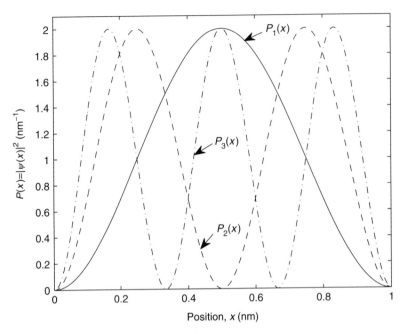

Figure 9.9 Probably distribution as a function of position for the first three wave functions in a 1 nm wide one-dimensional infinite square well.

$$\psi(x) = \sqrt{\frac{2}{d}} \sin \frac{n\pi}{d} x, \ n = 1, 2, 3 \dots \tag{9.36}$$

The probability that an electron is found at a point between x and $x + dx$ is given by

$$P(x)dx = |\psi(x)|^2 dx = \frac{2}{d} \sin^2 \left(\frac{n\pi}{d} x\right) dx. \tag{9.37}$$

The probability distribution $P(x)$ for an electron in the first three infinite square well states with $d = 1$ nm is shown in Figure 9.9. We observe that the electron in the $n = 1$ state is most probably found at $x = 0.5$ nm, while the electron in the $n = 2$ state has zero probability of being observed at $x = 0.5$ nm where the wave function has a node. Nevertheless, despite the fact that it has zero probability of being observed in the center of the well, the electron in the $n = 2$ state has equal probability of being found at $x = 0.25$ nm and $x = 0.75$ nm. The ability of a quantum electron to be observed at either side of the well without ever being observed in the center of the well is one of the non-intuitive results of quantum mechanics. The $n = 3$ state has two nodes at 0.33 nm and 0.66 nm and maxima at 0.167 nm, 0.5 nm, and 0.833 nm.

Following our development of what information we can find from the wave function, we can ask for the expectation value of the x-momentum:

$$\langle p_x \rangle = \int\limits_{x=0}^{d} \psi^*(x) \left[-i\hbar \frac{d}{dx} \right] \psi(x) dx = \frac{2}{d} \int\limits_{x=0}^{d} \sin kx \left[-i\hbar \frac{d}{dx} \right] \sin kx \, dx$$

$$= \frac{-2i\,\hbar k}{d} \int\limits_{x=0}^{d} \sin kx \, \cos kx \, dx = \left[\frac{-2i\,\hbar k}{2kd} \sin^2 kx \right]_{x=0}^{d} = 0 \qquad (9.38)$$

since $kd = \underline{n\pi}$. The infinite square wave states are *standing waves* and do not have any average momentum.

If we were to consider a finite square well – where U_o is finite but still greater than E – we would need to apply, in addition to the condition that the wave function is continuous at $x = 0$ and $x = d$, the continuity of the first derivative of the wave function at $x = 0$ and $x = d$. The solution is a sine function in the well (Region II) with a phase shift so the wave function joins smoothly onto extensions of $\psi(x)$ that decay exponentially into Region I and Region III. The energy eigenvalues are modified from the infinite square well with the deviation increasing as the state quantum number $\underline{n}$ increases. When the energy of the particle exceeds U_o, the solutions become traveling waves that reflect and transmit across the well.

In this chapter we have introduced the fundamental postulates of quantum mechanics and applied them to a few test cases in one dimension: free-particle traveling waves, quantum tunneling through a barrier, and an infinite square well. In the next chapter, we will apply the postulates to real three-dimensional systems of electrons in atoms and solids.

9.4   Numerical solution of Schrödinger's equation

If we break up the region over which we want to solve for the wave function into a set of discrete positions $x = [x_1, x_2, x_3, \ldots, x_N,]$, we can convert the continuous-space, time-independent Schrödinger's integral equation into a discretized matrix equation of the form $[A] \vec{V} = E \vec{V}$ where $\vec{V}$ is a vector representing the wave functions $\psi_n(x_i)$ and E is the corresponding energy. This is just a set of simultaneous linear equations that can be solved by the techniques of linear algebra. For example, the MATLAB `eig()` function can be used to find the eigenvalues E and eigenvectors $\vec{V}$ for any matrix $[A]$ which includes the discretized form of the second derivative for the $-\frac{\hbar^2}{2m} \frac{d^2}{dx^2} \psi(x)$ term and any x-dependent potential $V(x)$.

We will consider first the MATLAB solution for the infinite square well of width $d = 1$ nm which we can compare with the exact solutions we found in Section 9.3. Then we will consider how to find the numerical solution for the non-infinite square well. For the infinite square well, we have $U(x) \to \infty$ outside the well with the only consistent solution $\psi(x) = 0$. Inside the well, the potential energy $U(x) = 0$, with the added conditions that the continuity of the wave function requires $\psi(x_1) = \psi(x_N) = 0$. Using the discretized form for the first derivative of a function

$$\frac{df}{dx} = \frac{f(x_{[i+1]}) - f(x_{[i]})}{\Delta x} \tag{9.39}$$

we can find an expression for the second derivative

$$\frac{d^2 f}{dx^2} = \frac{\frac{df}{dx}\big|_{x_{[i+1]}} - \frac{df}{dx}\big|_{x_{[i-1]}}}{\Delta x} = \frac{f(x_{[i+1]}) - 2f(x_{[i]}) + f(x_{[i-1]})}{(\Delta x)^2}. \tag{9.40}$$

Since we only have N values of $x_{[i]}$, the second derivative can't be calculated for the two end points x_1 and x_N.

In converting from an integral equation such as Schrödinger's equation to a matrix equation, the accuracy of the solution depends on how finely the region of interest is divided into discrete intervals. For the infinite square well case, we will sample the wave function at only 10 points $N = 10$ giving nine discrete intervals from $x = 0$ to $x = 10$ nm. When $U_o \to \infty$ the only solution outside the well is $\psi(x) = 0$, so we only need to consider the solutions inside the well. Note that since we only have 10 points, we can find the second derivative at only 8 positions. The extra two equations necessary come from the conditions that $\psi(x_1) = \psi(x_{10}) = 0$ which can be put in the matrix form

$$\begin{bmatrix} A_{11} & \cdots & 0 \\ \vdots & \ddots & \vdots \\ 0 & \cdots & A_{NN} \end{bmatrix} \begin{pmatrix} \psi(x_1) \\ \vdots \\ \psi(x_N) \end{pmatrix} = E \begin{pmatrix} \psi(x_1) \\ \vdots \\ \psi(x_N) \end{pmatrix} \tag{9.41}$$

by selecting $A_{11} = A_{NN} = R$ (where R is a constant) and all other elements of the first and last row of [A] to be zero. This gives the first and last equation in the form

$$R\psi(x_1) = E\psi(x_1) \tag{9.42}$$

with the only consistent solutions of either (1) $E = R$ or (2) $\psi(x_1) = 0$. By selecting the value of R so that it will not be confused with any real energy solutions, we can guarantee that the wave function vanishes at x_1 and x_N.

The MATLAB code to set up the matrix [A] and solve for the eigenvalues and eigenvectors is below:

```
a=1e-9; N=10; del_x=a/(N-1); x=(0:(N-1))*del_x;   % Set up discrete
                                                      grid
A1=-2*eye(N);   % Diagonal terms -2* f(x_n) in second derivative
A2=zeros(N,N); A2(1:(N-1),2:N)=eye(N-1);   % Off-diagonal
                                              terms +f(x_n+1)
A3=zeros(N,N); A3(2:N,1:N-1)=eye(N-1);   % Off-diagonal
                                            terms +f(x_n-1)
A=A1+A2+A3;
C=(1.054e-34)^2/(2*9.11e-31*1.6e-19);
A(1,1)=-30;   %Can be any value; will create eigenvalue
                 -C*A(1,1)/del_x^2
A(1,2)=0;
```

```
A(N,N)=-30;   %Can be any value; will create eigenvalue
                -C*A(N,N)/del_x^2
A(N, N-1)=0;
A=-C*A/del_x^2

[V,d]=eig(A);    % Finds eigenvalues[d] and eigenvectors[V]
                    of[A]
E=diag(d)';       % Creates a vector of the eigenvalues on
                    diagonal of[d]
[Y,I]=sort(E);   % Sorts E from smallest to largest in Y, I is
                    index of positions
disp('Eigenvalues: '); disp(Y(1:10))

n_v=input('To plot eigenfunction, enter number of eigenvalue
(-1 to end): ');
while (n_v ~= -1)
    disp('Eigenvalue='); disp(Y(n_v))
    n_v2=I(n_v);
    plot(x, V(:, n_v2), 'o')
    n_v=input('To plot eigenfunction, enter number of
        eigenvalue (-1=end): ');
end
```

Running this code gives the following output:

```
A =
92.6017        0        0        0        0        0        0        0        0        0
-3.0867   6.1734  -3.0867        0        0        0        0        0        0        0
     0   -3.0867   6.1734  -3.0867        0        0        0        0        0        0
     0        0   -3.0867   6.1734  -3.0867        0        0        0        0        0
     0        0        0   -3.0867   6.1734  -3.0867        0        0        0        0
     0        0        0        0   -3.0867   6.1734  -3.0867        0        0        0
     0        0        0        0        0   -3.0867   6.1734  -3.0867        0        0
     0        0        0        0        0        0   -3.0867   6.1734  -3.0867        0
     0        0        0        0        0        0        0   -3.0867   6.1734  -3.0867
     0        0        0        0        0        0        0        0        0  92.6017

Eigenvalues:

 0.3723  1.4443  3.0867  5.1014  7.2455  9.2602  10.9026  11.9746  92.6017  92.6017

To plot eigenfunction, enter number of eigenvalue (-1 to end):
```

We observe the distinctive "[1 -2 1]" pattern of the discretized second derivative in each row of [A] except the first and last rows which introduce the two spurious solutions $E = R = 92.601$ eV. The other solutions track remarkably well with the exact solutions for an infinite square well with $d = 1$ nm, as shown in Table 9.1. The agreement is improved to nearly exact agreement within round-off error by increasing the sampling to $N = 500$, as indicated in Table 9.1.

Table 9.1 The numerically calculated energies (for $N = 180$) of the first six infinite square-well states for increasing number of points included in the simulation, compared to the exact solution as derived in Section 9.3

Model	Energies (eV)					
	E_1	E_2z	E_3	E_4	E_5	E_6z
Exact solution: $E = \frac{\hbar^2}{2m}\left(\frac{n\pi}{d}\right)^2$	0.376	1.504	3.385	6.018	9.403	13.54
Numerical result ($N = 10$)	0.372	1.444	3.087	5.101	7.245	9.260
Numerical result ($N = 50$)	0.376	1.502	3.374	5.985	9.322	13.37
Numerical result ($N = 100$)	0.376	1.504	3.382	6.010	9.383	13.50
Numerical result ($N = 500$)	0.376	1.504	3.385	6.017	9.402	13.54

The eigenfunctions are also very close to the exact analytic solution, with the two provisions that the normalization condition is not incorporated in the MATLAB code so the magnitude of the wave functions need to be matched, and also the numerical results occasionally give a wave function which is multiplied by −1 compared to the exact value. Since all physical predictions depend on the magnitude squared of the wave function, the factor of −1 is not physically meaningful. The first three wave functions ψ_1, ψ_2, and ψ_3 are shown for the exact solution and the numerical solution with $N = 10$ and $N = 50$ in Figure 9.10.

In the case of the finite square well we expect from our treatment of quantum tunneling that the wave function will not go to zero at the edge of the well, but will decay exponentially into the surrounding barriers. We will thus need to define the region of the analysis to extend beyond the well far enough that we can safely set the wave function equal to zero at the edges of the region considered. We will arbitrarily select to have our region of analysis be twice the size of the well with one-quarter of the points inside the barrier on each side of the well. This can be accomplished by adding to the [A] matrix a diagonal [U] matrix with the center half of the diagonal elements equal to zero and the first and last quarter of the diagonal elements equal to the height of the barrier, selected to be 10 eV in the units we have used for the $-\frac{\hbar^2}{2m}\frac{d^2}{dx^2}\psi(x)$ kinetic energy term. This can be accomplished by adding the following lines in the code above where the blank line is between between the lines A=−C*A/del_x^2; and [V,d]=eig(A);

```
% for finite quantum well, a=2* (size of well)
% 1/4 of the solution space in inside the barrier for both +/− x
U=zeros(N,N);
kmax=N/4;
for kk=1:kmax
    U(kk,kk)=10.;
end
for kk=N-kmax:N
    U(kk, kk)=10.;
end
A=A+U;
%end finite quantum well
```

Table 9.2 Numerically calculated energies of the first five quantum states for finite square wells with barrier potentials U_o as shown, compared to the exact solution for the infinite square well

Model	Energy (eV)				
	E_1	E_2	E_3	E_4	E_5
Infinite square well	0.376	1.504	3.385	6.018	9.403
Finite square well (U_o = 50 eV)	0.3406	1.3615	3.0603	5.4326	8.472
Finite square well (U_o = 30 eV)	0.3305	1.3208	2.9668	5.2623	8.1932
Finite square well (U_o = 10 eV)	0.3003	1.1960	2.6747	4.7006	7.1966

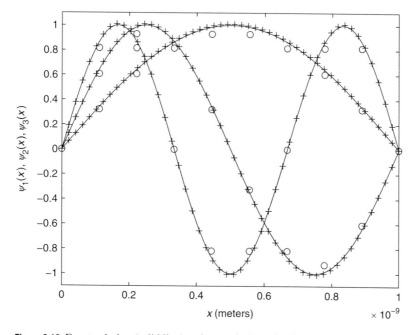

Figure 9.10 Exact solution (solid line) and numerical result with $N = 10$ ($\circ$) and $N = 50$ (+) for first three wave functions for an infinite square well of width 1 nm. The exact solutions are plotted without the normalization factor $\sqrt{\frac{2}{d}}$ and the numerical results are multiplied by -5 (for the $N = 50$ results) and by -2 (for the $N = 10$ ψ_1 and ψ_2 results) and by $+2$ (for the $N = 10$ ψ_3 result).

The values of the first four lowest energy quantum states of the finite square well with $U_o = 10$ eV, numerically calculated with $N = 180$, are shown in Table 9.2, and the first two wave functions $\psi_1(x)$ and $\psi_2(x)$ are plotted, along with the corresponding wave functions for the infinite square well, in Figure 9.11. We note that for the finite square well the wave functions extend, exponentially damped, into the barrier region. The energy of the finite square well states is reduced as the particle is not as tightly confined as in the infinite square well.

Clearly, the quantum mechanical wave functions and eigenvalue energies can be numerically calculated for any time-independent potential $U(x)$; two examples of this process are discussed in the chapter problems. The extension of these techniques to

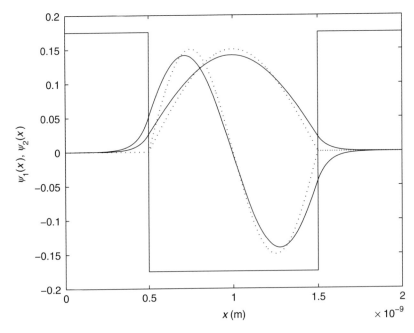

Figure 9.11 The first two wave functions $\psi_1(x)$ and $\psi_2(x)$ numerically calculated for a finite square well with a barrier potential $U_o = 10$ eV (solid lines) compared with the infinite wave functions $\psi_n(x) = A \sin \frac{n\pi}{d}$ with $A = 0.15$ (dotted lines). The extension of the well is indicated in the figure.

three dimensions and extended periodic systems such as crystal lattices has been extensively explored by quantum physicists and chemists since the development of digital computers in the 1950s, and some of the results will be surveyed in Chapters 10 and 11. The application of numerical solutions of Schrödinger's equation in artificially fabricated nano-structures to produce new and potentially useful device structures is a focus of much current scientific and engineering effort.

Chapter 9 problems

1. If the uncertainty in the position of a proton is 0.010 nm, what is the minimum uncertainty in the measurement of its velocity in the same direction?

2. Compare the probability of an electron tunneling through a barrier to that of a proton, if the energy of each particle is 5 eV and the barrier thickness and height are 0.01 nm and 6 eV, respectively.

3. What is the probability of a person tunneling through a garden wall? (Make reasonable necessary assumptions to answer this question.)

4. By how much does the tunneling probability increase or decrease, if
   a.   we double the width of the barrier?

    b.   we double the difference between the strength of the barrier and the energy of the particle?

5. What happens to the energy levels of an infinite quantum well if its width is reduced by a factor of two? Compare the difference in the first two energy levels in these two cases.

6. Consider an infinite quantum well of width L. What is the energy difference ΔE between its n and $n+2$ quantum levels?

7. **a.** If we approximate the atomic nucleus as an infinite potential well of width $L_1 = 1.5 \times 10^{-14}$ m with a proton trapped in it, calculate the first three energy levels and the wavelength of the photons emitted by the transitions between them.

    b. If we make a similar approximation for the atom as a whole and consider it to be represented by an infinite quantum well of width $L_2 = 10^{-10}$ m, calculate the first three energy levels for an electron trapped in it and the wavelength of the photons emitted by the transitions between them.

    c. Compare and discuss your results in (a) and (b).

8. What should be the range of width of an infinite well if we want the electronic transition from its first excited state to its ground state to result in a photon in the visible spectrum?

9. Consider the finite quantum well shown below. Find the first three wave functions (two even and one odd) and energy levels inside the well. To solve the transcendental equation, assume the particle is an electron with mass $m_e = 9.11e - 31$ kg, that $L = 0.2$ nm, and the height of the confining potential is $V_o = 20$ eV. Compare your results with the case for an infinite square well.

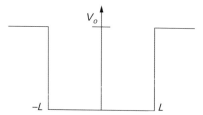

10. If the width of the quantum well in Problem 9 were doubled,

    a. How much energy is needed to knock the electron out of the well from its ground and first excited states?

    b. What is the maximum wavelength of the photon that can accomplish that?

    c. Is the ground state of the electron lower or higher than before?

11. Describe what happens when a quantum particle is traveling from the left in the figure in Problem 9 with energy $E > V_o$. Write down Schrödinger's equation for all three regions and find the equations that will allow you to determine all constants in the wave functions by applying the proper boundary conditions. (Do NOT solve the equations.)

12. Describe what happens when a quantum particle is traveling from left to right with energy $E > V_o$ (V_o is the strength of the barrier), as shown in the diagram below. Write down Schrödinger's equation for all three regions and find the equations that will allow you to determine all constants in the wave functions by applying the proper boundary conditions. (Do NOT solve the equations.)

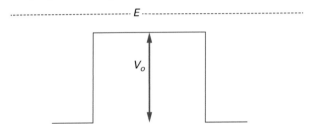

13. Consider an electron in a double well as shown below. Find the allowed energy levels. The outside walls are infinite, the width of the well L (= 10 nm), the inner barrier U_o (= 5 eV), and the barrier width d (= 0.02 nm). Consider all possible regions.

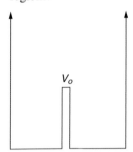

14. Coupled quantum wells.
   a. Modify the MATLAB code in Section 9.4 to numerically solve the one-dimensional Schrödinger equation for an electron in a potential of two quantum wells of width $a = 1$ nm separated by a potential barrier of height $U_o = 150$ eV and width $d = 0.5$ nm. You may assume the electron is confined within the region $0 < x < 2.5$ nm and the wave function is zero outside this region. Create a potential energy profile with $U = 0$ for $x < 1$ nm and $x > 1.5$ nm and $U = 150$ eV for $x < 1.5$ nm and add it to the diagonal elements of the [A] matrix as in the finite quantum well example.

   b. Modify this program so that you can study what happens when the potential barrier width d and height U_o are changed. Set $d = 0.1$ nm, the well widths to 1 nm, and reduce the barrier height to zero so the electron is in a single well of width 2.1 nm. How do the energy eigenvalues you find compare with what you would calculate from our closed-form solution of Schrödinger's equation in an infinite square well of width 2.1 nm?

c.   Let the barrier height increase to $U_o = 1.0$ eV with the width still fixed at 0.1 nm. How are the single-quantum-well states (energies and wave functions) modified by a thin, low barrier? Are the states with even quantum number n affected differently than those of odd quantum number n? Can you explain this by looking at the wave functions?

d.   Let the barrier height increase to $U_o = 20$ eV. How are the wave functions changed and how do your energy eigenvalues compare with the closed-form expression for a 1.0 nm wide infinite square well?

e.   Plot the energy of the lowest six energy levels as a function of the barrier height U_o for several values of U_o between 0 and 20 eV with $d = 0.1$ nm. Indicate on your plot the infinite quantum well energies for a well width of 1.0 nm and also for a well width of 2.1 nm.

15.   What are the allowed energy levels for energies $E > V_o$ in Problems 7 and 9? Are they quantized? Explain your answer.

16.   Harmonic oscillator quantum states.

   Another one-dimensional potential that is commonly solved in quantum mechanics textbooks is the *harmonic oscillator* potential. A harmonic oscillator is a system like the spring–mass system in which the restoring force is linearly dependent on the displacement from equilibrium: $F = -kx$. Integrating the force to get the potential energy, we find $U = \int F dx = \frac{1}{2}kx^2$. This is a reasonable approximation of interatomic forces in a molecule, and the vibrational states of molecules can be well approximated by a harmonic oscillator model.

a.   Modify the MATLAB code in Section 9.4 to model an electron in a harmonic oscillator potential with $k = 1 \times 10^{20}$ eV/m². You can assume that the wave function goes to zero at $|x| > 2$ nm (boundary conditions: for $\psi(x) = 0$ for $x = \pm 2$ nm). Find the lowest 10 allowed energies. What is the energy spacing between the energy levels? Does it change significantly as the electronic energy increases (higher quantum levels)? What do the wave functions look like? Plot out the wave functions for the five lowest energy states.

b.   On his Physics PhD oral exam, the first author was asked how the allowed wave functions and energies would change if the harmonic oscillator potential was modified so that for $x > 0$ the potential went to infinity (i.e., $\psi(x) \to 0$) – a sort of "half-harmonic-oscillator" potential. Modify your boundary conditions to make the wave function go to zero at $x = 0$ and $x = -2$ nm. How do the wave functions and energies change? Does this make sense to you?

10 Quantum electrons in atoms, molecules, and materials

The justification for the formalism of quantum mechanics developed in the last chapter is the rich variety of observed phenomena and real systems that are explained by quantum mechanics and are completely inexplicable by classical Newtonian mechanics and Maxwellian electromagnetic theory. As one example, once physicists developed the picture of an atom composed of a small positively charged nucleus surrounded by a cloud of electrons, they were left with an unanswerable puzzle. While Newtonian physics could explain electrons orbiting around the nucleus under the influence of the attractive Coulomb force, the electrons would be experiencing a constant centripetal acceleration and Maxwell's theory would predict that accelerating electrons should radiate electromagnetic radiation, lose energy, and spiral into the nucleus in a fraction of a second. So, classical physics cannot even account for one of the most fundamental truths: the stability of the atom and matter! The quantum solution of the hydrogen atom, presented later in this chapter, not only gave a satisfactory solution to this puzzle, but it predicted with exquisite accuracy the actual spectroscopic data for its electronic energy levels in the hydrogen atom. In this chapter, we will look at several examples of quantum systems and devices, using quantum mechanical concepts.

10.1 Schrödinger's equation in three dimensions

Since the real world is three dimensional, it will be necessary to extend the quantum mechanics concepts developed in the last chapter to three dimensions. In the three-dimensional Schrödinger's equation, the wave function $\Psi(x, y, z, t)$ follows by generalizing our momentum operator from $-i\hbar \frac{\partial}{\partial x}$ to $-i\hbar \vec{\nabla}$, where $\vec{\nabla} = \hat{x} \frac{\partial}{\partial x} + \hat{y} \frac{\partial}{\partial y} + \hat{z} \frac{\partial}{\partial z}$ is the gradient operator. The kinetic energy operator becomes $\frac{p^2}{2m} = \frac{\vec{p} \cdot \vec{p}}{2m} = -\frac{\hbar^2}{2m} \nabla^2$ where ∇^2 is the (scalar) Laplacian operator, which in rectangular coordinates is $\nabla^2 = \frac{\partial^2}{\partial x^2} + \frac{\partial^2}{\partial y^2} + \frac{\partial^2}{\partial z^2}$. The three-dimensional Schrödinger's equation then becomes

$$-\frac{\hbar^2}{2m} \nabla^2 \Psi(x, y, z, t) + U(x, y, z, t) \, \Psi(x, y, z, t) = i\hbar \frac{\partial}{\partial t} \Psi(x, y, z, t). \tag{10.1}$$

When the potential energy term $U(x, y, z, t)$ is a function only of x, y, and z and is independent of time t, we can solve the differential equation by separation of variables

$\Psi(x, y, z, t) = \psi(x, y, z)e^{-i\omega t}$ where $\psi(x, y, z)$ is the solution of the *time-independent Schrödinger's equation,*

$$-\frac{\hbar^2}{2m}\nabla^2\psi(x, y, z) + U(x, y, z)\,\psi(x, y, z) = E\psi(x, y, z) \qquad (10.2)$$

and $E = \hbar\omega$ is the total energy.

10.2 Free electrons in three dimensions (3D) (free-electron metals)

The simplest solution of the 3D Schrödinger's equation would be $U(x, y, z) = 0$, a free particle. The solution to the time-independent Schrödinger's equation is

$$\psi(x) = C_1 e^{i(k_x x + k_y y + k_z z)} + C_2\, e^{-i(k_x x + k_y y + k_z z)} \qquad (10.3)$$

with $\frac{\hbar^2}{2m}(k_x{}^2 + k_y{}^2 + k_z{}^2) = \frac{(p_x{}^2 + p_y{}^2 + p_z{}^2)}{2m} = E$. In vector notation, this is the same as

$$\psi(x) = C_1 e^{i\vec{k}\cdot\vec{r}} + C_2 e^{-i\vec{k}\cdot\vec{r}} \qquad (10.4)$$

where $\vec{k} = k_x\hat{x} + k_y\hat{y} + k_z\hat{z}$ is the *wave vector* and $\vec{r} = x\hat{x} + y\hat{y} + z\hat{z}$ is the position vector. **The solution represents *plane waves* traveling in the direction $\vec{k}$ (the first term with amplitude C_1) or in the direction $-\vec{k}$ (the second term with amplitude C_2).** The wavelength is $\lambda = \frac{2\pi}{|\vec{k}|}$ and planes of constant phase are perpendicular to $\vec{k}$.

We can consider electrons that are free to move in any direction in an isotropic metal to be approximated by this solution – if we address two problems. First, a collection of electrons will repel each other and fly off into space. If we assume that the electrons are moving in a background of uniform positive charge, which balances the negative charge of the electrons, they will remain in the region under consideration due to the attraction of the positive charge. We will assume that the sample is infinite in size, so as not to worry about edge effects. This model, referred to as *jellium*, is a reasonable first-order approximation for electrons in a metal. While the discrete nature of the ion cores is neglected, the fundamental nature of the positive background charge in balancing out the negative charge of the electrons is modeled. A second problem is that our model does not consider the effect of the electron–electron interactions in the definition of U. We will assume that we can set $U = 0$, switching off the electron–electron interactions, and make an argument for why this is reasonable later.

As we have seen, the Pauli Exclusion Principle does not permit more than two electrons being in the same quantum state, defined in three dimensions as a volume of *phase space* such that $\Delta x\,\Delta y\,\Delta z\,\Delta p_x\,\Delta p_y\,\Delta p_z = h^3$ where $h = 6.63 \times 10^{-34}$ J s is Planck's constant. Only two electrons (one spin up and one spin down) can be in the state around $p_x = p_y = p_z = 0$. After that state is filled, the next electrons must go into a state of higher kinetic energy. At $T = 0$ K (absolute zero) all the states with $|p| = \sqrt{p_x{}^2 + p_y{}^2 + p_z{}^2} < p_f$ must be filled and all the states with $|p| > p_f$ must be empty, where p_f is the *Fermi momentum*. If the electron density $n = \frac{N}{\Delta x\,\Delta y\,\Delta z} = \frac{N}{V}$ (# of

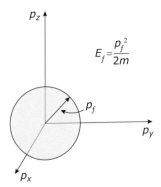

Figure 10.1 The lowest energy state of a free-electron metal has all states with momentum p less than the *Fermi momentum* p_f filled and all states with $p > p_f$ empty.

electrons/m³), then $n = \frac{1}{h^3} 2 \frac{4\pi}{3} p_f^3$, as shown in Figure 10.1. Substituting the *Fermi energy* $E_f = p_f^2/2m$ and using the more commonly used constant $\hbar = h/2\pi$, we find

$$n = \frac{1}{(2\pi\hbar)^3} 2 \frac{4\pi}{3} (2mE_f)^{3/2} \Rightarrow E_f = \frac{\hbar^2}{2m} (3\pi^2 n)^{2/3}. \tag{10.5}$$

Example 10.1 What are the Fermi energy, Fermi momentum, Fermi velocity, and Fermi wavelength of copper?

If we consider copper to have one free electron per atom, density $\rho = 8.94$ g/cm³ and atomic weight $AW = 63.55$ g/mole, and recall that Avogadro's number $N_A = 6.022 \times 10^{23}$ atoms/mole, we find that its electron density is

$$n = \frac{8.94 \left(\frac{g}{cm^3}\right)}{63.55 \left(\frac{g}{mole}\right)} \left(6.022 \times 10^{23} \left(\frac{atoms}{mole}\right)\right) \left(\frac{10^6 \ cm^3}{m^3}\right) (1 \ electron/atom)$$

$$= 8.47 \times 10^{28} \ electrons/m^3$$

from which we find the Fermi energy to be

$$E_f = \frac{\hbar^2}{2m} (3\pi^2 n)^{2/3} = 1.13 \times 10^{-18} \ J = 7.03 \ eV.$$

This value corresponds to a value of the momentum of an electron at the Fermi level $p_f = 1.43 \times 10^{-24}$ kg · m/s, an electron velocity at the Fermi level $v_f = p_f/m = 1.57 \times 10^6$ m/s, and an electron wavelength $\lambda_f = 2\pi/k = 2\pi\hbar/p_f = 4.62 \times 10^{-10}$ m $= 0.462$ nm.

At absolute zero, $T = 0$ K, all states in the "Fermi sphere" with $p < p_f$ are filled and all states with $p > p_f$ are empty. When we consider events where one electron – electron A – interacts with ("scatters off") another electron – electron B – with energy and momentum conserved, so that $\Delta E_A = -\Delta E_B$ and $\Delta \vec{p}_A = -\Delta \vec{p}_B$, Figure 10.2 shows that the only scattering events where each electron starts from an occupied state and ends at an unoccupied state are those where the electrons exchange states. Since electrons are identical particles there is no way to distinguish "electron A" from "electron B." Thus, this is not a meaningful scattering process, since the final state is indistinguishable from the initial state. At $T = 0$, therefore, the number of electron–electron scattering events approaches zero and our approximation to neglect the inter-electron potential is rigorously true – there isn't any phase space available for electron–electron scattering.

At $T > 0$ K, there is a band of states of width on the order of $\Delta E \approx k_B T$ around the Fermi surface where the electrons states are partly occupied, as we shall see later when we consider thermal occupancy. The key point here is that, while there are scattering events possible when $T > 0$ K, at room temperature $k_B T = 0.026$ eV and the band of states available for scattering is negligible compared to the size of the Fermi energy in metals ($E_f \sim 7$ eV). The conclusion that the number of electron–electron scattering events will be small holds true even at room temperature, and our non-interacting electron picture is still a good approximation. This is formalized in *Fermi Liquid Theory* where electrons in a metal are considered weakly interacting but with an increased mass due to the "exchange hole" created by the exclusion of other electrons of the same spin from the same volume of phase space.

Question 10.1

Explain why the electron–electron scattering event (I) in Figure 10.2(a) doesn't conserve energy. Does it conserve momentum? How do you know? Does the scattering event (II) conserve energy and momentum? Why is it forbidden? Explain how these restrictions are lifted in the event for $T > 0$ shown in Figure 10.2(b).

Since many of the properties of electrons in a metal depend on the number of electrons near the Fermi surface, the *density of states* at the Fermi energy is an important concept. The density of states $N(E) dE$ is defined as the number of states available between E and $E + dE$. As shown in Figure 10.3, one can find $N(E) dE$ by finding the volume of a spherical shell of width dp around the Fermi momentum p_f, dividing by $h^3/2$ to find the number of spin-up and spin-down electron states per unit volume in that volume, and then convert from p_f and dp to E_f and dE using the appropriate equation for electronic kinetic energy $E = \frac{p^2}{2m}$. Alternately, one can take a derivative with respect to E of Eq. (10.5) above for the total number of states with energy less than E. The result of either calculation is:

$$N(E) = \frac{1}{2\pi^2} \left(\frac{2m}{\hbar^2} \right)^{3/2} (E)^{1/2} \left(\frac{\text{electrons}}{\frac{m^3}{J}} \right). \tag{10.6}$$

Question 10.2

Can you derive Equation (10.6) (a) from Eq. (10.5), and (b) from Figure 10.3?

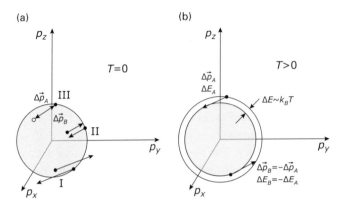

Figure 10.2 (a) Electron scattering events at $T = 0$ K that are forbidden by conservation of energy and momentum (I), the Pauli Exclusion Principle (II), or because the final state is identical to the initial state (III). (b) Allowed scattering events for $T > 0$ have initial and final states within $\sim k_B T$ of the Fermi energy $E_f = \frac{p_f^2}{2m}$.

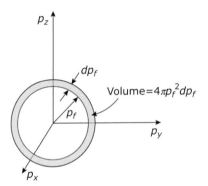

Figure 10.3 The volume of "phase space" between p_f and $p_f + dp_f$.

10.3   Application: scanning tunneling microscope (STM)

Metals are materials in which electrons are free to move in response to an applied voltage (electric field). The electrons in a metal are reasonably well-described by free non-interacting electrons moving in a region of uniform positive charge – the "jellium" model described above. The electrons must be confined in the metal by an attractive potential since they do not stream out of the metal into the surrounding air or vacuum except under unusual circumstances. The potential that confines electrons in a metal is referred to as the metal *work function*. A simplified plot of the potential experienced by an electron in a metal near the surface is shown in Figure 10.4. The work function, Φ, measured from the highest energy electron at the Fermi surface and the "vacuum level" of a free electron in the vacuum directly outside the metal, is usually between 1 and 10 eV.

If a probe made of the same metal is brought up very close to the free metal surface in a vacuum, as shown in Figure 10.5(a), the potential diagram would resemble the

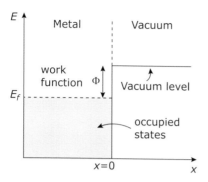

Figure 10.4 Quantum model of electron energy in a metal with a "work function" Φ. The surface of the metal is at $x = 0$ and the Fermi energy of electrons in the metal is Φ below the energy of an electron at rest immediately outside the metal.

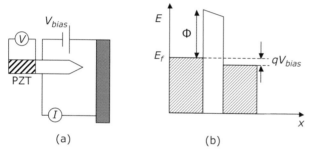

(a)   (b)

Figure 10.5 (a) Scanning tunneling microscope (STM) experimental apparatus for measuring the tunneling of electrons from a probe needle to a metal surface. The distance between the probe needle and the surface is controlled by the piezoelectric transducer (PZT) and the voltage difference is controlled by the bias voltage source V_{bias}. (b) An energy diagram of filled electron states (shaded) in the probe needle and in the metal surface. The Fermi energy in the metal surface is reduced by the bias voltage, creating empty states that electrons can tunnel into from the probe needle.

diagram in Figure 10.5(b). The Fermi levels in the two metals are offset by the bias voltage between the probe and the metal surface as shown on the right. As can be seen, the electrons in the probe can quantum mechanically tunnel from the probe to the unoccupied states within qV_{bias} above the Fermi level in the sample.

This observation is the basis of the scanning tunneling microscope (STM), developed in the early 1980s by Gerd Binnig and Heinrich Rohrer at IBM Zürich. The position of the probe is controlled very sensitively by the voltage V on a PZT (lead-zirconium-titanate) piezoelectric translator, so that the distance between the probe and the surface can be adjusted to make the tunneling current I between the probe and surface constant. As the probe is moved laterally across the surface, the map of the probe voltage V against the x–y position, gives a map of the topography of the surface. If the probe is driven against the surface and brought back again until the tunneling tip is effectively only one atom, the STM can give an atomic-scale map of the sample surface.

> ### Question 10.3
>
> What are the characteristics of PZT that make it useful for the STM? What is the purpose of the bias voltage V_{bias} and how is it different from the voltage V on the PZT? How do you think the tunneling tip is moved in the x–y direction across the surface to be studied?

We can find an estimate of the STM tunneling current as follows. Assume that the tunneling tip and the surface are both tungsten with a density $\rho = 19.3$ g/cm^3 and an atomic weight $AW = 183.35$ g/mole. Using Avogadro's number $N_A = 6.022 \times 10^{23}$ atoms/mole, we find the number of *tungsten atoms per volume* to be $N = \frac{\rho}{AW} N_A = \frac{19.3 \text{ g/cm}^3}{183.35 \text{ g/mole}} 6.022 \times 10^{23} \frac{\text{atoms}}{\text{mole}} \left(\frac{10^6 \text{ cm}^3}{\text{m}^3} \right) = 6.34 \times 10^{28}$ atoms/m^3. Tungsten is a transition metal with an outer electron configuration $6s^2 5p^6 5d^6$. If we assume that the two $6s$ electrons per atom comprise the conduction band, we find the *number of conduction electrons per volume* to be $n = 12.7 \times 10^{28}$ electrons/m^3. Assuming the conduction band is isotropic with an effective mass equal to the free-electron mass, we find the Fermi energy

$$E_f = \frac{\hbar^2}{2m} (3\pi^2 n)^{2/3} = 1.47 \times 10^{-18} \text{ J} = 9.2 \text{ eV}.$$

The velocity of an electron at the Fermi energy is then $v_f = \frac{p_f}{m} = \frac{\sqrt{2m E_f}}{m} = 1.8 \times 10^6$ m/s. The number of electrons per second traveling in the tip toward the vacuum barrier can be estimated by one-sixth of the number of electrons in a cylinder of area equal to the area of the single atom on the tunneling tip and length equal to v_f multiplied by one second. (The factor of 1/6 arises because only those electrons moving to the right in the diagram above – and not those moving to the left, moving up or down, or moving into or out of the page – are incident on the vacuum through the tunneling tip.) The only electrons from the tip that can tunnel into unoccupied states in the surface are those within qV_{bias} of the Fermi energy with a density $N(E)\Delta E = \frac{1}{2\pi^2} \left(\frac{2m}{\hbar^2} \right)^{3/2} (E_f)^{1/2} (qV_{bias})$ (using the results of the last section for the density of states at the Fermi level). If we assume that the tunneling tip is a single atom of diameter a, the number of electrons incident on the surface in time Δt that can tunnel into an unoccupied state in the surface is

$$n_{inc} = \frac{1}{6} v_f \Delta t \, \pi \left(\frac{a}{2} \right)^2 \frac{1}{2\pi^2} \left(\frac{2m}{\hbar^2} \right)^{3/2} (E_f)^{1/2} (qV_{bias}). \tag{10.7}$$

With a tunneling probability given in the last chapter $P_{tun} = e^{-2\sqrt{\frac{2m}{\hbar^2}(E_{vac} - E_f)} \, d} = e^{-2\sqrt{\frac{2m}{\hbar^2} \Phi} \, d}$, the *tunneling current* is

$$I_{tun} = q n_{tun} / \Delta t = q \frac{1}{6} v_f \, \pi \left(\frac{a}{2} \right)^2 \frac{1}{2\pi^2} \left(\frac{2m}{\hbar^2} \right)^{3/2} (E_f)^{1/2} (qV_{bias}) \, e^{-2\sqrt{\frac{2m}{\hbar^2} \Phi} \, d}. \tag{10.8}$$

The interatomic spacing in tungsten, a good approximation for the diameter of an atom, can be found from $a \cong \sqrt[3]{1/N} = \sqrt[3]{1/6.34 \times 10^{28} \text{ atoms/m}^3} = 2.5 \times 10^{-10}$ m. The *work function*, the energy difference between the Fermi energy and the vacuum level, is measured experimentally in tungsten to be $\Phi = 4.5$ eV. The tunneling current is thus

$$I_{tun} = 4.9 \times 10^{-5} \, V_{bias} \, e^{-2.2 \times 10^{10} \, d}.$$

For a bias voltage $V_{bias} = 0.2$ V and a tunneling distance $d = 0.25$ nm, we find the tunneling current $I_{tun} = 40$ nA. The tunneling current is highly sensitive to the distance d, decreasing to 0.16 nA if d increases to 0.5 nm. *By adjusting the distance d with the PZT to keep the tunneling current constant, the PZT voltage generates a very sensitive map of the surface topology.*

10.4   Quantum dots

The simplest three-dimensional confined electron system can be produced by fabricating nanoscale metal "dots" on an insulating substrate as shown below. The conduction electrons in the metal approximate free electrons with $U = 0$ confined ("trapped") in the structure by the electron work function. This is the three-dimensional analog of our quantum well. The energy levels of the cylindrical electron trap shown can be found by solving Schrödinger's equation in cylindrical coordinates with the requirement that the wave function be periodic if the azimuthal angle ϕ changes by 360°. For our purposes, it will be easier to consider a rectangular trap with dimensions L_x, L_y, and L_z, as shown in Figure 10.6(b).

For the rectangular box, the time-independent Schrödinger's equation in rectangular coordinates can be solved by separating $\psi(x, y, z) = \alpha(x)\beta(y)\gamma(z)$ into product functions which depend only on x, on y, or on z. Requiring the wave function to vanish at each edge of the box, results in a wave function

$$\psi(x, y, z) = A \, \sin k_x x \, \sin k_y y \, \sin k_z z \tag{10.9}$$

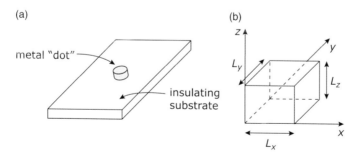

(a)   (b)

metal "dot"

insulating substrate

Figure 10.6 (a) A "quantum dot" produced by sputtering a thin film onto an insulating material and etching to create a free-standing metal dot of dimensions comparable to the electron wavelength. (b) An approximation of a rectangular "quantum box" for which energy levels can be found in rectangular coordinates in terms of infinite square-well wave functions in the x-, y-, and z-directions.

with the electron energy given by $E = \frac{\hbar^2}{2m}(k_x^2 + k_y^2 + k_z^2)$ and the quantization conditions

$$
\begin{aligned}
k_x L_x &= \underline{n}_x \pi, \\
k_y L_y &= \underline{n}_y \pi, \\
k_z L_z &= \underline{n}_z \pi
\end{aligned}
\tag{10.10}
$$

with $\underline{n}_x$, $\underline{n}_y$, and $\underline{n}_x$ required to be integers: 1, 2, 3, 4, ... The energy is thus quantized:

$$
E = \frac{\hbar^2 \pi^2}{2m}\left(\left(\frac{\underline{n}_x}{L_x}\right)^2 + \left(\frac{\underline{n}_y}{L_y}\right)^2 + \left(\frac{\underline{n}_z}{L_z}\right)^2\right).
\tag{10.11}
$$

In the case where the dimensions of the box are equal in each direction so that $L_x = L_y = L_z = L$, the energy is given by $E = \frac{\hbar^2 \pi^2}{2mL^2}(\underline{n}_x^2 + \underline{n}_y^2 + \underline{n}_z^2)$. There is only one "ground state" (state with the lowest energy) corresponding to $\underline{n}_x = 1$, $\underline{n}_y = 1$, $n_z = 1$ with an energy $E_1 = 3\frac{\hbar^2 \pi^2}{2mL^2}$. In contrast, at the next higher energy, $E_2 = 6\frac{\hbar^2 \pi^2}{2mL^2}$, there are three possible states with the same energy (degenerate states):

$$
\underline{n}_x = 2, \; \underline{n}_y = 1, \; \underline{n}_z = 1
$$

$$
\underline{n}_x = 1, \; \underline{n}_y = 2, \; \underline{n}_z = 1
$$

$$
\underline{n}_x = 1, \; \underline{n}_y = 1, \; \underline{n}_z = 2.
$$

Similarly, there are three states at energy $E_3 = 9\frac{\hbar^2 \pi^2}{2mL^2}$ corresponding to the values:

$$
\underline{n}_x = 2, \; \underline{n}_y = 2, \; \underline{n}_z = 1
$$

$$
\underline{n}_x = 1, \; \underline{n}_y = 2, \; \underline{n}_z = 2
$$

$$
\underline{n}_x = 2, \; \underline{n}_y = 1, \; \underline{n}_z = 2.
$$

On the other hand, there is only one state at $E_4 = 12\frac{\hbar^2 \pi^2}{2mL^2}$ with $\underline{n}_x = 2$, $\underline{n}_y = 2$, $\underline{n}_z = 2$.

The energy spectrum of the cubic electron trap is show in the diagram in Figure 10.7(a) with the degeneracy of the states indicated. Assuming that the electrons were non-interacting, if the trap was populated with multiple electrons, each of these states could have up to two electrons according to the Pauli Exclusion Principle, one with spin up and one with spin down. The diagram in Figure 10.7(b) shows the ground state of a cubic trap with eight electrons (assumed non-interacting). The energy of the eight-electron system is $42\frac{\hbar^2 \pi^2}{2mL^2}$.

10.5   The hydrogen atom

The cubic electron trap provides a background to study the first triumph of quantum mechanics, the solution of the hydrogen atom and the prediction of the observed hydrogen spectral lines. The system of one quantum electron in the potential of a single positively charged proton can be described as a particle moving in a Coulomb potential

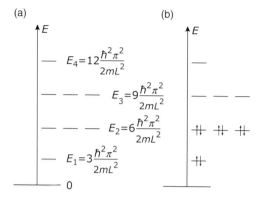

Figure 10.7 (a) Allowed energy levels of a cubic ($L_x = L_y = L_z = L$) electron trap modeled using infinite square-well potentials. (b) Occupancy of such a cubic electron trap by eight electrons in their lowest energy state allowed by the Pauli Exclusion Principle (assuming that the electron–electron interactions can be neglected).

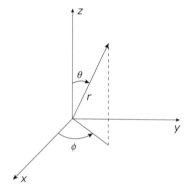

Figure 10.8 Spherical coordinate system with polar angle θ, azimuthal angle ϕ, and radius vector r shown.

$$U(r) = -\frac{1}{4\pi\epsilon_o}\frac{q^2}{r} \qquad (10.12)$$

where $\epsilon_o = 8.85 \times 10^{-12}$ F/m is the *permittivity of free space*. The problem is then to solve Schrödinger's equation

$$-\frac{\hbar^2}{2m}\nabla^2\psi - \frac{1}{4\pi\epsilon_o}\frac{q^2}{r}\psi = E\psi. \qquad (10.13)$$

The appropriate coordinate system is spherical coordinates (see Figure 10.8), with the wave function a function of the distance from the origin r, the polar angle θ, and the azimuthal angle ϕ: $\psi(r, \theta, \phi)$. In spherical coordinates, the Laplacian operator is

$$\nabla^2\psi = \frac{1}{r^2}\frac{\partial}{\partial r}\left(r^2\frac{\partial\psi}{\partial r}\right) + \frac{1}{r^2\sin\theta}\frac{\partial}{\partial\theta}\left(\sin\theta\frac{\partial\psi}{\partial\theta}\right) + \frac{1}{r^2\sin^2\theta}\frac{\partial^2\psi}{\partial\phi^2}. \qquad (10.14)$$

The partial differential equation that results by substituting the spherical Laplacian into Schrödinger's equation is *separable*. This means it can be separated into equations that just involve r, θ, or ϕ by assuming that the wave function ψ is the product of a function $R(r)$ that depends only on r, a function $\Theta(\theta)$ that depends only on θ, and a function $\Phi(\phi)$ that depends only on ϕ: $\psi(r, \theta, \phi) = R(r)\Theta(\theta)\Phi(\phi)$.

The equations that the three functions must satisfy are

$$\frac{1}{r^2}\frac{d}{dr}\left(r^2\frac{dR}{dr}\right) + \left[\frac{2mE}{\hbar^2} + \frac{mq^2}{\hbar^2 4\pi\epsilon_o}\frac{2}{r} - \frac{\ell(\ell+1)}{r^2}\right]R = 0 \tag{10.15a}$$

$$\frac{1}{\sin\theta}\frac{d}{d\theta}\left(\sin\theta\frac{d\Theta}{d\theta}\right) + \left[\ell(\ell+1) - \frac{m^2}{\sin^2\theta}\right]\Theta = 0 \tag{10.15b}$$

$$\frac{d^2\Phi}{d\phi^2} + m^2\Phi = 0. \tag{10.15c}$$

The third equation Eq. (10.15c) has the solution $\Phi(\phi) = Ae^{im\phi} + Be^{-im\phi}$ (or $\Phi(\phi) = A\sin m\phi + B\cos m\phi$). Since the wave function must have the same value if $\phi \rightarrow \phi + 2\pi$, it is required that $m = 0, \pm 1, \pm 2 \ldots$ The solutions for the second equation are the *Legendre functions* $\Theta(\theta)$ with the requirement that ℓ is a positive integer with $\ell \geq |m|$. For $\ell = 0$ $\Theta = 1$, for $\ell = 1$ $\Theta = \cos\theta$, for $\ell = 2$ $\Theta = (3\cos\theta - 1)$, etc.

The radial functions $R(r)$ are also known functions, the *Laguerre functions*. The equation requires that $n = \sqrt{\frac{mq^4}{2\hbar^2 E(4\pi\epsilon_o)^2}}$ is also a positive integer with $n \geq \ell + 1$. This implies that the values of energy E are quantized

$$E = -\frac{mq^4}{32\epsilon_o^2\pi^2\hbar^2}\frac{1}{n^2} = -\frac{13.6\text{ eV}}{n^2}. \tag{10.16}$$

Using terminology from the spectroscopic community, the states with $\ell = 0$ are referred to as s states, those with $\ell = 1$ are p states, those with $\ell = 2$ are d states, and $\ell = 3$ are called f states. In hydrogen, all the states with the same value of n all have the same energy: they are degenerate states. The states with their energy, notation, and total occupancy (one spin up and one spin down electron per state) are indicated in Table 10.1.

Table 10.1 Electron quantum states of hydrogen

Orbital quantum number	Angular momentum (L) quantum number	L_z quantum number	Orbital name	Energy	Occupancy
$n = 1$	$\ell = 0$	$m = 0$	1 s	−13.61 eV	2
$n = 2$	$\ell = 0$	$m = 0$	2 s	−3.40 eV	2
$n = 2$	$\ell = 1$	$m = -1, 0, 1$	2 p	−3.40 eV	6
$n = 3$	$\ell = 0$	$m = 0$	3 s	−1.51 eV	2
$n = 3$	$\ell = 1$	$m = -1, 0, 1$	3 p	−1.51 eV	6
$n = 3$	$\ell = 2$	$m = -2, -1, 0, 1, 2$	3 d	−1.51 eV	10

Lyman and Balmer series

The transitions between these levels had been spectroscopically observed in hydrogen plasmas before quantum mechanics was invented. The explanation of the spectral lines from first principles and the accurate prediction of the observed emission/absorption wavelengths from fundamental constants was one of the exceptional results of the new quantum theory and Schrödinger's equation. The electronic transitions involved in the *Balmer series* of spectral lines in the visible wavelengths and the *Lyman series* in the ultraviolet are indicated in Figure 10.9. The wavelengths that correspond to emission from these transitions, from the relation that $\Delta E = \hbar\omega = \frac{h2\pi c}{\lambda} = \frac{hc}{\lambda}$ are shown in Table 10.2, calculated from the equation

Table 10.2 Energy differences and emission wavelengths from the Lyman and Balmer hydrogen series

Series	Final State	Initial State	ΔE (eV)	$\lambda = hc/\Delta E$ (nm)
Lyman	$n = 1$	$n = 2$	10.21	121.5
Lyman	$n = 1$	$n = 3$	12.09	102.5
Lyman	$n = 1$	$n = 4$	12.76	97.20
Lyman	$n = 1$	$n = \infty$	13.61	91.13
Balmer	$n = 2$	$n = 3$	1.900	656.1
Balmer	$n = 2$	$n = 4$	2.551	486.0
Balmer	$n = 2$	$n = 5$	2.857	433.9
Balmer	$n = 2$	$n = \infty$	3.402	340.2

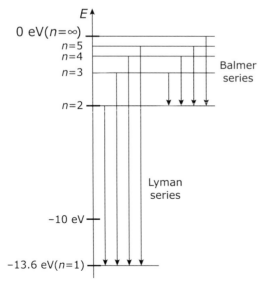

Figure 10.9 Electron transitions between quantum states in the hydrogen atom, indicating the Lyman and Balmer series transitions.

$$\frac{hc}{\lambda} = \frac{mq^4}{32\epsilon_o^2\pi^2\hbar^2}\left(\frac{1}{n_i^2} - \frac{1}{n_f^2}\right). \tag{10.17}$$

The table entry for initial state $n = \infty$ represents the onset of a continuum of emission from free electrons in the vacuum. There is no column to compare the calculated wavelength from the equation above with the experimentally measured values because the spectroscopically measured Lyman and Balmer transitions are now accepted as one of the primary ways to determine the values of the fundamental constants m, q, $\hbar$, and ϵ_o in the equation. So, by definition, the experimentally measured wavelengths are identical to the theoretical results from the equation.

Question 10.4

What are the wavelengths of the first three lowest energy transitions which have a final state $n = 3$? What spectral region are these transitions in?

10.6   Multi-electron atoms (electronic configuration of the elements in the periodic table)

The solution of the Schrödinger equation for an electron in the complicated potential of the Coulomb attraction with the nucleus combined with the Coulomb repulsion from the other electrons in a multi-electron atom is very challenging – no closed-form solution exists even in an atom as simple as helium, where a single extra electron is present. The difficulty of solving the many-body problem of multiple electrons in the attractive force of the nucleus is what led the eminent physicist Philip Morrison to quip, "I know a lot about the physics of hydrogen and the chemistry of helium, but not a lot about the physics of helium or the chemistry of hydrogen."

Nevertheless, although lacking an exact quantum mechanical solution, a great deal is known about the quantum states of multiple-electron atoms from the electronic properties of the atoms and the observed electronic transitions. *It is known that in multi-electron atoms, the degeneracy of the s, p, d, and f states for the same value of n is lifted by the electron–electron repulsion.* This is shown (not to scale) in the schematic diagram in Figure 10.10(b). Most noticeably, the 3d states are at a higher energy than the 4s states. Practically, this means that as the atomic number increases above argon with atomic number $= 18$ (with two electrons, one spin up and one spin down, in the 1s, 2s, 2p, 3s, and 3p states), the next two elements, potassium and calcium, have an electron configuration with 1 and 2 electrons respectively in their 4s states. The next ten *transition elements,* from scandium (atomic number $= 21$) to zinc (atomic number $= 30$), correspond to the element-by-element filling of the 3d states before the 4p states become occupied in elements 31 (gallium) through 36 (krypton). The 3d transition metals display a wide range of magnetic properties – including the ferromagnetic elements iron, nickel, and cobalt – due to the filling of

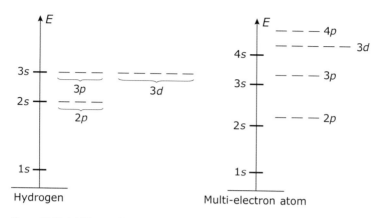

Figure 10.10 (a) Energy levels in single electron (hydrogen) atom. (b) The energy levels in a multi-electron atom where the degeneracy of the states is lifted by the electron–electron interactions.

the $3d$ states, along with metallic conductivities due to the occupancy of between one and two electrons in the $4s$ states.

Similarly, the filling of the $4f$ ($\ell = 3$) states occurs after the $6s$ states are filled. The elements with partially filled $4f$ states are the *rare earth elements*.

Question 10.5

List at least two rare earth elements that have important engineering applications and the devices that you use that contain those elements.

10.7 Atomic wave functions, chemical bonds, and hybridization

The atomic s states with the quantum number $\ell = 0$ are characterized by having no net angular momentum. The s-state wave functions are spherically symmetrical with a number of radial nodes equal to one less than the n quantum number: the $1s$ state has no radial nodes, the $2s$ state has one radial node (the wave function goes to zero at one value of radius r), the $3s$ state has two radial nodes, etc.

The p-state wave functions with $\ell = 1$ have one angular nodal plane. The p_x wave function has a y–z nodal plane and thus consists of two lobes in the $\pm x$-directions, etc. The angular dependence of the p wave functions is shown in Figure 10.11.

By the Pauli Exclusion Principle, each p-state orbital can accommodate two electrons, one spin up and one spin down. One should resist the temptation to assume that one electron is in one of the wave function lobes and the other is in the other lobe. In fact, each electron occupies both lobes equally – even though there is zero probability that the electron is in the nodal plane separating the two lobes.

The d-state wave functions with $\ell = 2$ have two angular nodal planes. Four of the d-state orbitals have four lobes directed along or at $45°$ angles to the Cartesian x–y–z

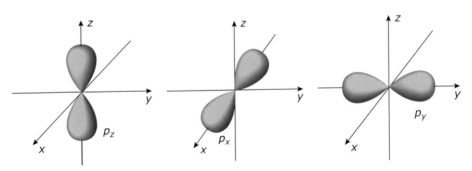

Figure 10.11 Distribution in space of the angular components of the *p*-state wave functions.

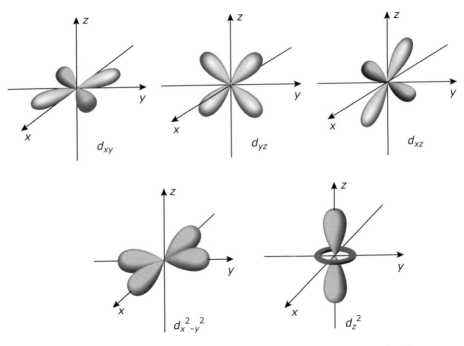

Figure 10.12 Distribution in space of the angular components of the *d*-state wave functions.

axes; the fifth *d*-state orbital has lobes in the $\pm z$-direction oriented through a ring in the *x–y* plane, as shown in Figure 10.12.

Recall that the $\ell = 2$ *d*-states are allowed only in the $n = 3$ or higher shells (and, in fact, the 3*d* states are filled only after the 4*s* states). Thus the 3*d* wave functions, with no radial nodes, are located closer to the nucleus than the 3*s* and 3*p* states in the same shell. The result is that the *d*-electrons play relatively little part in the chemical bonding of the atoms compared with the more spatially extended outer 4*s* and 3*p* states.

We will next consider chemical bonding concepts involving the *s* and *p* states. Chemical bonding involves a conceptual puzzle: we know that electrons have negative charge and therefore repel each other. But we also know that, in fact, atoms with

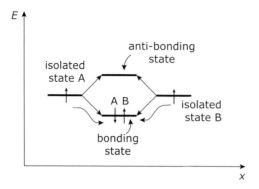

Figure 10.13 Splitting of single-electron state on two adjacent atoms into hybrid "bonding" and "anti-bonding" states as the interatomic distance decreases.

unfilled electron shells attract other atoms with the appropriate *valence* to provide enough electrons to complete the shell. Thus, in practice, electrons act as if they attract other electrons to produce chemical bonds. This paradox is explained through quantum mechanics. When two electron quantum states interact, they can form hybrid states which are a linear combination of the two component states. Each electron, resulting combination state one spin up and one spin down can contain in accordance with the Pauli Exclusion Principle. There are, in general, two of these combination states: a bonding state (with increased amplitude between the two atoms) at a lower energy than the component states, and an anti-bonding state (with reduced amplitude between the atoms) at a higher energy than the component states as shown in Figure 10.13. If the component states each have a single electron, both electrons can go into the lower energy. *Even allowing for an increase in energy due to the electron Coulomb repulsion, the decrease in energy resulting from the reduction in the electron spatial confinement makes it energetically favorable for the atoms to remain close together with both electrons in the bonding state.*

Chemical bonding states can be created from atomic states that are directed along the line between the two atoms (*sigma bonds*), such as an s state and a p_x state between two atoms which approach each other along the x-direction. *Pi bonds*, on the other hand, are formed between atomic states that are directed perpendicular to the direction of approach of the atoms. An example of pi bonds would be a bonding state between two p_z or between two p_y orbitals on atoms which approach each other along the x-axis. *In general, sigma bonds allow for more overlap between the component states than pi bonds, and therefore are stronger.*

It can sometimes be energetically favorable for the electrons in atomic orbitals to rearrange themselves into hybrid states to optimize the bond energies. For example, we have observed that in a multi-electron atom the $2p$ states have a higher energy than the $2s$ states. Thus the ground state of the carbon atom (Element 6, with 6 electrons) should have 2 electrons in the $1s$ state, 2 electrons in the $2s$ state, and 2 electrons in the $2p$ states. This is indicated by the electron configuration $1s^2 2s^2 2p^2$ where we have used the standard chemical notation where the exponent following the s, p, d, f notation represents the number of electrons in that state. From the previous discussion, the $2s$ state is spherically symmetrical, while the $2p$ states are perpendicularly located at $90°$ from each other. This, however, is not the optimum configuration for bonding geometries. It is energetically favorable for the electron states on carbon to rearrange themselves into higher energy *hybrid* states that are linear combinations of the s and p orbitals. The extra energy in forming the hybrid atomic orbitals is recovered from the bonding energy of the compounds that can be formed.

For example, solid carbon at room temperature and pressure has an equilibrium form of *graphite,* which has a structure with atoms strongly bonded in layers with relatively weak bonds between the layers. The sliding of the layers of atoms over each other gives graphite its lubricant properties. *Graphene,* created by peeling single layers of carbon atoms off graphite crystals using Scotch tape, has extremely interesting mechanical and electronic properties, which have just begun to be investigated and exploited in new carbon technologies. The three hybrid states created from a combination of an s state and two p states as illustrated below are notated as sp^2 hybrid states and have the trigonal configuration as shown in Figure 10.14.

The ground-state energy of the $1s^2 2p^1 (sp^2)^3$ state is higher than the ground state as shown in Figure 10.15, but the extra energy is recovered in the bonding of the sp^2 states on adjacent atoms in the graphite planes. (Note the dual use of exponents to represent both the composition of the hybrid state (sp^2) and also the occupancy of the hybrid states by three electrons.) The structure of atoms in each plane is the trigonal geometry of the sp^2 hybrid states. The planes are more weakly bonded together by bonding involving the unpaired $2p_z$ electrons.

Carbon also has another crystal structure in its high-temperature/high-pressure form of *diamond.* Diamond is characterized by bonds in tetrahedral directions with all the bond angles at the $109°$ tetrahedral angles. The bonding in diamond involves hybrid atomic orbitals composed of one $2s$ state and three $2p$ states: sp^3 *hybrid orbitals as*

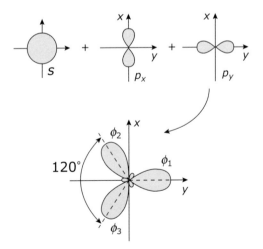

Figure 10.14 Hybridization of the $2s$ state with the $2p_x$ and $2p_y$ atomic states to create three planar sp^2 states.

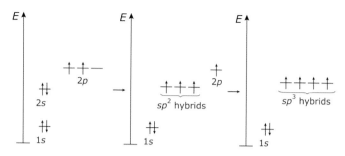

Figure 10.15 Energy of sp^2 and sp^3 hybrid states compared to the electron $1s$, $2s$, and $2p$ states.

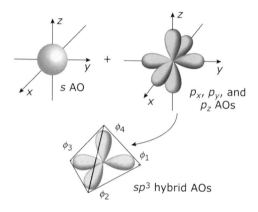

Figure 10.16 s-state, p-state, and hybrid sp^3 state atomic orbitals. Hybridization of the $2s$ state with the $2p_x$, $2p_y$, and $2p_z$ atomic states can create four tetrahedral sp^3 states.

shown in Figure 10.16. The ground state of the carbon atom in the sp^3 configuration is $1s^2 (sp^3)^4$ as in Figure 10.15. Each of the four sp^3 states is half-filled with one electron. If we bring up four hydrogen atoms, each with one electron, to form sigma bonds with each sp^3 hybrid orbital, we end up with the molecule of methane CH_4. The bond angles of methane are experimentally determined to be exactly the 109° tetrahedral angles.

We can use the concept of the sp^3 hybrid orbitals to understand other simple related molecules. If we imagine removing a proton from the center of one of the hydrogen atoms of methane and adding it to the carbon nucleus (also adding a chargeless neutron), we end up with a nucleus of Element 7, nitrogen, with three hydrogen atoms as shown in Figure 10.17. The molecule is now NH_3, *ammonia*. The two electrons, which remain in the sp^3 orbital that is no longer bonded to a hydrogen atom, form a *lone pair* (*lp*) of electrons. The bond angle in ammonia, measured to be 107°, is reduced by the somewhat greater repulsion of the lone pair electrons compared with electrons in the bonding states with a proton in them. If a second proton is removed from methane and added to the central nucleus (adding another neutron), we end up with an oxygen nucleus bonded to two hydrogen atoms with two lone electron pairs, the molecule of *water*, H_2O, with a bond angle reduced to 104.5° by the repulsion from the two lone-pair electrons. One more such transformation results in a HF *hydrogen fluoride* or

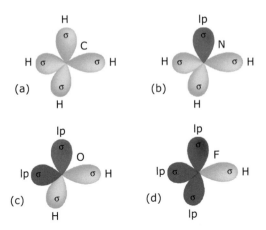

Figure 10.17 Tetrahedral sp^3 orbitals in the bonding of methane (CH$_4$), ammonia (NH$_3$), water (H$_2$O), and hydrofluoric acid (HF). As the number of valence electrons in the central atom increases from 4 in carbon, to 5 in nitrogen, 6 in oxygen, and 7 in fluorine, the σ-bonds with hydrogen atoms are replaced by lone pairs of electrons in un-bonded sp^3 orbitals. The proton repulsion from the bonded hydrogen atoms increases the H-X-H bond angles.

hydrofluoric acid molecule with three lone pairs and one bonded hydrogen, which can dissociate into a tetrahedral F− ion and a bare proton H+ ion. Each of these molecules can be understood as a base structure of the tetrahedral sp^3 orbitals.

The carbon tetrahedral sp^3 orbitals can be understood as the basis of carbon double bonds (sharing an edge of the tetrahedron with two bonded orbitals) and triple bonds (sharing a face of the tetrahedron with three bonded orbitals). In compounds with other elements, hybridization can be less important and bonds with predominant p-orbital contributions will lead to bond angles closer to the 90° between the p_x, p_y, and p_z orbitals.

Chemical molecules and chemistry can thus be understood by the solutions of Schrödinger's equation for atomic electrons in the potential of positive nuclei and their interactions with the electrons on other nearby atoms. When we move to solids where the atoms are closely packed, the translational symmetry of the potential that the electrons see due to the lattice of positively charged ion cores causes a new set of solutions to Schrödinger's equation that we will explore in the next chapter. The properties of these solutions are the subject of the field of *solid state physics* and are exploited in the wide range of solid state electronic and electro-mechanical devices.

Chapter 10 problems

1. Consider electrons in a three-dimensional quantum well with size 1 nm × 1 nm × 0.5 nm. Calculate the energy in eV of the lowest five energy levels for this quantum well. How many electrons can be in each of these energy levels?

2. Assume there were 12 electrons in the 3D quantum well in Problem 1. Allowing for the Pauli Exclusion Principle (and ignoring any interaction between the electrons!),

what is the total energy of the electrons in the ground state (lowest energy state) of the system?

3. Considering the 12 electrons in Problem 2, find the frequency and wavelength of the three lowest energy photons that would cause allowed excitations of the system in its ground state.

4. Aluminum has 13 electrons per atom, a density of 2.7 g/cm^3, and an atomic weight of 26.98 g/mole.

   a. What is its electronic configuration – how many electrons are in each atomic state?

   b. Given that aluminum is a metal and that each atom contributes three conduction electrons, find the Fermi energy in eV and the density of states at the Fermi level for aluminum.

   c. What is the velocity of an electron at the Fermi energy in aluminum?

5. What photon wavelength would cause an excitation from the $n = 3$ to $n = 4$ electronic energy levels in hydrogen? Is this a photon of visible/infrared/microwave frequency?

6. Referring to a periodic table, and assuming that only the outer electrons – and not the d-state electrons – contribute to the electron conduction, what density of conduction electrons, Fermi energy, and Fermi velocity would you predict for potassium, calcium, scandium, vanadium, cobalt, and gallium from the table below?

Element	Atomic number	Electron configuration	Atomic weight (g/mole)	Density (g/cm³)
Potassium (K)	19	$4s^1$	39.10	0.901
Calcium (Ca)	20	$4s^2$	40.08	1.53
Scandium (Sc)	21	$3d^1\ 4s^2$	44.96	2.99
Vanadium (V)	23	$3d^3\ 4s^2$	50.94	6.09
Cobalt (Co)	27	$3d^7\ 4s^2$	58.93	8.90
Gallium (Ga)	31	$3d^{10}\ 4s^2\ 4p^1$	69.72	5.91

7. Rules for the order of filling of atomic orbitals are given by *Hund's Rules*. Predict the occupancy of the 5 d-states in the transition metals from element 21 to 30 as predicted by Hund's Rules. Copper (Element 29) is usually considered to be *monovalent* (has only one conduction electron per atom), similar to the *noble metals* silver and gold directly below it in the periodic table. Can you explain from the outer electron configuration how copper could have only a single conduction electron?

8. The compound benzene (C_6H_6) is usually drawn as a hexagon of carbon atoms, alternately single and double bonded, with hydrogen atoms bonded to the outside of the ring. In terms of *sp* hybridization, how would you explain the bonding of benzene? In particular, where does the double bond come from and how is it

determined that, for example, carbon (1) is double bonded to carbon (2) but only single bonded to carbon (6)?

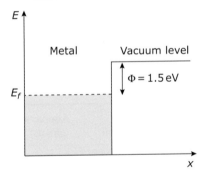

9. Monochromatic light is directed onto a clean metal surface in a vacuum. Light at wavelengths shorter than 410 nm causes electrons to be emitted from the surface. Light at wavelengths longer than 410nm does not produce any photoelec-trons emitted from the surface, no matter how intense the source is. What is happening here?

10. A metal has a work function $\Phi = 1.5$ eV as shown below.
    a. A photon of wavelength $\lambda = 550$ nm is absorbed by an electron at the Fermi level which is photo-emitted from the surface. What is the maximum velocity that the electron could have in the vacuum?

11. Consider a lithium atom. If two of the three electrons have quantum numbers (n, ℓ, m_ℓ, m_s) of $(1,0,0, +\frac{1}{2})$ and $(1,0,0, -\frac{1}{2})$, what are possible quantum numbers for the third electron, if the atom is (a) in the ground state, (b) in the first excited state and (c) in the second excited state?

11 Quantum electrons in solids

In Chapter 10, we looked at quantum solutions for electrons in three-dimensional systems including quantum dots, tunneling electrons, atoms, and molecules. In this chapter we will go further in studying electrons in solids, introducing the concept of crystal symmetry and the resulting *band states* that underlie the fundamental electronic properties of solids, including the distinction between metals, insulators, and semiconductors and the origin of the optical properties of solids which we will re-examine from an electromagnetic perspective in Chapter 14.

The concepts of Bloch states, crystal momentum, reciprocal lattice vectors, electron/hole states, and effective masses form the basis of solid state devices and thus all modern electronics and opto-electronics. We will introduce these concepts in one-dimensional systems first and then generalize to three-dimensional systems. Finally, we will examine low-dimensional systems to understand how these systems modify the behavior of quantum electrons in ways that can be useful to many applications.

11.1   Kronig–Penney potential, Bloch's theorem, and energy bands

Electrons in solids are interacting with a periodic lattice of Coulomb attractive potentials due to the ion cores. We have seen the solution of Schrödinger's equation for an electron in a single Coulomb attractive potential in the previous chapter where we looked at the quantum solution of the hydrogen atom. When we consider the electron in a lattice of periodic potentials, the symmetry of the problem puts additional constraints on the electron wave function. We can explore these constraints by looking at a one-dimensional potential $U(x)$ consisting of a one-dimensional lattice of square wells. This potential, shown in Figure 11.1, is known as the Kronig–Penney potential. It is assumed that the potential continues infinitely in the negative x- and positive x- directions with a repeating lattice constant of a.

The first conclusion we draw is that, whatever the electron wave function, the electron probability of being at any value x should be the same as the probability of being found at $x + a$, $x + 2a$, or $x + na$, where n is an integer. Since the lattice is invariant under a translation of na, there is no physical reason why an electron should be more probably located in the proximity of one well as opposed to any other. This implies that the electron wave function at $x + na$ can differ at most by a phase factor $e^{i\delta}$ from the wave function at x: $\psi(x + na) = e^{i\delta}u(x)$ where $u(x)$ has the periodicity of the lattice

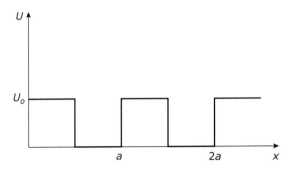

Figure 11.1 Kronig–Penney potential, a one-dimensional approximation of a crystal lattice potential.

$u(x) = u(x + \underline{n}a)$. This follows since the electron probability of being at $x + a$ is given by $P(x + a) = |\psi(x + a)|^2 = e^{i\delta} u(x) e^{-i\delta} u^*(x) = |u(x)|^2 = P(x)$, since the squared magnitude of the phase factor $e^{i\delta}$ gives just 1 when multiplied by its complex conjugate $e^{-i\delta}$.

Bloch's theorem states that the electron wave functions in a periodic potential are of the form

$$\psi_k(x) = e^{ikx} u_k(x) \tag{11.1}$$

where $u_k(x) = u_k(x + \underline{n}a)$. Bloch's theorem satisfies the requirement that the wave function can only deviate from the periodicity of the lattice by a phase factor and says that the phase factor is of the form of a plane wave e^{ikx}. The wave function thus consists of a part $u_k(x)$ that varies rapidly with x, repeating every lattice constant, modulated by a plane wave that for small k (long wavelength) can vary slowly through the lattice. The plane wave vector k plays a critical part in the electron properties: $\hbar k$ is the *crystal momentum,* which describes how the electron carries momentum through the crystal and contributes a term proportional to $(\hbar k)^2$ to the electron energy, as we will see later.

The periodicity of the lattice leads to an important property of the plane wave vector k. We note

$$\psi(x + a) = e^{ik(x+a)} u_k(x + a) = e^{ika} e^{ikx} u_k(x) = e^{ika} \psi(x). \tag{11.2}$$

From this we can conclude that changing the value of k by integer multiples of $2\pi/a$ will not affect the wave function: $e^{i\left(k + \frac{n2\pi}{a}\right)a} = e^{ika} e^{i\underline{n}2\pi} = e^{ika}$ (since $e^{i\underline{n}2\pi} = \cos \underline{n}2\pi + i \sin \underline{n}2\pi = 1$). This implies that changing the wave vector k by an integral number of *reciprocal lattice vectors* $K = \frac{2\pi}{a}$ does not change the wave function. This means that we can completely describe the wave function by just specifying the wave function between $k = -\frac{\pi}{a}$ and $k = +\frac{\pi}{a}$ (the region of "k-space" known as the *first Brillouin zone*). For values of k outside this region the wave function can be transformed back into the region $-\frac{\pi}{a} < k < \frac{\pi}{a}$ by adding or subtracting integral numbers of reciprocal lattice vectors.

Physically, changing k by a reciprocal lattice vector corresponds to a reflection of the electron off the crystal lattice with a transfer of momentum $\hbar K$ to the lattice. The electron plane wave propagating through the crystal can undergo partial reflection at

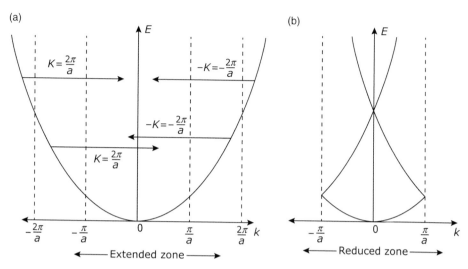

Figure 11.2 Energy vs. wave vector for a one-dimensional Bloch wave function in an "empty lattice" (a Kronig–Penney potential with $U_o = 0$), shown in (a) an extended zone and in (b) in the reduced zone created by imposing the Bloch theorem condition $\psi_k(x) = \psi_{k \pm \frac{2\pi n}{a}}(x)$: the wave function is not changed if the wave vector is shifted by an integral number of reciprocal lattice vectors $K = \frac{2\pi}{a}$.

every lattice site. When $\lambda = 2a$ (when $k = \frac{2\pi}{\lambda} = \frac{\pi}{a}$), the reflected waves reinforce each other creating a wave propagating in the opposite x-direction with a momentum change $\Delta p = \hbar \Delta k = \pm \left[\left(\hbar \frac{\pi}{a} \right) - \left(-\hbar \frac{\pi}{a} \right) \right] = \pm \hbar \frac{2\pi}{a} = \pm \hbar K$.

Let us try to make these concepts clear by an example. We will consider a Kronig–Penney potential with $U_o \to 0$, the so-called "empty lattice." In effect, we have just a free particle, but considered to be in a lattice with a periodic potential and the corresponding reciprocal lattice vectors. The wave function is just a Bloch function with $u_k(x) = 1$: $\psi_k(x) = e^{ikx}$. We can find the energy from Schrödinger's equation; the result is the familiar result for a free particle: $E = \frac{\hbar^2 k^2}{2m}$.

Plotting the energy E vs. wave vector k, we get the familiar parabola for a plane wave as shown in Figure 11.2(a). If we now apply the conditions on the wave vector that are a consequence of the periodicity of the lattice, we can transpose the state energies for $|k| > \pi/a$ by adding or subtracting a reciprocal lattice vector $K = 2\pi/a$ to reduce the solutions to the first Brillouin zone, as shown Figure 11.2(b).

So far, we have not included any effects of the actual periodic potential except for the equivalence of the wave function solutions that differ by a reciprocal lattice vector. If we consider the effect of the actual potential when we let U_o become a non-zero small positive number, we will find that at the zone boundary at $k = \pm \pi/a$ we have $\lambda = 2\pi/k = 2a$. For this wavelength we have strong reflection of the lattice and our wave functions will be a combination of a plane wave moving in the positive x-direction and a wave moving in the negative x-direction $\psi(x) = Ae^{ikx} + Be^{-ikx}$. When the amplitudes of the two waves are nearly equal, a standing wave – either a sine or cosine wave – is produced. One of these solutions represents an electron probability with a maximum in

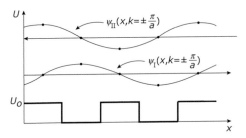

Figure 11.3 The two standing-wave solutions of the one-dimensional Schrödinger equation with $k = \pm\frac{\pi}{a}$. One solution ψ_I corresponds to a sine wave with maximum amplitude over the regions with $U(x) = U_o$; the other solution ψ_{II} corresponds to a cosine wave with maximum amplitude over regions with $U(x) = 0$.

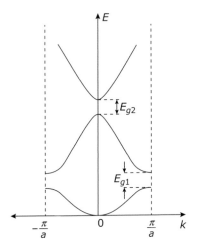

Figure 11.4 Energy vs. wave vector in reduced zone with degeneracy at the zone edge removed by the effect of the finite U_o.

the potential energy wells ($\psi_I(x, k = \pm\pi/a)$ as shown in Figure 11.3), while the other represents an electron probability with a maximum on the potential energy peaks $\psi_{II}(x, k = \pm\pi/a)$. The energy of ψ_I will be less than the energy of ψ_{II} because the electron probability is greater in the lower energy regions of the potential.

The effect of the potential compared with the empty lattice is to remove the degeneracy in the $E(k)$ plots at $\pm\pi/a$, $\pm 2\pi/a$, etc. The resulting E vs. k plot in the reduced zone is shown in Figure 11.4. The effect is to open up gaps in the energy spectrum (E_{g1} and E_{g2}) where there are no electron states. This partition of the energy spectrum into *bands*, where there are states at some value of k, and *energy gaps*, where there are no states is, characteristic of all solids. The plot of E vs. k is referred to as a *band diagram* and illustrates the *band structure* of the solid. While the real potentials of the ion cores are more complex than the simple square wells we have considered, the band shapes we have found for the (nearly) empty lattice can be recognized in the calculated band diagrams for real materials.

> **Question 11.1**
>
> Consider a lattice where the separation between lattice sites is $a = 0.5$ nm. What is the wavelength of the electron Bloch state at $k = 0$? What is the wavelength of the state with $k = \frac{\pi}{a}$? Does the answer depend on which band the electron is in?

Each electron Bloch state in a solid $\psi_k(x) = e^{ikx} u_k(x)$ is characterized by its band index and by the value of crystal momentum $\vec{k}$, and, in accordance with the Pauli Exclusion Principle, can only be occupied by one electron (per unit volume) of each spin. The usual phase space volume associated with a single state of a free electron, $\Delta x \, \Delta p_x \, \Delta y \, \Delta p_y \, \Delta z \, \Delta p_z = h^3$, applies to band states with the crystal momentum $\hbar \vec{k}$ playing the role of the momentum $\vec{p}$. So in three dimensions there are two spin states/m^3 in a $(2\pi)^3$ volume of k-space in each band.

> **Question 11.2**
>
> Verify that $\Delta x \, \Delta p_x \, \Delta y \, \Delta p_y \, \Delta z \, \Delta p_z = h^3$, corresponding to one electron state, gives a volume of $(2\pi)^3$ of k-space for one electron state/m^3.

If we consider the bands in the x-direction, for example, we find $\Delta x \, \Delta p_x = \Delta x \, \hbar \Delta k_x = h$ and thus an interval $\Delta k_x = 2\pi/\Delta x$ corresponds to a single state. If we fill up a one-dimensional band at $T = 0$ K with n_L electrons/unit length, for example, we get a picture like the one shown in Figure 11.5, with each dot representing one spin up and one spin down electron in a length of k-space $= 2\pi$. The states fill up from the lowest energy up to a maximum energy (at $T = 0$) E_f, the *Fermi energy*. The momentum of an electron at E_f is k_f, where $\hbar k_f$ is the *Fermi momentum*. E_f and $\hbar k_f$ are, respectively, the energy and momentum, of the highest occupied state in the material at absolute zero $T = 0$ K. For this one-dimensional system, we have

$$n_L = 2 \frac{2k_f}{(2\pi)}$$

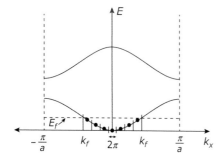

Figure 11.5 Lowest energy electron filling of one-dimensional Bloch states consistent with the Pauli restriction that each state corresponds to an interval $\Delta k = 2\pi$.

where the prefactor of 2 comes from the two spin states and the (2π) in the denominator is the length of linear k-space corresponding to one state.

> **Question 11.3**
>
> For a one-dimensional system with $a = 0.5$ nm how many electrons per meter (both spin up and spin down) could fit into the lowest energy band in Figure 11.5? How many electrons per meter could fit into the second band? For the band diagram illustrated in Figure 11.5, describe the filled electron states at $T = 0$ K if $n_L = 3 \times 10^9$ electrons/m. What is the Fermi wave vector k_f? What Fermi momentum does this correspond to? What states would be filled if $n_L = 6 \times 10^9$ electrons/m? What would the Fermi wave vector be in this case?

11.2   Metals, insulators, and semiconductors

This simple band picture allows us to understand the difference between solids that are *metals*, *insulators*, or *semiconductors*. If the Fermi energy lies within the band, as in Figure 11.5, then the material is a metal. Under the influence of an electric field $-\mathcal{E}$ in the negative x-direction, the electrons will experience a force $F = (-e)(-\mathcal{E})$ in the positive x-direction. The electrons are free to change their momentum to states at larger values of k_x giving more electron states moving in the $+x$-direction than in the $-x$-direction: the metal is carrying a net current of electrons in the $+x$-direction. Since conventional current is defined as the direction that *positive* charges move, we have a net current parallel to the electric field in the negative x-direction.

In a perfect metal, electrons would continue to move to higher values of k indefinitely (or at least until they reach a value $k = \pi/a$ at the edge of the first Brillouin zone and reflect off the lattice to the corresponding state at $k = -\pi/a$). (See Figure 11.6.) In practice, the electrons that have moved to states with $k > k_f$ will scatter off impurities in the lattice or, at temperatures $T > 0$, from thermal vibrations (*phonons*) to unfilled states

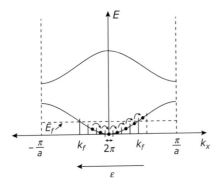

Figure 11.6 Shifting of occupied electron states in a material with a partially filled band under the influence of an electric field $\mathcal{E}$ directed in the negative x-direction. More electron states with positive k_x are occupied, leading to electrical conductivity in metals.

below the Fermi energy. In equilibrium, there will be some filled states with energy greater than E_f with positive k and some unfilled states with energy less than E_f and negative k. Since there are more electrons moving in the positive x-direction than in the negative x-direction, there is a net current which flows in response to the electric field.

We note that even with no applied electric field, there are electrons in states with non-zero k. Except for the two electrons in the state in the (2π) interval of k-space around $k = 0$, the electrons occupy states with finite crystal momentum $\hbar k$. With no applied electric field, there are just as many electrons moving toward positive x as moving toward negative x and no net current is carried.

In contrast with metals where the electrons only partially fill a band and the Fermi energy is within the continuum of states in the band, consider a material where the number of electrons is sufficient to just completely fill a band. (Because of the relation of bands to atomic states and the effect of chemical bonds in filling of atomic shells, this condition is much more common than it might appear at first sight.) In this case, the Fermi energy at $T = 0$ is located halfway between the highest filled state and the lowest empty state, i.e. in the middle of the band gap. This situation is shown in Figure 11.7. The electrical properties of this material can be deduced by first noting that the filled band carries no current because it has just as many electrons traveling in the negative x-direction (negative k_x) as in the positive x-direction. In addition, in the filled band, the electrons have no vacant states to transition to in response to an electric field. Unless the electric field is very strong (electric breakdown), an electron cannot acquire enough energy from the electric field to hop across the energy gap to the empty states in the higher band. This material conducts no current in response to an electric field and it is called an *insulator*.

Insulators with small band gaps – less than or on the order of 2.5 eV – are called *semiconductors*. A typical semiconductor band structure is shown in Figure 11.8. We note that, in common with insulators, the Fermi energy in a semiconductor at $T = 0$ K is located in a band gap. And, in fact, at low temperatures semiconductors, like insulators, do not conduct electricity for the same reason: there are no states at the Fermi level in which the electrons can rearrange themselves in response to an electric field to carry current preferentially in one direction. However, in semiconductors, unlike insulators, the band gap is relatively small and there are electron states associated with impurities at

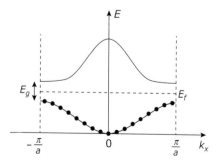

Figure 11.7 Lowest energy occupancy of electron band in material with a filled band. No empty states are available for electrons to transition to under the influence of an electric field, leading to insulating electrical behavior.

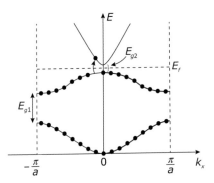

Figure 11.8 A semiconductor-like band structure with a small enough energy gap E_{g2} that electrons can be thermally excited from an occupied to an unoccupied band.

energies close to the band edges. In silicon, the quintessential semiconductor, the band gap E_{g2} is 1.1 eV and impurities of phosphorus, with an atomic valence one more than silicon, have a shallow electron bound state that is only 0.044 eV below the edge of the empty band above the band gap. The result of the relatively small energy gap and presence of *shallow impurity states* is that, at room temperature or above, electrons can be thermally excited from the highest filled band (the *valence band*) or from the shallow impurity states into the lowest empty band (the *conduction band*). The electrons that are thermally excited into the conduction band can move freely in response to an electric field and carry a current.

In addition, the vacancy that is left in the valence band provides a state for electrons in the valence band to adjust themselves in response to an electric field, which results in the valence band electrons producing a net current. If the electrons in the valence band move to the plus k_x-direction in response to the electric field, they leave vacancies in the minus k_x-direction. The vacancies appear to move in the opposite direction to the electrons: they behave as though they were *positively charged*. These vacancies in the valence band are referred to as *holes*. Current is carried in a semiconductor at temperatures above absolute zero by a combination of electrons in the conduction band and holes in the valence band. A semiconductor which is deliberately doped with impurities that have electron states close to the conduction band edge (such as phosphorus in silicon) has free electrons in the conduction band at room temperature and is called *n-type*. Semiconductors containing dopant atoms, such as boron in silicon, that create empty states close to the valence band edge so that electrons can be thermally promoted from the valence band – creating more holes in the valence band – are called *p-type*.

Question 11.4

Consider a semiconductor with both electrons and holes that has an electric field directed in the positive x-direction $\vec{\mathcal{E}} = \mathcal{E}_o \hat{x}$. What direction would the electrons move? What direction is the current due to the electron movement? What direction would the holes move? What direction is the current due to the hole movement?

11.3   Effective masses in semiconductors

If we look near the band minimum or maximum at the band baps, we can expand the band energy $E(k)$ in a Taylor series:

$$E(k) = E_o + \frac{dE}{dk}(k - k_o) + \frac{1}{2}\frac{d^2 E}{dk^2}(k - k_o)^2 + \ldots \tag{11.3}$$

At a maximum or minimum, the first derivative $\frac{dE}{dk}$ is zero. Neglecting the terms higher than $(k - k_o)^2$, we write

$$E(k) = E_o + \frac{1}{2}\frac{d^2 E}{dk^2}(k - k_o)^2 = E_o + \frac{\hbar^2}{2m^*}(k - k_o)^2 \tag{11.4}$$

where we have defined the *effective mass* $m^* = \frac{\hbar^2}{\frac{d^2 E}{dk^2}}$. The effective mass, inversely proportional to the band curvature, determines the energy E vs. crystal momentum $\hbar k$ relation of an electron in the band just as real mass does for a free particle. The acceleration of an electron in the band due to an electric field, for example, is given by $\vec{a} = \frac{\vec{F}}{m^*} = \frac{-e\vec{\mathcal{E}}}{m^*}$. *Due to large band curvature, effective masses can be small fractions of the free-electron mass.*

The band structures of silicon (Si), germanium (Ge), and gallium arsenide (GaAs) are sketched in Figure 11.9. We note that, while the minimum of the conduction band in GaAs is at $k = 0$, the minimum of the conduction band in Si (indicated by the value of k_o) is about 85% of the way between $k = 0$ and the Brillouin zone edge in the [100] direction. Since there are six equivalent [100] directions corresponding to the direction of the faces of a cube ([100], [−100], [010], [0−10], [001], [00−1]), Si has six equivalent *electron pockets*. The curvature of the Si bands is less along the [100] directions than perpendicular to the [100] directions, implying that the mass in each pocket is *aniso-tropic* and can be represented by a *mass tensor,* as we saw in Chapter 4.

The minimum of the conduction band in Ge is at the Brillouin zone edge in the [111] directions – corresponding to the directions from the center toward the corner of a cube. There are eight corners of a cube and the electron pockets each extend into two Brillouin zones. Thus there are eight half-pockets (or four full pockets) extending in the direction from the center of a cube to the corners (in the [111] directions). The Ge electron

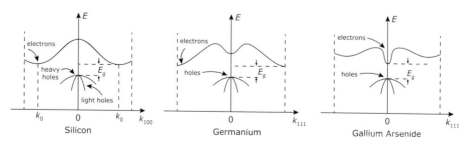

Figure 11.9 The highest occupied and lowest unoccupied bands for intrinsic (undoped) silicon, germanium, and gallium arsenide illustrating the location in the Brilliouin zone of the electron and hole pockets.

Table 11.1 Effective mass parameters for semiconductors

	E_g (eV)	# of pockets	k_o	$m_\parallel^*/m_o$	$m_\perp^*/m_o$	m^*_{cond}/m_o	m^*_{dos}/m_o
GaAs electrons	1.43	1	0	0.067	0.067	0.067	0.067
GaAs holes	1.43	2	0			0.34	0.47
Si electrons	1.11	6	0.85[100], etc.	0.98	0.19	0.26	1.08
Si holes	1.11	2	0			0.39	0.81
Ge electrons	0.67	4	[111], etc.	1.64	0.082	0.12	0.56
Ge holes	0.67	2	0			0.21	0.29

pockets are also highly anisotropic with a smaller mass perpendicular to the [111] direction than parallel to the [111] direction. The longitudinal and transverse effective mass for the electron bands in Si, Ge, and GaAs are indicated in Table 11.1. For each of these semiconductors, the maximum of the valence band below the energy gap E_g is at the center of the band at $k = 0$. The average curvature of the valence bands at $k = 0$ is isotropic and the value of the "hole" effective mass is given in Table 11.1.

The electron pockets in silicon and germanium are anisotropic with different curvature along the axis of the pocket and perpendicular to the axis of the pocket and can be described by a mass tensor

$$m^* = \begin{bmatrix} m_\perp & 0 & 0 \\ 0 & m_\perp & 0 \\ 0 & 0 & m_\parallel \end{bmatrix}.$$

The *conductivity mass* m^*_{cond} describes how the electrons accelerate in response to an electric field $\vec{a} = \frac{-e\vec{\mathcal{E}}}{m^*_{cond}}$, and can be calculated by adding the reciprocals of the mass tensor of each pocket in a common coordinate system, as we saw in Chapter 4. Since the crystals are cubic the conductivity mass is isotropic with a value given in the table. The *density of states mass* m^*_{dos} describes the filling of the bands with electrons $n = \frac{2}{(2\pi)^3} \frac{4}{3}\pi \left(\frac{2m^*_{dos}E_f}{\hbar^2}\right)^{3/2}$ (as in Eq. (10.5)) and is found by a different averaging process – it is approximately the conductivity mass times the number of pockets.

Question 11.5

Derive the result $n = \frac{2}{(2\pi)^3} \frac{4}{3}\pi \left(\frac{2m^*_{dos}E_f}{\hbar^2}\right)^{3/2}$ for three dimensions by calculating the volume of a sphere in k-space of radius k_f and assuming two electron states per a volume of $(2\pi)^3$ of k-space with $E_f = \frac{p_f^2}{2m^*_{dos}} = \frac{(\hbar k_f)^2}{2m^*_{dos}}$. Show that this equation is the same as Eq. (10.5) with $m = m^*_{dos}$.

11.4   Optical absorption in solids

Electrons can transition from one quantum state to another under the influence of an electromagnetic field. The quantum model for this process is that an electron transitions from a lower energy state to a higher energy state by absorbing a photon or photons. An electron can also transition from a higher to a lower energy state by emitting a photon or photons. Both processes are described by a quantum *matrix element* which involves an integration over the wave functions of the initial and final states with a perturbing potential related to the electromagnetic field. To lowest order, the matrix element for the transition between the quantum states n and m is

$$H_{nm} = \int \Psi_n^* q\vec{\mathcal{E}}_o \, e^{i(\vec{q}\cdot\vec{r}-\omega t)} \cdot \vec{r} \, \Psi_m \, d\vec{r} \tag{11.5}$$

where Ψ_m and Ψ_n are the wave function of the initial and final states, and $q\vec{\mathcal{E}}_o e^{i(\vec{q}\cdot\vec{r}-\omega t)} \cdot \vec{r}$ is the electric field energy associated with the electromagnetic wave $\vec{\mathcal{E}}_o e^{i(\vec{q}\cdot\vec{r}-\omega t)}$.

Factors that affect optical absorption by electrons in quantum states include (1) the initial electron state must be occupied and the final state unoccupied, (2) the energy difference between the initial and final states must be equal to the photon energy, and (3) the momentum difference between the initial and final electron states $\Delta\vec{p} = \hbar\vec{k}_f - \hbar\vec{k}_i$ must be equal to the photon momentum $\hbar\vec{q}$. In addition, (4) the transition needs to be "allowed" – the matrix element can't be zero due to symmetry effects, for example – and (5) the absorption strength depends on having a large number of states available for the transition – the *joint density of states* must be large. We will discuss each of these effects in more detail below.

For optical transitions in solids, Ψ_m and Ψ_n are Bloch states $e^{i\vec{k}\cdot\vec{r}}u_k(\vec{r})$ with Bloch wave vectors $\vec{k}_m$ and $\vec{k}_n$. The integration of the exponentials leads to delta functions on the electromagnetic wave frequency ω and vector $\vec{q}$ and such that $E_m - E_n = \hbar\omega$ and $\vec{k}_m - \vec{k}_n = \vec{q}$. We recognize these conditions as the conservation of photon energy $\hbar\omega$ and momentum: the photon energy must equal the energy difference between the initial and final band states and the photon momentum $\hbar\vec{q}$ must equal the difference in crystal momentum between the initial and final band states.

We note that the wave vector at an optical wavelength of 500 nm is of the order of $|\vec{q}| = 2\pi/\lambda = 2\pi/500\times10^{-9}$ m $\cong 10^7$ m^{-1}. In contrast, the first Brillouin zone spans a wave vector range of $2\pi/a$ where a, the lattice constant, is on the order of 0.5 nm. Thus, $2\pi/a = 2\pi/0.5\times10^{-9}$ m $\cong 10^{10}$ m^{-1}, three orders of magnitude greater than the optical photon wave vector. Thus the optical wave vector is essentially zero on the scale of the band state wave vector, and the photon momentum conservation requirement implies that the initial and final band state wave vectors $\vec{k}_m$ and $\vec{k}_n$ must be nearly equal. The optical transition is a *vertical* transition in the band k-space, as shown in Figure 11.10. Photon momenta are small compared to the Brillouin zone until wavelengths become short enough that they are comparable to the lattice spacing, a regime well out of the optical range, into the X-ray spectral region. At these short wavelengths, the photon energies are so large that electronic transitions are primarily from atomic core states and not valence-band-to-conduction-band transitions.

A band gap optical absorption or emission process in a *direct-gap* material like GaAs (where the conduction band minimum and the valence band maximum occur at the same k-value in the Brillouin zone as shown in Figure 11.11(a)), therefore, is much

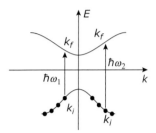

Figure 11.10 Optical transitions in a solid from an initial occupied state at $k = k_i$ to a final unoccupied state at $k = k_f$. The energy difference between the initial and final states is related to the optical photon frequency by $\Delta E = \hbar\omega$.

(a)   (b)

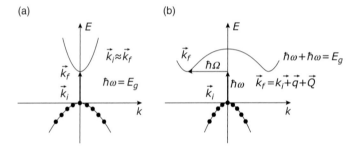

Figure 11.11 (a) Band gap absorption in direct-gap GaAs compared with (b) indirect-gap silicon where a lattice vibration (phonon) must provide the momentum $\hbar\Omega$ to make up the difference between the initial state $\hbar k_i$ and final state $\hbar k_f$.

more probable than an optical band gap transition in an *indirect-gap* material, like silicon, where the valence band maximum is at the center of the zone ($\vec{k} = 0$), while the conduction band minima are at a different value of $\vec{k}$, close to the zone edges as shown in Figure 11.11(b) $\hbar\omega = E_g$. For a photon absorption or emission process to occur near in silicon, another mechanism – usually a lattice vibration or *phonon* – must be present to conserve momentum. The probability of this *three-particle interaction* (photon–electron–phonon) is many orders of magnitude less than the probability of a direct absorption or emission process where the initial and final band wave vectors are the same.

Question 11.6

What is the longest wavelength of light that can be absorbed by a valence-band-to-conduction-band electron transition in GaAs, Si, and Ge? Do you think that these semiconductors would absorb any light at shorter wavelengths? Would any of these materials be useful for infrared detection in the 3 μm $< \lambda$ < 5 μm or the 8 μm $< \lambda < 12$ μm atmospheric "windows" of low infrared absorption? What materials are used for infrared detectors at these wavelengths?

The reduced absorption probability for indirect-gap materials implies that a silicon solar cell needs to be considerably thicker than in a comparable direct-gap material like GaAs. Silicon is the current material of choice for solar cells only because silicon material and processing costs for large-area solar arrays are a fraction of those for any other material, and not because silicon is an efficient optical material. The relative weak optical absorption of silicon is why present high efficiency solar cells use a *multi-layer* structure with two or more stacked p-n junctions to maximize the collection of photo-generated electrons.

The same arguments hold for emission: the radiative lifetime in silicon is so long that non-radiative processes dominate the recombination of conduction-band electrons and valence-band holes, and thus light emission is weak from indirect semiconductors like silicon and germanium. Light-emitting-diodes (LEDs) are typically made from direct-gap III-V materials like aluminum gallium arsenide (red), gallium phosphide (green), or indium gallium nitride (blue).

Similarly, higher-order *two-photon* processes can lead to *non-linear* optical absorption when the photon energy is less than the band gap $\hbar\omega < E_g$. But the absorption due to these higher-order processes is orders of magnitude less than the absorption due to one-photon direct transitions.

In addition to the requirements of wave-vector/momentum conservation and energy conservation that result from the $e^{i\vec{k}\cdot\vec{r}}$ part of the Bloch band function, the matrix element for the optical transition also depends on the short-range part of the potential. If the matrix element integral $\int u_n(\vec{r})^* \, q\vec{\mathcal{E}}_o \cdot \vec{r} \, u_m(\vec{r})d\vec{r}$ is zero due to symmetry reasons, for example, the transition is described as *dipole-forbidden*. Absorption or emission from dipole-forbidden transitions can occur due to higher-order processes, but they are weak compared to *dipole-allowed* transitions.

The total optical absorption depends on the sum of the transitions that are described by the matrix elements. Adding up all the possible transitions that satisfy the conditions on conservation of photon energy and momentum gives a *joint density of states*: the product of the total number of occupied valence-band initial states between E and $E + dE$ and unoccupied conduction-band final states between $E + \hbar\omega$ and $E + \hbar\omega + dE$. For direct optical transitions, the states that should be included are those at the same value of $\vec{k}$. There are two cases that lead to a large joint density of states. One is a transition to or from a band minimum or maximum, where the slope of the E vs. k curve is nearly flat and there is a large number of states within a narrow dE range. The second is when an occupied valence band is nearly parallel with an unoccupied conduction band.

11.5   Stimulated and spontaneous optical emission

All of our discussion so far has been about optical absorption processes, but the same factors also affect emission processes where, under the influence of an electromagnetic field, an electron jumps from a higher energy state to a lower energy state and emits a photon in the process. In fact, because the integral in the matrix element above is not affected if the states Ψ_n and Ψ_m are interchanged, there is a perfect symmetry between absorption and the mirror process of emission in the presence of an electromagnetic wave. Absorption, a transition from a lower energy state to a higher energy state, results in the energy of the electromagnetic field being converted into energy of the electron. Emission is a transition from a higher energy state to a lower energy state with energy being carried off in the electromagnetic field. If this emission process is precipitated by an existing electromagnetic field, the process is known as *stimulated emission* and results in a single photon with a given $\vec{q}$ and ω causing the emission of a second photon of the same $\vec{q}$ and ω and the same *polarization* (the electric field $\vec{\mathcal{E}}_o$ of the emitted photon is in the same direction as the electric field of the photon that precipitates the transition). In addition to stimulated emission, there is a process of *spontaneous emission*, where an electron transitions from a higher to a lower energy state and emits a photon without a precipitating photon being present.

Since absorption and stimulated emission are described by the same matrix element – depending on whether Ψ_n is at a higher or lower energy than Ψ_m – the rate of optical absorption and the rate of stimulated emission differ only by differences in the occupancy of the higher and lower energy states. In thermal equilibrium the occupancy of higher energy states is exponentially smaller than lower energy states, and therefore stimulated emission is usually much smaller than absorption. But in non-equilibrium situations where a *population inversion* exists with more electrons in a higher energy state, stimulated emission can lead to an intense emission chain reaction that can be exploited in lasers. The relations between the absorption and stimulated emission strength and the spontaneous emission strength were calculated by Einstein and are known as the *Einstein A and B coefficients*.

11.6 Low-dimensional systems: 2D electron gas (GaAs/GaAlAs heterostructure)

The modification of energies and density of states caused by quantum effects in low-dimensional systems, and the ability to tune these effects through fabrication techniques, have been extensively exploited in nanotechnology. Quantum dots, quantum wires, and two-dimensional quantum systems have been fabricated by a variety of deposition and etching techniques, and their properties have been extensively studied and exploited in devices. One of the first low-dimensional systems to be fabricated was the two-dimensional electron gas in III-V thin-film heterostructures. For example, by the process of *molecular beam epitaxy* GaAs can be alloyed with aluminum to form $Ga_{1-x}Al_xAs$ where x is the atomic percentage of aluminum that is substituted for gallium in the GaAs lattice. The band gap of GaAs (1.43 eV) can be continuously increased by substituting aluminum for gallium $Ga_{1-x}Al_xAs$ and for $x < 0.45$ is given by $E_g = 1.43 + 1.25x$ (eV). Thus, a layer of pure GaAs can be grown between two layers of GaAlAs creating a band structure like the one shown in Figure 11.12 as a function of z, the distance perpendicular to the growth surface.

The electrons in the conduction band will collect in the low-band-gap GaAs layer where they will be confined in the z-dimension but still be able to move freely in the

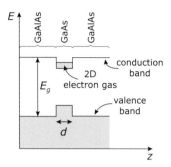

Figure 11.12 Conduction and valence band energies at $k = 0$ as a function of distance from the device surface in GaAlAs-GaAs-GaAlAs heterostructure.

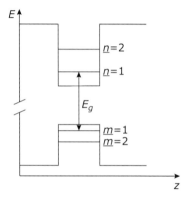

Figure 11.13 One-dimensional electron square-well energies in the GaAs region of a heterostructure. The lowest energy transition is from the $\underline{m} = 1$ state in the valence band to the $\underline{n} = 1$ state in the conduction band.

transverse x- and y-directions. If the dopant atoms are restricted to the GaAlAs layers – a technique called *modulation doping* – the electrons in the GaAs layer can propagate without scattering off impurity atoms, a property that is exploited in *high-electron mobility transistors*.

In the z-direction, the electrons are in a one-dimensional quantum well, as shown in Figure 11.13, while in the x- and y-directions the electrons are free electrons with an effective mass characteristic of the GaAs conduction band ($m_c{}^* = 0.067m_o$), and the electron energy is given by:

$$E = \frac{\hbar^2 \pi^2}{2m^*} \frac{n^2}{d^2} + \frac{\hbar^2 k_x^2}{2m^*} + \frac{\hbar^2 k_y^2}{2m^*} \tag{11.6}$$

where $n = 1, 2, 3\ldots$ is the quantum number.

The two-dimensional nature of the electrons in the heterostructure has two useful effects. First, the band gap can be tuned by varying d, the thickness of the GaAs layer. Including the quantum energy of the holes in the quantum well, the band gap from the $\underline{m} = 1$ valence band state to the $\underline{n} = 1$ conduction band state becomes

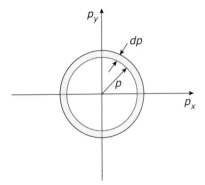

Figure 11.14 Momentum-space diagram of two-dimensional ring of states between p and $p + dp$ used to calculate the density of states for a 2D electron gas.

$$E_g = 1.43 \text{ eV} + \frac{\hbar^2 \pi^2}{2m_c^*} \frac{1}{d^2} + \frac{\hbar^2 \pi^2}{2m_v^*} \frac{1}{d^2} \tag{11.7}$$

where the valence band effective mass m_v^* is approximately a factor of five larger than the conduction band effective mass. The band gap can thus be adjusted to correspond to a photon energy of interest in photodetectors or semiconductor lasers by adjusting the thickness of the film d.

Second, since the electron gas with a given square well state $\underline{n}$ depends only on $p_x = \hbar k_x$ and $p_y = \hbar k_y$ (and not on p_z) as shown in Figure 11.14, the density of states can be calculated by calculating the two-dimensional phase space between p and $p + dp$, dividing by $(h)^2$ – the phase space area corresponding to one state in two dimensions – and multiplying by two spin states as below:

$$N(E)dE = \frac{2}{(h)^2} 2\pi p dp. \tag{11.8}$$

If we use $E = \frac{p^2}{2m^*} \Rightarrow dE = \frac{2p\,dp}{2m^*} = \frac{p\,dp}{m^*}$, we find:

$$N_{2D}(E) = \frac{2}{h^2} 2\pi p \frac{m^*}{p} = \frac{m^*}{\pi \hbar^2} \left(\frac{\text{electrons}}{\text{m}^2}{\text{J}} \right) \tag{11.9}$$

(where we have converted from h to $\hbar$).

This result tells us that the density of states in two dimensions does not depend on the energy! Within each sub band corresponding to $n = 1, 2, 3 \ldots$, the density of states is a constant $\frac{m^*}{\pi \hbar^2}$. The total density of states for a two-dimensional system is the sum of $\frac{m^*}{\pi \hbar^2}$ for each sub band,

$$N_{2D}(E) = \frac{m^*}{\pi \hbar^2} \sum_n \theta(E - E_n) \left(\frac{\text{electrons}}{\text{m}^2}{\text{J}} \right) \tag{11.10}$$

where $\theta(E - E_n)$ is a step function, as shown in Figure 11.15, compared with the three-dimensional density of states which we found to be

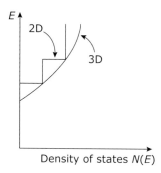

E

2D

3D

Density of states $N(E)$

Figure 11.15 Density of states for 2D electron gas compared with 3D density of states.

$$N_{3D}(E) = \frac{1}{2\pi^2}\left(\frac{2m}{\hbar^2}\right)^{3/2}(E)^{1/2}\left(\frac{\text{electrons}}{\text{m}^3}{\text{J}}\right).$$ (11.11)

The effect of this step-like density of states is that the onset of optical transitions between two-dimensional states is more abrupt than the corresponding transitions in three-dimensional systems. The combination of an onset energy (frequency) that can be tuned by changing the thickness of the layer and a more abrupt onset of absorption or emission due to the step-like two-dimensional density of states (along with the direct band gap) makes GaAs/GaAlAs and other III-V heterostructures ideal systems for optical detectors or laser diodes.

The unusual properties of low-dimensional systems are just beginning to be exploited in device applications. Carbon, for example, can exist in the layered two-dimensional form of graphene, the one-dimensional form of nanotubes where the graphite layers are rolled into a tube, and the "zero-dimensional" form of C_{60} "fullerene" molecules or "Bucky-balls," named for their resemblance to the geodesic domes designed by noted architect and futurist Buckminster Fuller.

We will study the occupancy of the quantum states in solids as a function of temperature in Chapter 13, where we consider the statistics of quantum particles at finite temperatures.

Chapter 11 problems

1. How many conduction electrons are in 1 cm^3 of copper? What about in 1 cm^3 of magnesium?

2. What is the momentum of a visible light photon with wavelength 500 nm? What wavelength of photon would have a momentum corresponding to the momentum of an electron at the first Brillouin zone edge $\left(\frac{\hbar K}{2} = \frac{\hbar\pi}{a} \simeq \frac{\hbar\pi}{0.5 \times 10^{-9}\text{ m}}\right)$?

3. What energy would the photons in Problem 2 carry?

4. Calculate the acceleration of an electron in silicon, germanium, and gallium arsenide due to an electric field $\mathcal{E} = 1$ V/m. Assuming that the electron began at rest at zero energy (bottom of the conduction band), how much energy would it

have 5×10^{-14} s later? (This is a typical *electron scattering time* – the time between electron scattering events.) How fast would the electron be going at this time?

5. Calculate the longest wavelength of light that will cause a band-to-band electronic transition in silicon, germanium, gallium arsenide, lead telluride, diamond, silicon dioxide (SiO_2), and aluminum oxide (Al_2O_3).

6. Find the Fermi energy in eV and the density of states at the Fermi energy in number of electrons/(m^3/eV) for copper and silver (assumed to have one conduction electron per atom), for tungsten and calcium (assumed to have two conduction electrons per atom), and for aluminum (assumed to have three conduction electrons per atom).

7. Explore additional applications of GaAs/GaAlAs and other quantum-well-hetero-structures and their use in devices.

Problems 8 and 9

By calculating the volume of k-space and using the relation that one electron state occupies a volume of $(2\pi)^3$ in three dimensions, we calculated in the text the Fermi energy E_f in terms of electron density n and the density of states $D(E) = \frac{dn}{dE}$ for a 3D free-electron metal. Problems 8 and 9 below extend the analysis to two- and one-dimensional systems.

AlGaAs   GaAs

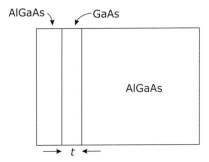

8. An electron in the GaAs layer of a modulation-doped GaAs/GaAlAs quantum well can be modeled as a particle-in-a-box in the direction perpendicular to the interface with energy $E_\perp = \frac{l^2 \pi^2 \hbar^2}{2m^* t^2}$ (where t is the thickness of the quantum well and $l = 1, 2, 3, \ldots$), and as a free particle parallel to the interface with energy $E_{||} = \frac{\hbar^2 k_x^2}{2m^*} + \frac{\hbar^2 k_y^2}{2m^*}$. The total energy is $E = E_\perp + E_{||}$ and the effective mass in GaAs is $m^* = 0.067\, m_o$ where m_o is the free-electron mass.
   a. Find the Fermi energy at $T = 0$ for a two-dimensional electron gas in a GaAs layer of thickness $t = 20$ nm with an area electron density of 1×10^{16} m^{-2}. (Hint: Consider the diagram below of the occupied states in 2D k-space for the $l = 1$ sub band.) How many sub bands are occupied?

b. What is the Fermi energy if the electron density is increased to 3×10^{16} m^{-2}? How many sub bands are occupied?

c. Graph the density of states $D(E)$ vs. energy from $E = 0$ to $E = 0.150$ eV.

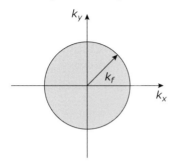

9. Electrons circulating in a strong magnetic field in the z-direction are quantized in their motion perpendicular to the magnetic field and move as free particles parallel to the magnetic field with energy levels ("Landau levels") given by:

$$E = \left(l + \frac{1}{2} + s\right)\hbar\omega_c + \frac{\hbar^2 k_z^2}{2m^*}$$

where $l = 0, 1, 2, \ldots$ is the Landau level index, $s = \pm 1/2$ is the electron spin, and ω_c is the electron cyclotron resonance frequency, given by $\omega_c = eB/m^*$ where B is the magnetic field, e is the electron charge, and m^* is the electron effective mass.

a. For a semiconductor with $m^* = 0.02\ m_o$ in a magnetic field $B = 5$ Tesla, plot the energy in eV vs. k_z for first five Landau levels ($l = 0\ s = -1/2$; $l = 0$ $s = +1/2$; $l = 1\ s = -1/2$; $l = 1\ s = +1/2$; $l = 2\ s = -1/2$) on the same plot over the range $k_z = -1 \times 10^8$ to $+1 \times 10^8$ m^{-1}.

b. For the semiconductor in part (a) in a field of 5 tesla, calculate and plot the density of states $D(E)$ vs. E over the range from $E = 0$ to 0.100 eV. You will need to know that, due to the number of electron orbits in the x–y plane, each interval $\Delta k_z = 1/(2\pi)$ corresponds to a number $\frac{eB}{2\pi\hbar}$ of electron states, neglecting spin. Note that the lowest Landau level ($l = 0\ s = -1/2$) has only one spin state.

c. What magnetic field applied to the semiconductor with $m^* = 0.02\ m_o$ and an electron density of $n = 1 \times 10^{23}$ m^{-3} at $T = 0$ would result in only the lowest Landau level being occupied (the so-called "extreme quantum limit")? What field applied to a metal with $m^* = m_o$ and $n = 1 \times 10^{28}$ m^{-3} would result in only the lowest Landau level being occupied? (Magnetic fields large enough to bring free electrons into the extreme quantum limit are believed to exist on some neutron stars!)

12 Thermal physics: energy, heat, and thermodynamics

Which has more internal energy: a pot of boiling water or the air in the room? This question is a good way to begin to talk about thermal phenomena. First, it introduces the concept that heat is a form of energy – the equivalence of energy and heat is the foundation of the *First Law of Thermodynamics*. Second, the answer depends on understanding the difference between *average* kinetic energy of molecules – which is what we measure by *temperature* – and the *total* internal energy of a system of molecules – which is changed by adding or removing *heat*. The water molecules in the pot have a higher average energy, but there may be fewer molecules in the pot than in the room air. We will be developing quantitative techniques to address this problem in this chapter, but first we need to consider the concepts of thermal equilibrium and heat capacity.

12.1 Thermal equilibrium

The first scientists who developed thermometers in the sixteenth and seventeenth centuries were surprised to find that different objects in their labs were all at the same temperature. When we touch objects that feel cold, what we are sensing is that they conduct our body heat away faster. Thus metal objects feel cooler than wood or cloth objects, but, in fact, to the first approximation everything in the room is at the same temperature, "room temperature." This is because the room is an example of a system that is in – or near – *thermal equilibrium*.

The temperature is a measure of the average kinetic energy of the molecules in the system. If an object at a higher temperature comes in contact with a lower temperature object, energy will flow from the higher temperature object to the lower temperature object. This process goes on until the two objects are at the same temperature. All the objects in the room are in contact with the air and, through the air, in indirect contact with all the other objects in the room. The definition of thermal equilibrium is a system in which all the component objects are at the same temperature.

In a gas, intermolecular collisions drive the kinetic energy of a molecule toward the average kinetic energy characterized by the temperature of the system. Consider injecting a fast-moving molecule into a collection of gas molecules in equilibrium at a given temperature. Through collisions that conserve momentum and energy (elastic collisions) with the molecules already present, the injected molecule will lose energy to the gas until it has energy close to the average energy. In collisions that follow, the molecule may gain or lose energy, but in the gas as a whole the distributions of energies will conform closely to a statistical distribution that is characterized, both in its average and in the distribution around the average, by functions of a single parameter which is the temperature.

12.2 The equipartition theorem

The equipartition theorem is a consequence of classical mechanics, which states that in thermal equilibrium the average energy of a molecule is equally divided over each *degree of freedom* available to the molecule with $\langle E_{int} \rangle = \frac{1}{2} k_B T$ per degree of freedom. Here $k_B = 1.381 \times 10^{-23}$ J/K $= 8.617 \times 10^{-5}$ eV/K is *Boltzmann's constant*, T is the temperature in kelvin, and a degree of freedom refers to all independent quadratic terms in the energy. In thermal equilibrium, the internal energy of a system will be equally distributed over all accessible modes – we would not expect, for example, the energy in the x-direction motion to be significantly different than the energy in the y-direction motion. There are several important modifications of the equipartition theorem that are required by quantum theory, and we will discuss them as we encounter them.

As an example of equipartition, consider a monatomic atom (helium, for example). The energy is given by just the kinetic energy

$$E_{int} = KE = \frac{1}{2}mv_x^2 + \frac{1}{2}mv_y^2 + \frac{1}{2}mv_z^2 = \frac{p_x^2}{2m} + \frac{p_y^2}{2m} + \frac{p_z^2}{2m}. \tag{12.1}$$

Since the x-, y-, and z-velocities or momenta can vary independently, there are thus three independent quadratic terms, and the equipartition theorem predicts that the average energy (average kinetic energy) of each helium molecule is $\langle E \rangle = \frac{3}{2} k_B T$.

What are the implications of this prediction? First, it predicts that as the temperature decreases toward $T = 0$ K the molecular kinetic energy goes to zero. Since we cannot have a negative kinetic energy, this implies that zero degrees on the Kelvin scale (-273 °C) is the lowest possible temperature, *absolute zero*. Second, this predicts the change in temperature of a system to which energy is added. If an amount of heat energy dQ added to a system results in a temperature change dT, the *heat capacity* of a system is defined as $C = dQ/dT$. For a monatomic gas at constant volume, the equipartition theorem predicts the heat capacity should be $C = \frac{dQ}{dT} = \frac{d\langle E_{int} \rangle}{dT} = \frac{3}{2} k_B$ per molecule.

Heat capacity depends on the size of the system – the number of molecules. If the heat capacity is normalized by the system mass or number of particles, for example, this yields the material-dependent quantity *specific heat*. Specific heat, represented by a lower-case c, depends on the material considered, but not on the size of the system.

The *molar specific heat* c_{mol} is the amount of heat needed to raise one mole of molecules – defined as the number of C^{12} atoms in a mass of 12 grams – one degree kelvin. If we multiply $\frac{3}{2} k_B$ by the number of molecules in a mole – Avogadro's number $N_A = 6.022 \times 10^{23}$ – we find the molar specific heat, the heat capacity of Avogadro's number of molecules, to be: $c_{mol} = \frac{3}{2} k_B N_A = \frac{3}{2} R$ where $R = k_B N_A = 8.314$ J/mole K is

Table 12.1 Specific heat at constant volume for various gases

	Gas	c_{mole} (J/mole·K)	c_{mole}/R	Atomic wt. (g/mole)	c_V (J/kg·K)
Monatomic gases	Helium	12.5	1.5	4.00	3120
	Neon	12.47	1.5	20.18	618
	Argon	12.5	1.5	39.95	313
Diatomic gases	Hydrogen (H_2)	20.4	2.45	2.016	10,160
	Nitrogen (N_2)	20.6	2.49	28.01	736
	Oxygen (O_2)	21.1	2.54	32.00	659
	Nitric oxide (NO)	20.9	2.51	30.00	696
Polyatomic gases	Hydrogen sulfide (H_2S)	25.4	3.06	34.08	745
	Water vapor (H_2O)	26.6	3.20	18.02	1476
	Carbon dioxide (CO_2)	28.2	3.40	44.01	641

the *universal gas constant*. The first three rows in Table 12.1 compare the molar specific heats at constant volume of monatomic gases with the predictions of the equipartition theorem, $c_{mole} = \frac{3}{2}R = 12.47\,\text{J/K}$. The results are all quite close to the equipartition prediction. The last column expresses the specific heat per kilogram instead of per mole – the apparent differences in the specific heats per kilogram are due to the different atomic masses of the gases.

> **Question 12.1**
>
> How is it possible for a much heavier atom, such as argon with an atomic weight of 39.95 g/mol, to have the same energy as a light atom like helium with an atomic weight of 4 g/mol?

Things are quite different when we look at the next four diatomic gases, which have a molar specific heat more like $2.5R$ than $1.5R$. What is different here? The difference is that diatomic molecules have degrees of freedom related to rotation. One could argue that a monatomic atom can also rotate – just with a smaller moment of inertia – but, quantum mechanically, the rotation angle of the atom does not enter into the Hamiltonian. Quantum mechanically, a rotated monatomic atom is indistinguishable from a non-rotated atom, and states that are indistinguishable are identical in quantum mechanics. Rotation does not change the state and therefore does not contribute to the independent degrees of freedom.

For diatomic molecules, on the other hand, the total energy includes both translational kinetic energy and rotational kinetic energy. For a diatomic molecule with its axis along the z-direction, for example,

$$E = \frac{1}{2}mv_x^2 + \frac{1}{2}mv_y^2 + \frac{1}{2}mv_z^2 + \frac{1}{2}I_x\omega_x^2 + \frac{1}{2}I_y\omega_y^2 \tag{12.2}$$

which gives five independent quadratic terms, three for the x-, y-, and z-components of the translational velocity and two for the x- and y-component of the rotational velocity, ω_x and ω_y. We note that the rotation around the z-axis of symmetry does not create an

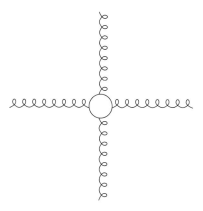

Figure 12.1 An atom in a crystal lattice modeled as held in position by spring restoring forces.

independent degree of freedom for the same reason that rotation does not contribute to the monatomic atoms: quantum mechanically the molecule is indistinguishable following rotation of the molecule around its z-axis and thus rotation around the z-axis does not result in an independent term in the energy.

Thus, with five independent quadratic terms in the energy, equipartition predicts an average energy $\langle E_{int} \rangle = \frac{5}{2} k_B T$ and a molar specific heat of $2.5R$. This expectation is verified in the table, where the four diatomic gases all have a molar specific heat close to $2.5R$. In molecules with three atoms, on the other hand, the molar specific heats generally exceed $3.0R$ which would be expected for a non-linear rotating molecule with six degrees of freedom – three translational and three rotational. In addition, we do not observe the difference we might expect between a linear molecule like CO_2 and non-linear molecules like H_2S and H_2O. The problem here is that we have not yet considered vibrational degrees of freedom.

To consider how vibrations add to the degrees of freedom let us begin by considering vibrations in solids. Consider first a single atom in a lattice as shown in Figure 12.1. As the atom vibrates in three dimensions, it will be driven back to its equilibrium position by interatomic elastic forces, which are modeled by the springs shown in the diagram. In this case, in addition to the kinetic energy of motion the energy of the atom will have contributions from the potential energy of the springs as they compress and restore. Recalling from our discussion of the spring–mass system, the energy associated with the compression of a spring is found from the work required to do the compression

$$U = -\int \vec{F} \cdot d\vec{x} = -\int (-kx)dx = \frac{1}{2} k x^2. \tag{12.3}$$

Considering all three dimensions, we find the energy of an atom in a solid will be given by

$$E = KE + U = \frac{1}{2}mv_x^2 + \frac{1}{2}mv_y^2 + \frac{1}{2}mv_z^2 + \frac{1}{2}kx^2 + \frac{1}{2}ky^2 + \frac{1}{2}kz^2 \tag{12.4}$$

assuming the spring constant k is the same in all directions. This expression for the energy has six independent quadratic terms, three related to the kinetic energy involving the square of the velocity (or momentum) and three related to the potential energy involving the square of the particle positions x, y, and z. We would not expect to find all the atom's energy in the kinetic energy with no potential energy component, nor would we expect to find all the energy in the potential energy. Instead, the atom's energy is exchanged back and forth between the kinetic and potential energy in all three directions and the equipartition theorem tells us that each component will have the same average energy $\frac{1}{2} k_B T$ with a total average energy $\langle E_{int} \rangle = 6 \cdot \frac{1}{2} k_B T = 3 k_B T$.

Question 12.2

What is the average energy in eV of gaseous argon atoms at room temperature (300 K)? What forms of energy does this include? Are there other ways that an argon atom could absorb energy? Can you estimate how much energy it would take to excite these other forms of energy?

Since it is difficult to constrain solids or liquids to a constant volume as they are heated, the specific heats are given at *constant pressure* (typically atmospheric pressure) rather than constant volume. We note that, although there is no particular pattern in the specific heat per kilogram (last column), the specific heat per atom (or per mole of atoms) is remarkably close for most elemental solids, ionic solids (NaCl), and glassy solids (As$_2$Se$_3$) at about $3k_B$ per atom or $3R$ per mole. This corresponds closely to the expected six degrees of freedom – three dimensions of velocity and three dimensions of elastic potential energy.

Table 12.2 Molar specific heats at constant pressure for several solids and liquids

	Material	c_{mole} (J/mole·K)	$2\,c_{mole}/R$ (degrees of freedom per atom)	Atomic wt. (g/mole)	c_p (J/kg·K)
Elemental solids	Aluminum	24.4	5.9	26.98	904
	Copper	24.5	5.9	63.54	386
	Diamond	6.1	1.5	12.01	508
	Graphite	8.5	2.0	12.01	708
	Lead	26.5	6.4	207.2	128
	Silver	25.5	6.1	107.9	236
	Tungsten	24.8	5.9	183.8	134
	NaCl	50	6.0	58.44	855
	As$_2$Se$_3$	121.6	5.9	387.0	314
	Al$_2$O$_3$	89.72	4.3	102.0	880
	SiO$_2$	44	3.5	60.08	730
	Ice (H$_2$O @ $-10°$C)	38.1	3.1	18.01	2110
Liquids	Mercury	28.1	6.8	200.6	140
	Water	75.4	6.1	18.01	4190
	Ethyl alcohol (C$_2$H$_6$O)	112	3.0	46.0	2430

However, there are some striking exceptions. Carbon in the diamond lattice has a molar specific heat only one-quarter as large as the expected six degrees of freedom would suggest. The structural form of the lattice clearly has something to do with this, since carbon in the form of graphite has distinctively different and higher specific heat per atom. In addition, strongly covalently bonded solids, such as silica (SiO_2), sapphire (Al_2O_3), and ice (H_2O), also have significantly lower specific heats per atom.

12.3   Quantum deviations from equipartition theorem

To resolve these discrepancies from the classical equipartition prediction, we need to look at how the classical picture is modified by quantum theory. Quantum theory says that electrons, atoms, and systems exist in *quantum states* of a particular energy. At $T = 0$ K, the system exists in its lowest energy "ground state." As energy is added to the system, increasing the temperature, higher energy states begin to be occupied. As we will see later, if energy can be exchanged between various states of the system each system state with the same available energy will be occupied with equal probability. But this implies that the energy will be roughly evenly distributed among all the particles in the system: for one atom to be in a state with all the extra added energy, while the other atoms all remain in their ground state, is so relatively improbable that it never happens in thermal equilibrium.

Thus thermal equilibrium is characterized by the energy of a system evenly distributed over the component particles of the system, with each particle having close to the average energy k_BT. At relatively low temperatures, therefore, the component atoms and electrons are largely confined to quantum states with energies around k_BT or below. Atomic and electronic states with energies $E_i \gg k_BT$ are *frozen out* and occupied with vanishingly small probability.

Kinetic energy states exist with essentially vanishing small energy (each state occupying an area h^3 of phase space), so at the lowest temperatures only states of molecular motion are occupied to any significant extent. When the temperature is high enough that k_BT is comparable to the first of the rotational energy levels $E_\ell = \frac{\hbar^2 \ell(\ell+1)}{2I}$ (where I is the molecule's moment of inertia and ℓ is an integral quantum number) the energy levels associated with the rotational degrees of freedom become significantly occupied. For diatomic gases, such as hydrogen at room temperature, the energy levels associated with the translational and rotational degrees of freedom are occupied, but the vibrational energies $E_n = \left(n + \frac{1}{2}\right)\hbar\omega$ are still only occupied at the ground level $n = 0$. There are only five degrees of freedom which are significantly occupied, each contributing $\frac{1}{2}k_BT$ to the average energy, leading to a molar specific heat $c_V = \frac{5}{2} R$. At higher temperatures, around 1000 K, the vibrational modes of the H_2 molecule, with $\hbar\omega = 0.547$ eV, will start to be excited, opening up another degree of freedom. Electronic degrees of freedom will only come into play at even higher temperatures, just before dissociation when the specific heat will return to 2 atoms $\times$ 3 degrees of freedom $\times \frac{1}{2} R = 6/2\ R$ (see Figure 12.2).

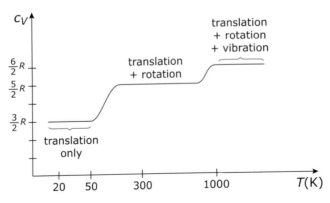

Figure 12.2 Specific heat at constant volume as a function of temperature for diatomic hydrogen gas.

Vibrational modes in other gases may occur at lower frequencies and be significantly involved in the thermal response even at room temperature. In solids, vibrational modes are expressed in extended modes called *phonons* that usually have a fairly low energy spectrum and contribute to the specific heat at low temperatures. Exceptions to this include diamond, which has a high energy phonon spectrum, due to the high hardness/elastic constant and low mass. Recall that the vibrational frequency of an isolated oscillator is given by $\omega = \sqrt{\frac{k}{m}}$ where k is the spring constant and m is the mass. Even as a solid, diamond has high energy vibrational modes that are not fully occupied near room temperature. Not until temperatures near 1000 K does diamond achieve a molar specific heat approaching that of other solids with six degrees of freedom.

With the concepts of specific heat and heat capacity, we can answer the question posed at the beginning of this chapter: which has more internal energy, a pot of boiling water or the air in the room? Considering first the water, if we assume that the pot holds about a quart of water and "a pint's a pound the world around," we can say we have about 2 pounds of water – call it 1 kilogram. With a molecular weight of 18 g/mole (16 for oxygen and 1 each for the two hydrogen atoms) this corresponds to 1 kg/0.018 kg mole^{-1} = 55.5 moles of water molecules. If we consider water to be a collection of weakly interacting molecules, each with three translational degrees of freedom and three rotational degrees of freedom, equipartition would give us a molar specific heat $c_m = 6 \times \frac{1}{2} k_B N_A = 3R = 24.9\,\text{J/mole} \cdot \text{K}$. However, for reasons related to the strong intermolecular electronic forces – "hydrogen bonds" – in water, the specific heat of water is almost three times as large, 75.4 J/mole · K. Thus, the heat capacity of one kilogram of water is $C_V = c_m m/AW = 75.4\,\frac{\text{J}}{\text{mole} \cdot \text{K}} \cdot 1\,\text{kg}/0.018\,\frac{\text{kg}}{\text{mole}} = 4190\,\text{J/K}$. The molar specific heat of the air in the room, on the other hand, is given approximately by five degrees of freedom (three translational + two rotational for the diatomic N_2 molecule) times $R/2$: $c_m = 5 \times 8.31/2 = 20.8$ J/mole · K. The total heat capacity of the air is given by multiplying the molar specific heat by the volume of the room and the density of air and then dividing by the atomic weight.

$C_V = c_m V \rho / AW = 20.8 \, \frac{J}{mole \cdot K} \cdot (6 \, m \times 4 \, m \times 3 \, m) \cdot 1.2 \frac{kg}{m^3} / 0.028 \frac{kg}{mole} = 64,000 \, J/K.$

Equating the heat lost from the boiling water (initial temperature $= 100 + 273$ K) with the heat gained by the air in the room (initial temperature $= 20 + 273$ K), we find a final temperature of the room as follows:

$$Heat \; lost = (373 - T_f) \cdot 4190 \frac{J}{K} = Heat \; gained = (T_f - 293) \cdot 64000 \frac{J}{K}$$
$$(64000 + 4190) \cdot T_f = 373 \cdot 4190 + 293 \cdot 64000$$
$$T_f = 298 \; K.$$

The water drops 70 degrees, while the room air heats up less than 5 degrees! In fact, since the air will transfer its heat to the walls and contents of the room, the actual increase in temperature will be even less.

Temperature, in short, is a measure of the average internal energy of the particles of a system. **Heat**, on the other hand, is a measure of the total energy added to or transferred away from a system. A larger system can have more total energy available than a smaller system at a higher temperature.

12.4   Equation of state

In a molecular picture of materials, a complete description of a system would be to specify the exact position and momentum of each molecule. Such a specification would be mind-boggling complex, and in addition it would need to be updated every instant of time as the molecules of the material interact and are accelerated by intermolecular forces and interactions with the system container. Instead, we seek to find a set of macroscopic parameters that can describe the properties of the system in a state of thermal equilibrium. We understand that there will be statistical fluctuations of the system around the equilibrium values – and we can even calculate how large these fluctuations will be – but the implicit assumption is that these fluctuations will become vanishingly small as the system becomes large. This assumption is, in fact, confirmed both by calculation and by empirical measurements.

If the system we are considering is a gas or liquid which conforms to the shape of the container, the set of parameters that we need to describe the system would include the number of molecules n, the volume of the system V, the pressure the molecules exert on any internal or container surface P, and the temperature of the system (expressed in degrees kelvin) T. These parameters are not all independent – they are related by an *equation of state* which depends on the details of the intermolecular forces as well as any external forces, such as the gravitational force on the molecules. If the system is a solid, we would also have to specify the crystal structure of the molecules and any internal shear stresses. The *phase change* or change of state associated with the system when it changes from solid to liquid, from liquid to gas, or changes crystal structures in a solid – and the dependence of the phase change on macroscopic parameters such as temperature and pressure – is a particularly important and complex topic which depends not only on the details of the intermolecular forces, but often on the details of collective modes of motion such as vibrational waves or *phonons*. For the time being we neglect these details and consider liquids and, especially, gases that are characterized by the macroscopic variables n, P, V, and T.

The forces between molecules of a gas are a combination of a short-range repulsion and a longer-range attraction. The short-range forces arise because the molecules cannot pass through each other due to the repulsion between the outer electrons of each molecule which becomes infinite as the distance between the electrons goes to zero. The longer-range attraction can be understood because fluctuations in the center of charge of the positively charged nucleus and the surrounding negatively charged electron cloud can spontaneously create an *electric dipole*: a positive charge q centered at a point a distance d away from the center of a negative charge $-q$. A molecule with a spontaneous electric dipole qd can induce a charge separation on a neighboring molecule – the displaced negative charge will tend to repel the negative electrons and attract the positive nucleus on the neighbor molecule. The oppositely directed dipoles on two molecules result in a net attractive force because the positive side of one dipole will be closer to the negative side of the neighbor dipole. This attractive potential due to the induced electric dipoles of nearby molecules can be calculated to depend on the intermolecular distance r as $-\frac{1}{r^6}$ (the minus sign indicating that the potential is attractive). The repulsive force due to the Pauli exclusion forces preventing the overlap of the outer electrons of the two molecules is harder to calculate, but it is often approximated as depending on the inter-molecular distance as $+\frac{1}{r^{12}}$. The resultant *Leonard-Jones 6–12 potential* is expressed as

$$V(r) = V_m \left[\left(\frac{r_m}{r}\right)^{12} - 2\left(\frac{r_m}{r}\right)^6 \right]$$

where V_m is the depth of the potential well and r_m is the distance at which the potential reaches its minimum as shown in Figure 12.3(a). The distance r_m is close to the intermolecular distance in a liquid.

Question 12.3

Can you show that the potential $V(r)$ above has a minimum at $r = r_m$ and that the potential $V(r) = -V_m$ at $r = r_m$?

The Leonard-Jones potential can be used as a calculational model, but the parameters V_m and r_m are usually determined empirically for a given gas or liquid. A simplification of the exact intermolecular potential is shown in Figure 12.3(b). In this approximation, the molecules are completely non-interacting until they come into contact and experience an infinite repulsive force at an effective radius of zero. This idealized potential can be shown to lead to the relation between the macroscopic parameters given by the *Ideal Gas Law*

$$PV = nk_BT \qquad (Ideal\ Gas\ Law) \qquad (12.5)$$

where k_B is the Boltzmann constant that gives the relation between the Kelvin temperature and the average internal energy per degree of freedom in the equipartion theorem.

Table 12.3 Van der Waals parameters for some gases

Gas	Chemical Formula	$a \left(\frac{m^6 \cdot Pa}{mol^2} \right)$	$b \left(\frac{m^3}{mol} \right)$
Argon	Ar	0.1355	0.0320×10^{-3}
Carbon dioxide	CO_2	0.3640	0.04267×10^{-3}
Chlorine	Cl_2	0.6579	0.05622×10^{-3}
Ethane	C_2H_6	0.5562	0.9638×10^{-3}
Helium	He	0.003457	0.0237×10^{-3}
Hydrogen	H_2	0.2476	0.02661×10^{-3}
Methane	CH_4	0.2283	0.04278×10^{-3}
Mercury vapor	Hg	0.8200	0.01696×10^{-3}
Neon	Ne	0.02135	0.01709×10^{-3}
Nitrogen	N_2	0.1408	0.03913×10^{-3}
Oxygen	O_2	0.1378	0.03183×10^{-3}
Sulfur dioxide	SO_2	0.6803	0.05636×10^{-3}
Toluene	$C_6H_5CH_3$	2.406	0.1463×10^{-3}
Water vapor	H_2O	0.5536	0.03049×10^{-3}

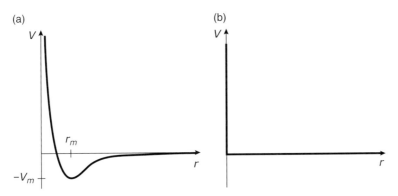

Figure 12.3 Interatomic potential energy as a function of the distance between two atoms. (a) The Leonard–Jones potential including the Pauli exclusion repulsion of the electrons at short distances and the attraction due to dipole–dipole interaction at greater distances. (b) An approximation of the actual potential that results in the ideal gas equation of state.

The ideal gas law is not exact for any gas, but it closely represents the behavior of most gases at low densities. For macroscopic systems it is useful to describe a system in terms of the number of moles of a gas N instead of the number of molecules. For N moles of gas, the Ideal Gas Law becomes

$$PV = N N_A k_B T = NRT \qquad (12.6)$$

where $N_A = 6.022 \times 10^{23}$ is Avogadro's number, which gives the number of molecules in a mole, and $R = k_B N_A = 8.314$ J/mole K is the universal gas constant.

Two effects of the real potential vs. distance profile are neglected in the Ideal Gas Law: (1) the weak attractive force between the molecules that reduces the pressure exerted by the gas, and (2) the strong repulsive core potential that decreases the effective volume that the gas molecules are free to move around in. These two corrections can be modeled using two empirical parameters that can be determined for the specific gas in question in the *Van der Waals equation of state*. In terms of moles of gas N, the Van der Waals equation is

$$\left(P + a\frac{N^2}{V^2}\right)(V - bN) = NRT. \tag{12.7}$$

The empirically determined constants a and b are listed for a few gases in Table 12.3.

Example 12.1 An air-filled cylindrical tank 3 meters in diameter and 10 meters long is sealed at a temperature of 35 °C. If the temperature drops to 0 °C, what is the total force on the curved cylindrical surface of the tank?

Solution: Before the tank is sealed, the pressure is atmospheric pressure, approximately 10^5 pascal. The volume of the tank is $V = \pi 1.5^2\, 10 = 70.7$ m^3. Assuming that the air is an ideal gas initially at 35 °C, we find the number of moles of air is $N = \frac{PV}{RT} = \frac{70.7 \times 10^5}{8.31 \times (273+35)} = 2760$ moles. Upon cooling to 0 °C, the pressure inside is reduced to $P = \frac{NRT}{V} = 2280 \cdot 8.31 \cdot \frac{273}{70.7} = 8.864 \times 10^4$ Pa. The total force on the tank is the difference in inside/outside pressure in pascals (1 Pa = 1 N/m^2) times the surface area of the tank $F = (10^5 - 8.86 \times 10^4)\, \pi \cdot 3 \cdot 10 = 1.07 \times 10^6$ N.

If we use the Van der Waals equation to get a better approximation for N and p we need to solve the equation

$$\left(P + a\frac{N^2}{V^2}\right)(V - bN) = NRT$$

with $a = 0.1408$ and $b = 0.03913 \times 10^{-3}$ for nitrogen which comprises about 78% of air. This equation can be solved numerically (using MATLAB's *fzero* function, for example) and gives $N = 2762$. Solving for the pressure inside the tank at 0 °C, we find $p = 8.861 \times 10^4$ Pa. The Ideal Gas Law is quite accurate for air at atmospheric pressure!

We note that the noble gases (He, Ne, Ar) have only small corrections in the Van der Waals a and b parameters; this is a reflection of their simple, monatomic structure and small size. Hydrogen gas has nearly as small a b parameter, since the molecule is also small, but a larger a parameter since the diatomic structure allows for much easier polarization. At the opposite limit, large, multi-atom molecules, such as ethane or toluene, have relatively large Van der Waals corrections from the ideal gas behavior.

At elevated pressures, the Van der Waals corrections become important and give information about the change of state from gas to liquid. Consider the Van der Waals equation for carbon dioxide, for example. In the state space around a molar volume of 1.3×10^{-4} m^3 and a pressure of 7.4×10^6 Pa (near 74 atmospheres), for temperatures less than 304.2 K the Van der Waals pressure/molar-volume curve has a region shown between points A and B in the dashed line in Figure 12.4 where increasing pressure results in *increasing* volume. This would imply a negative elasticity and a violation of Newton's Laws! In fact, what happens is that the volume decreases at constant pressure

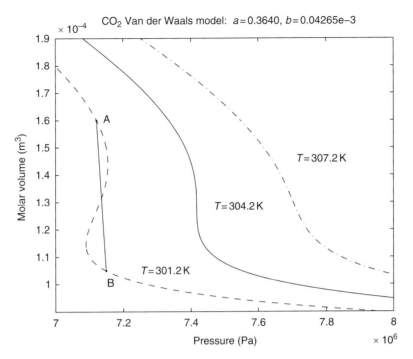

Figure 12.4 Molar volume as a function of pressure for carbon dioxide (CO_2) calculated from the Van der Waals equation for three temperatures, the critical temperature $T_c = 304.2$ K, $T = (T_c - 3$ K), and $T = (T_c + 3$ K). In the region between A and B below the critical temperature the material is in a mixed state with solid and liquid coexisting. Above the critical temperature the distinction between solid and liquid states disappears.

following the straight line between A and B as the gas phase at higher volume coexists in equilibrium with the liquid phase at lower volume.[3] Above A, only the gas phase exists; below B only the liquid phase exists. For temperatures above the critical temperature $T_c = 302.4$ K, the P–V curve always has a negative slope and there is no gas-to-liquid transformation. Above the critical temperature the properties of the gas and liquid phase become identical: CO_2 exists as a "supercritical fluid" completely filling a container like a gas, but at a density like a liquid.

12.5   Thermal expansion

Since temperature is a measure of the average kinetic and potential energy of a system and since potential energy in a harmonic system is defined in terms of atomic displacements, it is not surprising that increased temperature would be associated with increased

---

[3] At room pressure carbon dioxide notably does not have a liquid phase, converting directly from solid "dry ice" to a gas. This is not true at elevated pressures where the gas can be in equilibrium with a liquid phase.

atomic displacement and linear expansion. In a harmonic system, potential energy is given in terms of displacement in one dimension by

$$U_x = \frac{1}{2} k_s x^2. \tag{12.8}$$

Equipartition tells us that the average potential energy in one dimension is $\frac{1}{2} k_B T$, so the average displacement squared should be related to temperature by

$$\frac{1}{2} k_B T = \langle U \rangle = \frac{1}{2} k_s \langle x^2 \rangle. \tag{12.9}$$

An increase in temperature should be roughly related to an increase in the mean squared displacement. This is not always true – the most famous counter-example is water between 0 and 4 °C where, due to the incipient crystallization, increasing temperature causes the volume of water to decrease. But, in general, increase in temperature is related to an increase in linear dimensions and an increase in volume.

In general, we can write any expression for a change in linear dimension with temperature $x = x(T)$ as a Taylor series:

$$x(T) = x(T_o) + \frac{dx}{dT}\bigg|_{T_o} (T - T_o) + \frac{1}{2} \frac{d^2x}{dT^2}\bigg|_{T_o} (T - T_o)^2 + \ \cdots \tag{12.10}$$

If we assume that terms above the first order are small, we find

$$x(T) - x(T_o) = x - x_o = \frac{dx}{dT}\bigg|_{T_o} (T - T_o). \tag{12.11}$$

Multiplying by N, the number of atoms in a length L, we find

$$N(x - x_o) = (L - L_o) = \Delta L = \frac{dNx}{dT}\bigg|_{T_o} (T - T_o) \tag{12.12}$$

$$\Rightarrow \Delta L = L_o \, \alpha \, \Delta T \tag{12.13}$$

where we have introduced the *coefficient of linear expansion* $\alpha = \frac{1}{L_o} \frac{dL}{dT}\big|_{T_o}$ which gives the percent change in linear length with temperature. Note that α may well be a function of temperature $\alpha(T)$ which allows for effects beyond the first-order Taylor series term, as well as anomalous effects, such as the decrease in length with increasing temperature in water near 0 °C. Note that, since all linear dimensions increase by the same alpha, the linear dimension of a hole will get larger (for positive α) in the same way that a linear object increases.

For liquids or gases where the linear dimensions are fixed by the container, a more meaningful parameter is the *volume coefficient of expansion* $\beta = \frac{1}{V_o} \frac{dV}{dT}\big|_{T_o}$ with the change in volume given by

$$\Delta V = V_o \beta \, \Delta T. \tag{12.14}$$

Table 12.4 Coefficient of linear expansion α for different materials (note the large variation)

Material	α
Steel	11×10^{-6} /°C
Aluminum	23×10^{-6} /°C
Lead	29×10^{-6} /°C
Concrete	12×10^{-6} /°C
Invar	0.7×10^{-6} /°C
Fused quartz (SiO_2)	0.5×10^{-6} /°C

Example 12.2 Find the volume coefficient of expansion for an ideal gas.

For an ideal gas with an equation of state $PV = NRT$ we find, for fixed pressure P_o and temperature T_o, $P_o = NRT_o/V$, and

$$\beta = \frac{1}{V_o} \frac{dV}{dT}\bigg]_{T_o} = \frac{1}{V_o} \frac{NR}{P_o} = \frac{NR}{NRT_o} = \frac{1}{T_o}$$

which implies that $\frac{\Delta V}{V_o} = \frac{\Delta T}{T_o}$.

In a solid, the relation between the coefficient of linear expansion α and the volume coefficient of expansion β can be derived from the relationship that volume equals length cubed:

$$V(T) = [L(T)]^3 = L_o^3 (1 + \alpha \Delta T)^3. \tag{12.15}$$

Using the binomial expansion $(1 + \delta)^n \cong 1 + n\delta$, we find:

$$V(T) \cong L_o^3 (1 + 3\alpha \Delta T) = V_o (1 + 3\alpha \Delta T). \tag{12.16}$$

Comparing with $\Delta V = V(T) - V_o = V_o \beta \Delta T$, we see that $\beta \cong 3\alpha$.

Example 12.3 Mercury thermometer

Consider a thermometer with a large bulb of mercury approximated as a sphere of radius $r_1 = 3$ mm connected to a thin column with a radius $r_2 = 0.1$ mm. Mercury has a volume coefficient of expansion $\beta = 181 \times 10^{-6}$/°C. The thermometer is made of glass (fused quartz) – for simplicity we can ignore the linear coefficient of expansion of the glass body. How much does the height of the mercury in the column change in mm/°C?

Solution: Ignoring the much smaller volume of the mercury in the column, the change in volume is $\Delta V = V_o \beta \Delta T = \frac{4}{3} \pi r_1^3 \beta \Delta T$. This translates to a change in the column height $\Delta d = \Delta V / A$ where A is the area of the column of mercury $A = \pi r_2^2$. This gives

Continued

Example 12.3 (*cont.*)

$$\Delta d = \frac{\frac{4}{3}\pi\, r_1{}^3\, \beta\, \Delta T}{\pi r_2{}^2} = \frac{\frac{4}{3}\, 0.003^3\, 181 \times 10^{-6}\, \Delta T}{0.0001^2}$$

from which we find $\frac{\Delta d}{\Delta T} = 0.648\,\frac{mm}{°C}$.

Question 12.4

Inexpensive thermometers use alcohol dyed red instead of mercury as an expansion fluid in the thermometer. Look up the thermal expansion characteristics of alcohol to determine what differences this might mean for the thermometer. Does the thermal expansion coefficient depend on temperature?

12.6   The First Law of Thermodynamics: Work–Energy Theorem

We now have examined systems that change volume when their temperature is changed. From the definition of work, increasing volume under ambient air pressure requires work

$$W = \int \vec{F} \cdot d\vec{x} = \int \vec{F}/_A \cdot A\, d\vec{x} = \int P\, dV \tag{12.17}$$

where the last equality comes from the definition of pressure = force/area and the condition of equilibrium that a fluid always exerts a force which is perpendicular to a surface.

    We can ask how our picture of heat exchange, heat capacity, and change in temperature is modified in the case in which a system does work or has work done on it. A hint that some modification is necessary can be found by examining the specific heat of gases. Specific heat of gases measured at constant volume is always less than the specific heat measured at constant pressure (with the volume free to change). The difference is that at constant volume $dV = 0$ and the gas does no work while being heated, while at constant pressure (by a vial sealed by a constant force as in Figure 12.5, for example) the volume of the heated gas will change and the gas will do work.

    The concepts of work, energy, and heat are related by the First Law of Thermodynamics which can be viewed as a generalization of Conservation of Mechanical Energy to include the effects of heating (including by friction). The result is that the sum of mechanical *and* thermal energy is conserved. Mechanical energy can be lost or gained, but only through a corresponding increase or decrease in the thermal energy of the system. In short, the First Law of Thermodynamics tells you that "you can't win." The total amount of energy in a system cannot be increased (or decreased) by any means; energy may change form but the total will be constant.

(a)   (b)

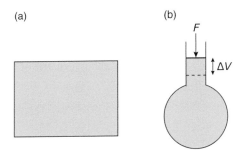

Figure 12.5 Experimental configuration for finding specific heat at (a) constant volume and (b) at constant pressure with the volume allowed to change.

In equation form, the **First Law of Thermodynamics** can be written as follows:

$$\Delta E_{int} = Q - W. \tag{12.18}$$

In this equation, ΔE_{int} **is the change in the internal energy of the system**, Q **is the heat added to the system**, and W **is the work that is done by the system**. The internal energy can be changed by:

1. Changing the temperature of the system as expressed by the heat capacity of the system.
2. Changing the state of the system (from solid to liquid or liquid to gas, for example, or other structural transformation such as a phase change to a different crystal structure in a solid).
3. Changing the chemical composition of the system (a change in the chemical bonding will release energy, if the new bonds are stronger, or absorb energy, if the resulting bonds are weaker than before).

The change in internal energy ΔE_{int} can be either positive or negative as the system gains or loses energy. The heat Q added to the system and the work W done by the system can be positive or negative as well. **Negative Q corresponds to heat *removed* from the system** and **negative W corresponds to work done *on* the system.**

12.7   Examples of the First Law of Thermodynamics

a.   Calorimetry

In a thermally isolated (*adiabatic*) system, 500 g of aluminum at 100 °C is dropped into 500 g of water with 15 g of ice in a 10 g aluminum cup at 0°C. What is the final temperature?

   In this problem, the system consists of the 500 g heated aluminum mass, the 500 g of water, the 15 g of ice, and the 10 g aluminum cup. For an adiabatic system, the heat added Q is zero, and we will neglect any change in volume by the melting ice, so we assume the work done by the system W is also zero. With $Q = 0$ and $W = 0$, the First

Law tells us that the change in internal energy ΔE_{int} is zero. Thus the decrease in internal energy by the aluminum mass when its temperature decreases from 100 °C must be balanced by the increase in internal energy by the aluminum cup and the water when their temperature rises from 0 °C, and by the increase in internal energy of the ice when it undergoes a change of state (when it melts) and then by the increase in temperature of the melted ice from 0 °C.

$$\Delta E_{int} \left(H_2O + \text{ice} + \text{Al cup} \right) = -\Delta E_{int} \left(\text{Al mass} \right).$$

We will consider each of these terms in turn and then solve the equation to determine the final temperature T_f. The decrease in internal energy of the aluminum mass in this temperature range is entirely due to the decrease in temperature. From Table 12.2, aluminum has a specific heat at constant atmospheric pressure $c_p = 904$ J/kg K $= 0.216$ cal/g °C in calorimetric units (1 calorie of heat $= 4.1868$ joules). This gives for the aluminum mass and cup

$$\Delta E_{int} \left(\text{Al mass} \right) = mc_p \, \Delta T = 500 \text{ g} \cdot 0.216 \frac{\text{cal}}{\text{g} \cdot {}^\circ\text{C}} \cdot (T_f - 100^\circ\text{C})$$

$$\Delta E_{int} \left(\text{Al cup} \right) = mc_p \, \Delta T = 10 \text{ g} \cdot 0.216 \frac{\text{cal}}{\text{g} \cdot {}^\circ\text{C}} \cdot (T_f - 0^\circ\text{C}).$$

By the definition of a calorie, water has a specific heat of 1 cal/g °C. So the change in internal energy of the 500 g of water is:

$$\Delta E_{int} \left(\text{water} \right) = mc_p \, \Delta T = 500 \text{ g} \cdot 1.0 \frac{\text{cal}}{\text{g} \cdot {}^\circ\text{C}} \cdot (T_f - 0^\circ\text{C}).$$

Finally, there is the 15 g of ice. The change in internal energy of the ice comes both from the energy required to change the state from solid ice to liquid water plus the energy to heat the resulting water to the final temperature. The *latent heat of fusion* of water, the energy to melt the ice, is experimentally determined to be $L_f = 80$ calories/g. The melted ice has the same c_p as water: 1 calorie/g °C. The total energy to bring the ice to the final temperature (assuming that the ice is all melted) is:

$$\Delta E_{int} \left(\text{ice} \right) = mL_f + mc_p \, \Delta T = 15 \text{ g} \cdot 80 \frac{\text{cal}}{\text{g}} + 15 \text{ g} \cdot 1.0 \frac{\text{cal}}{\text{g} \cdot {}^\circ\text{C}} \cdot (T_f - 0^\circ\text{C})$$

Putting this all together, we find:

$$500 \cdot 1.0 \cdot (T_f - 0) + 15 \cdot 80 + 15 \cdot 1.0 \cdot (T_f - 0) + 10 \cdot 0.216 \cdot (T_f - 0)$$
$$= -500 \cdot 0.216 \cdot (T_f - 100).$$

Collecting terms in T_f and solving, we get

$$(500 + 15 + 2.16 + 108)T_f = 10,800 - 1200$$

$$\Rightarrow T_f = 15.36\,^\circ\text{C}.$$

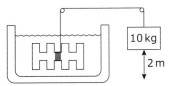

Figure 12.6 Experimental apparatus to turn the mechanical energy of a falling weight into thermal energy of a fluid.

b.   ## Work done on adiabatic system

Consider a system where a falling weight powers counter-rotating paddles that agitate 1500 grams of water in a thermally isolated container of negligible heat capacity as shown in Figure 12.6. If the paddles are driven by a 10 kg mass that falls a total of 2 m, how much is the temperature of the water raised?

Since the system here is adiabatic, no heat can enter or leave the system. Since the work here is done *on* the system, W the work done *by* the system is negative. The First Law gives us:

$$\Delta E_{int} = mc_p\,\Delta T = Q - W = 0 + mg\Delta h.$$

If we ignore the heat capacity of the container and the agitating mechanism, the change of temperature of the water can be found from

$$1500 \text{ g} \cdot 1\frac{\text{cal}}{\text{g} \cdot {}^\circ\text{C}} \cdot \Delta T = 10 \text{ kg} \cdot 9.8\frac{\text{m}}{\text{s}^2} \cdot 2\,\text{m} \cdot \left(\frac{1 \text{ cal}}{4.1868 \text{ J}}\right)$$

$$\Rightarrow \Delta T = 0.031\ {}^\circ\text{C}.$$

The relatively small increase in temperature due to the application of work to a system is the reason that the equivalence of heat and work/energy was overlooked for so long. In the late eighteenth century, Benjamin Thompson (later Count Rumford) observed that in the process of boring cannons heat was produced in the chips of brass produced that could be used to heat water. Thus heat – at that time believed to be a type of fluid *phlogiston* produced by heat sources, such as fire, that could flow into surrounding objects – was shown to be produced by the application of work, another form of energy.

c.   ## Energy in food

How many vertical feet could an 80 kg rock climber climb from the energy in one raisin with 2 calories?

Recognizing that food calories are actually *kilocalories* and assuming that the rock climber is 100 percent efficient in converting the food energy into mechanical energy of climbing, we have

$$2000 \text{ cal} = mg\Delta h \quad \Rightarrow \quad \Delta h = \frac{2000}{80 \cdot 9.8}\left(\frac{4.1868 \text{ J}}{1 \text{ cal}}\right) = 10.7\,\text{m}.$$

d.   ### Free expansion (Joule–Thompson cooling)

Consider a closed container of compressed nitrogen gas at initial pressure P_o. A movable diaphragm is released and the gas expands to twice its initial volume. How much is the temperature of the gas reduced?

Cooling by expansion is a commonly observed phenomenon that is also the basis for most refrigeration and air conditioning applications. In this case, since the pressure of the gas will change as the enclosure expands, we should use the differential form of the First Law of Thermodynamics

$$dE_{int} = dQ - dW.$$

Assuming that the container is insulated so the process is adiabatic, we have $dQ = 0$. The differential work is expressed in terms of the instantaneous pressure as $dW = P\,dV$, so we have

$$dE_{int} = -P\,dV.$$

For a diatomic gas, the change in internal energy with temperature is given by $dE_{int} = \frac{5}{2} NR\,dT$ and if we can approximate the gas equation of state by the Ideal Gas Law we have $P = NRT/V$, yielding

$$\frac{5}{2} NR\,dT = -\frac{NRT}{V}\,dV.$$

Separating the variables V and T to opposite sides of the equation we find,

$$\frac{dT}{T} = -\frac{2}{5}\frac{dV}{V}.$$

Integration gives us $\ln\left[\frac{T}{T_o}\right] = -\frac{2}{5}\ln\left[\frac{V}{V_o}\right]$, and if $V = 2V_o$ we find,

$$T = T_o\, e^{-\frac{2}{5}\ln\left[\frac{V}{V_o}\right]} = T_o\, e^{-\frac{2}{5}\ln 2} = (0.76)T_o.$$

Microscopically, we understand this decrease in temperature in terms of molecules of the gas bouncing off the receding diaphragm, by simple momentum considerations reflecting with decreased velocity and therefore decreased energy. In practical refrigeration applications, more complex molecular gases with longer-range interactions are used where the expansion against the intra-molecular forces and possibly a change of state from liquid to gas increase the change in the internal energy on expansion.

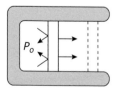

Figure 12.7 Adiabatic (no heat transfer) expansion of a gas against a restraining force or pressure.

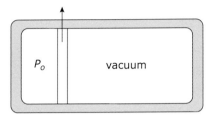

Figure 12.8 Free expansion of a gas into an evacuated space. The barrier is removed instantaneously in a transverse direction.

e.

Free expansion into vacuum

In this problem a compressed gas at initial pressure P_o in a thermally insulated chamber is separated from an evacuated space by a barrier that is rapidly removed in a lateral direction (see Figure 12.8). What is the change in the temperature of the gas, assuming an ideal gas with molecules that are non-interacting except for instantaneous point collisions?

> **Question 12.5**
>
> The reader is encouraged to apply the First Law to the problem to find a solution before reading on.

Since the chamber is insulated, the heat Q added to the system is zero. But in this case, since the vacuum does not exert a pressure for the gas to do work on the system, the work done by the system W is also zero. The First Law thus tells us that the change in internal energy,

$$\Delta E_{int} = Q - W = 0 - 0 = 0.$$

With no change in internal energy, the kinetic energy of the molecules and the temperature do not change. There is no change of temperature for an ideal gas expanding into vacuum!

This somewhat counterintuitive result also follows from microscopic arguments. Assuming the barrier is removed nearly instantaneously and the edge of the barrier has no appreciable area, the gas molecules do not interact with any object. The molecules continue into the evacuated space with no change of velocity, and therefore no change in temperature.

12.8   Molar specific heats, c_V vs. c_p for an ideal gas

In review, the equipartition theorem gave us that the internal energy for a monatomic gas with only translational degrees of freedom is $\frac{3}{2} k_B T$ per molecule. In differential form:

$$d E_{int} = n \frac{3}{2} k_B \, dT = \frac{3}{2} N N_A k_B \, dT = \frac{3}{2} N R \, dT = dQ - dW \tag{12.19}$$

where n is the number of molecules and N is the number of moles and N_A is Avogadro's number, giving the number of molecules per mole. At constant volume, no work is done, so $dW = 0$ and the specific heat per mole is

$$c_V = \frac{1}{N} \frac{dQ}{dT} = \frac{3}{2} R. \tag{12.20}$$

When the pressure is fixed but the volume can change, $\frac{dQ}{dT}$ will give us the specific heat at constant pressure. Assuming first a monatomic gas, we find

$$dE_{int} = \frac{3}{2} N R \, dT = dQ - dW \tag{12.21}$$

where at constant pressure $dW = P \, dV$. If we assume an Ideal Gas Law equation of state $PV = NRT$, at constant pressure $dV = \frac{NR}{P} \, dT$. Thus we get

$$dE_{int} = \frac{3}{2} N R \, dT = dQ - P \, dV = dQ - N R \, dT. \tag{12.22}$$

Rearranging, we find

$$dQ = \left(\frac{3}{2} NR + NR \right) dT \tag{12.23}$$

and the molar specific heat at constant pressure is

$$c_p = \frac{1}{N} \frac{dQ}{dT} = \frac{3}{2} R + R = \frac{5}{2} R \quad \text{(monatomic gas)}. \tag{12.24}$$

For a diatomic molecule with two rotational degrees of freedom, we have $c_V = \frac{5}{2} R$; if three rotational degrees of freedom are possible, $c_V = 3R$. **Generally, our analysis gives us** $c_p = c_V + R$, so for a diatomic gas (N_2, for example), $c_p = \frac{5}{2} R + R = \frac{7}{2} R$. Table 12.5 compares the measured molar specific heats c_V and c_p with the results calculated from the equipartition theorem and the First Law result $c_p = c_V + R$ for four gases.

Table 12.5 Measured specific heat at constant volume and constant pressure compared with predictions of equipartition theorem and First Law

Gas	c_V (J/mole·K)	c_p (J/mole·K)
He	$12.49 \left(\frac{3}{2} R = 12.5 \right)$	$20.7 \ (c_V + R = 20.8)$
N_2	$20.81 \left(\frac{5}{2} R = 20.7 \right)$	$29.1 \ (c_V + R = 29.1)$
NH_3	$28.23 \left(\frac{6}{2} R = 24.9 \right)$	$37.24 \ (c_V + R = 37.3)$
CH_4	$27.22 \left(\frac{6}{2} R = 24.9 \right)$	$35.54 \ (c_V + R = 35.5)$

Even in the cases where the equipartition result for c_V isn't a particularly good approximation, the relation between c_V and c_p is given quite well by the First Law of Thermodynamics and the ideal gas equation of state.

12.9   Enthalpy

We are familiar with the concept that a system will spontaneously transition to a state of lower potential energy. Given an initial push to overcome any small barrier holding it in place, a ball will naturally roll down a hill converting its (gravitational) potential energy into kinetic energy of movement. Similarly, we can compare the internal energy of a system before and after a chemical reaction by calculating how E_{int} compares before and after the reaction as the energy is changed by the creation or breaking of chemical bonds. However, a little thought will make it apparent that such a description is incomplete: in addition to the change in internal energy caused by the change in chemical bonding, the system has to do work against the ambient (usually atmospheric) pressure if the reaction products take up more volume than the reactants. If a reaction reduces the volume, work will be done on the system by the ambient pressure.

To take into account this work done by (or on) the system in expanding (or contracting) against the ambient pressure, a new state function called the system *enthalpy* is defined

$$H = E_{int} + PV. \tag{12.25}$$

As with total energy (kinetic plus potential energy), the zero point of enthalpy is not defined, and what is important is the change in enthalpy

$$\Delta H = \Delta E_{int} + \Delta(PV) \tag{12.26}$$

as the system undergoes a reaction or change in state.

Using the First Law, we can write $\Delta E_{int} = Q - P\Delta V$, leading to the relation

$$\Delta H = Q - P\Delta V + P\Delta V + V\Delta P = Q + V\Delta P. \tag{12.27}$$

In the most common case where the ambient pressure doesn't change, we find that the change in enthalpy just equals the heat added to the system

$$\Delta H = Q \quad \text{(at constant pressure)}. \tag{12.28}$$

A system that is transitioning at constant pressure to a lower energy state will make the transformation spontaneously with the generation of an amount of heat equal to the decrease in enthalpy. Such a reaction is described as *exothermic*. (Recall that Q is the heat added to a system; a negative Q is the heat generated by the system.) Similarly a reaction that requires heat to be supplied from the environment is *endothermic*. Tables of the enthalpy per mole of different reactants and products exist that can be consulted to determine whether the final state is at a net lower or higher energy. Enthalpy contours as a function of pressure will tell a chemical engineer or process chemist what the lowest energy state is for a chemical system at a given temperature and pressure.

However, enthalpy considerations alone do not determine the equilibrium state of a system. The heat absorbed or produced by the reaction will affect the environment of the system. We will need to consider the combined effect on the system and the environment to see if a reaction will occur spontaneously. And before we can do these calculations accurately, we need to consider the role of probability in the process. The combination of the system and its environment does not always end up in the lowest energy state – but it will always end up in the *most probable state*. To understand this distinction, we need to introduce the concept of entropy and the Second Law of Thermodynamics. This critically important topic is treated in the next section.

12.10   Entropy and the Second Law of Thermodynamics

Not all processes permitted by the First Law of Thermodynamics will occur spontan-eously. For example, consider the following scenario: A boat (see Figure 12.9) is to be propelled by taking on board water at ambient temperature (say 20 °C), energy is extracted from the water by cooling and freezing it and that energy is used to propel the ship. The ice remaining from the process is then tossed overboard, and the ship is propelled without the use of any energy source.

Can this process work? Clearly, if it could there would be no energy shortage! One is tempted to say that the process is forbidden by Conservation of Energy, but if we look at the work–energy theorem in the First Law of Thermodynamics, nothing appears to be a problem.

The First Law states that $\Delta E_{int} = Q - W$ where ΔE_{int} is negative (internal energy is extracted from the water to cool and freeze it), Q is zero (no heat is added to system), and W, the work done by the system is positive (propelling the ship through the water). The ice thrown overboard is ultimately melted and heated back to the ambient tempera-ture by the ocean water, presumably cooling the ocean temperature by an infinitesimal amount, but energy (and work) are conserved throughout the process. So, why doesn't this process work?

The impossibility of this "perpetual energy machine of the second type" is due to a violation of the *Second Law of Thermodynamics,* which says that the *entropy* of the universe cannot decrease in any real physical process. Entropy is usually described as "disorder," but it can be more accurately described as the tendency of a system to be in its *most probable state* consistent with its total energy being conserved.

As an example, we can ask: how is it that all of the air molecules in a room don't all collect on one side of the room – suffocating those unlucky people on the side with no

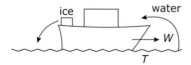

Figure 12.9 A scheme to power a boat by converting the heat removed from the sea water (by converting the water to ice) into work which drives the boat.

Table 12.6 Distribution of two molecules in room

Left side of room	Right side of room	Number of states
AB	–	1
–	AB	1
A	B	(2 indistinguishable states)
B	A	

air? One could say that it would presumably require some work to compress the air into one side of the room, but that could be balanced by a decrease in the temperature (internal energy) of the air in accordance with the First Law of Thermodynamics. The people on the side with air might be a little chilled, but they would certainly be better off than their colleagues gasping for air on the other side of the room.

So why don't we go around terrified of spontaneous suffocation due to the random motion of the air molecules? To answer this question, we need to look at some simple statistics of particles. Let us consider the case where there are only two molecules in the room, call them molecule A and molecule B. Table 12.6 indicates all the possible ways that the two molecules could distribute themselves between the right and left sides of the room.

As you can see, the (indistinguishable) situations where there is one molecule on each side of the room is twice as likely as the situation where both molecules are on either the right or left side.

If we consider the case where the number of molecules is four, A, B, C, and D, we can find the number of possible states either by listing all the possible distributions or by using elementary statistics that gives the number of configurations of N objects divided into two states with M in one state and $(N - M)$ in the other:

$$\Omega = \frac{N!}{M!(N - M)!} .$$
(12.29)

For example, there is only one state with all four molecules on the left side and none on the right:

$$\Omega_{4|0} = \frac{4!}{4!(0)!} = \frac{4 \cdot 3 \cdot 2 \cdot 1}{4 \cdot 3 \cdot 2 \cdot 1 \cdot (1)} = 1.$$

Similarly, there are four (indistinguishable) states with one molecule (A, B, C, or D) on the left and all the others on the right:

$$\Omega_{1|3} = \frac{4!}{1!(3)!} = \frac{4 \cdot 3 \cdot 2 \cdot 1}{1 \cdot (3 \cdot 2 \cdot 1)} = 4$$

and six (indistinguishable) states with two molecules on the left and two on the right:

$$\Omega_{2|2} = \frac{4!}{2!(2)!} = \frac{4 \cdot 3 \cdot 2 \cdot 1}{2 \cdot 1 \cdot (2 \cdot 1)} = 6.$$

Table 12.7 Distribution of four molecules in room

Left side of room	Right side of room	Number of states
XXXX	–	1
–	XXXX	1
X	XXX	4
XXX	X	4
XX	XX	6

The number of states in each configuration for four molecules is shown in the Table 12.7.

If we assume that each of these states is energetically equivalent and further assume that a system will occupy each of its energetically equivalent configurations with equal probability, we see that the system is six times as likely to have two molecules on each side of the room as to have all the molecules on one side or the other. As the number of molecules N increases the number of configurations with the molecules essentially equally divided increases as $\Omega_{\frac{N}{2}|\frac{N}{2}} = \frac{N!}{(N/2)! \cdot (N/2)!}$ while there remains only one configuration with all the molecules on one side or the other. If the system is equally likely to be found in any of its accessible configurations, the equally divided state is $\frac{N!}{(N/2)! \cdot (N/2)!}$ or nearly $N!$ times as likely as the state with all the molecules on one side. Since the factorial function is faster growing than even the exponential function, the probability of the equally (or nearly equally) divided state is so much more probable than a state with all (or mostly all) of the molecules on one side, that the molecules are virtually certain to be nearly equally divided.

Does this mean that probability is all that keeps half the class from suffocating? Essentially yes. There is some increase in energy ($\approx P \Delta V$) for the state with the smaller volume which we will learn how to deal with later through the concept of *free energy*, but the major reason for the equal division of molecules is the much larger number of available states with the molecules equally divided. When the number of molecules gets very large ($\sim 6 \times 10^{23}$ per mole of gas), the probabilities become so large they are virtual certainties.

Using the concept of statistical mechanics we can restate the Second Law of Thermodynamics: *a system with a large number of (energetically equivalent) available states will be in the most statistically probable state*. This can be made quantitative by **defining entropy as**

$$S = k_B \ln \Omega \tag{12.30}$$

where k_B is Boltzmann's constant and Ω is the *multiplicity of states*, the number of states accessible to system with given constraints (fixed energy, volume, number of particles, etc.). The greater number of states accessible to the system, the greater the entropy.

The logarithmic dependence on number of accessible states leads to a convenient result: **entropy is additive with the size of the system.** If we have two systems, one with Ω_1 states available to it and the other with Ω_2 states available to it, the total number of states available in the combined system is $\Omega_1 \Omega_2$ (because Ω_2 states are available in the second system for each state in the first system). The total entropy is thus $S_T = k_B$ $\ln(\Omega_1\Omega_2) = k_B \ln(\Omega_1) + k_B \ln(\Omega_2) = S_1 + S_2$.

The number of states available to a system increases exponentially as the energy of the system increases by adding heat. This leads to the differential relationship

$$dS = \frac{dQ}{T} \tag{12.31}$$

with dQ as the heat added and T the temperature in degrees kelvin.

With this definition for entropy S the **Second Law of Thermodynamics** can be written as: *in any physical process the entropy of the universe (the system under consideration and all systems that it interacts with) either increases or remains unchanged.* The total entropy cannot decrease in any physically realizable process. Another way of saying this mathematically is $\Delta S \geq 0$ for any process. **The case where** $\Delta S = 0$ **is called a reversible process**, indicating the system and its environment can be restored to their original state. **If $\Delta S > 0$ the process is irreversible**, and the system and its environment cannot be restored to their original state without increasing the entropy of some larger outside system.

If the First Law of Thermodynamics tells us "you can't win," the Second Law's prohibition against a decrease in entropy can be expressed as "you can't even break even."

As an example, consider the system shown in Figure 12.10, with a gas in a thermally insulated enclosure compressed by the weight of a mass M. The mass is removed, and the compressed gas expands and pushes the diaphragm upward until the internal pressure of the gas is balanced by the atmospheric pressure and the weight of the diaphragm. The compressed gas does work in moving the diaphragm upward, but no heat enters the gas through the insulated container – the process is adiabatic (no heat is added).

The First Law of Thermodynamics tells us that $\Delta E_{int} = Q - W$ and since the system does work, while there is no heat added, the internal energy (temperature) of the gas decreases. Since $dQ = 0$, there is no change in entropy $dS = dQ/T = 0$ and the process is *reversible*; if the weight is put back on the diaphragm the gas will return to its original volume and temperature. We note that **work done by the system (or on the system) does not change the entropy, even though the temperature changes**.

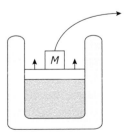

Figure 12.10 A reversible thermodynamic process: the weight is removed and the gas expands without any heat added (adiabatic expansion).

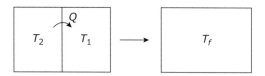

Figure 12.11 An irreversible thermodynamic process: two bodies at different temperatures T_1 and $T_2 > T_1$ are put in thermal contact and reach an equilibrium temperature T_f $(T_1 < T_f < T_2)$.

The entropy changes only if the internal energy is changed by adding or removing heat from the system.

As an example of an irreversible process, consider two systems at different temperatures T_1 and T_2, with $T_2 > T_1$, that are brought into contact with each other and allowed to come to thermal equilibrium at a temperature T_f that is between T_1 and T_2 (see Figure 12.11). The temperature of each system will be continuously changing toward the final temperature T_f, so to calculate the change in entropy we should integrate the differential relation $dS = dQ/T$ to find $\Delta S = \int \frac{dQ}{T}$. We can understand quantitatively what is going on by noting that the heat which leaves the gas at T_2 ($dQ = -Q$ is negative, because the heat is leaving the system T_2) is the same as the heat which enters the gas at T_1 ($dQ = +Q$ is positive because the heat is entering the system T_1). The temperature of the gas which starts at T_2 will always be between T_2 and T_f and the gas which starts at T_1 will be between T_1 and T_f. We can then approximate the integrals as: $\Delta S_1 = Q/T_{1+}$ and $\Delta S_2 = Q/T_{2-}$ where T_{1+} is some temperature between T_1 and T_f and T_{2-} is a temperature between T_2 and T_f. Since $T_{2-} > T_{1+}$, this means that $|\Delta S_1| > |\Delta S_2|$ so the total change in entropy $\Delta S = \Delta S_1 + \Delta S_2 = Q/T_{1+} - Q/T_{2-} > 0$. **The entropy of the entire system increases in this spontaneous process of heat transfer from a warmer to a colder object.** The process is *irreversible*, as the systems cannot be restored to their original temperatures without interacting with an external system, with a larger increase in entropy.

Question 12.6

Do the air conditioner and the refrigerator violate the Second Law of Thermodynamics since they cool the air inside the room and inside the refrigerator, thus decreasing the entropy inside those spaces? Justify your answer.

With this introduction, we can now understand what is wrong with our ambient-water-temperature-powered ship engine. As shown in Figure 12.12, the engine works by extracting heat Q from the water, reducing its temperature from T_2 to T_1. That heat is then used to provide work W that runs the ship. However, if we look at the change in entropy $\Delta S = \int \frac{dQ}{T}$ we see that $\Delta S < 0$ because dQ is negative since the heat Q is leaving the water. Since the work done W does not change the entropy, there is no corresponding increase in entropy and the process results in a decrease in the entropy of the universe. As such, the process is physically unrealizable, since it results in the system transitioning spontaneously from a more probable state to a less probable state!

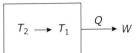

Figure 12.12 Schematic of the operation of "perpetual motion" drive of the boat in Figure 12.9. Heat is extracted from the ocean water, reducing its temperature to T_1. The removed heat is then converted to work to drive the boat. This process decreases the entropy of the universe and is thus forbidden by the Second Law of Thermodynamics.

12.11   Heat engines

So, if we can't simply extract heat from a system and convert it directly into work, how does a real heat engine work? As shown in Figure 12.13, a heat engine operates between two heat reservoirs at T_2 and T_1 with $T_2 > T_1$. It is assumed that the heat reservoirs are large enough that their temperatures are not affected by the heat Q_2 and Q_1 that are transferred from and to them. (Think of a fired furnace at T_2 and a cooling tower at T_1.) The work extracted by the heat engine is the difference between the heat Q_2 extracted from the hot reservoir and the heat Q_1 transferred to the cool reservoir.

The net change in entropy must be the sum of the component entropy changes. The change in entropies in the heat reservoirs are:

$$\Delta S_2 = - Q_2/T_2, \quad \Delta S_1 = + Q_1/T_1.$$

There is no entropy change associated with the work extracted.

Since $W = Q_2 - Q_1 \Rightarrow Q_1 = Q_2 - W$, we can use this relation to eliminate Q_1.

$$\Delta S_2 = - Q_2/T_2, \quad \Delta S_1 = + (Q_2 - W)/T_1$$

$$\Delta S_{TOT} = \Delta S_2 + \Delta S_1 = \frac{Q_2 - W}{T_1} - \frac{Q_2}{T_2} = Q_2 \left(\frac{1}{T_1} - \frac{1}{T_2} \right) - \frac{W}{T_1}.$$

From the Second Law, the *best* that you can do in a process is $\Delta S_{TOT} = 0$. From the equation above, $\Delta S_{TOT} = 0$ **gives a result for the *Carnot efficiency*, the work extracted divided by the heat extracted from the high temperature reservoir,** as

$$\frac{W}{Q_2} = T_1 \left(\frac{1}{T_1} - \frac{1}{T_2} \right) = 1 - \frac{T_1}{T_2}. \qquad (12.32)$$

The Carnot efficiency is the theoretical upper limit of a heat engine operating between temperatures T_2 and T_1.

As an example, the Carnot efficiency of a heat engine operating between 800 K and 300 K is $1 - \frac{300}{800} = 62.5\%$. If the high-temperature reservoir is at 1000 K and the low-temperature reservoir is at 300 K, the Carnot efficiency is $1 - \frac{300}{1000} = 70\%$. The efficiency of real heat engines is less than the Carnot efficiency, but approaches this limit in well-designed systems. The energy extracted from the high-temperature reservoir and transferred to the low-temperature reservoir is lost to the process, but this lost energy is required to ensure that the heat engine has a total entropy change which is not negative.

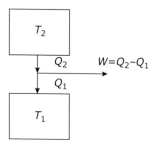

Figure 12.13 Schematic of the operation of a real heat engine. An amount of heat Q_2 is removed from the high-temperature reservoir at T_2 and a smaller amount of heat Q_1 is transferred to a low-temperature reservoir at T_2. Work equal to the difference between Q_2 and Q_1 is extracted from the engine.

12.12   The Third Law of Thermodynamics

Clearly there is a problem with the definition $dS = \frac{dQ}{T}$ if T goes to zero or becomes negative. There is an obvious problem if $T < 0$ since adding heat to a system would result in a decrease in entropy. So, negative Kelvin temperatures are physically unrealizable. The definition of dS is also a problem if $T \rightarrow 0$ K, unless dQ also equals zero: we can't extract any heat from a system at T. Thus, 0 K is the lowest temperature possible, *absolute zero*. Due to the quantum zero point energy – for example, for a harmonic oscillator $E = \left(n + \frac{1}{2}\right)\hbar\omega_0$ and the minimum energy is $\frac{1}{2}\hbar\omega_0$ – the energy of a system does not necessarily go to zero as $T \rightarrow 0$ K. However, the entropy of the system does go to zero as $T \rightarrow 0$ and the system approaches a single ground state of lowest energy. This result, that **zero degrees on the Kelvin scale is the lowest possible temperature and represents a state of zero entropy**, is referred to as the *Third Law of Thermodynamics*.

    If the First Law tells us "you can't win" and the Second Law tells us "you can't even break even," the Third Law can be thought of as telling us "you can't get out of the game." There is no temperature regime where the total change in entropy of the system plus the environment can be made negative in a physically realizable system.

> **Question 12.7**
>
> Consider what we know about quantum electrons in a metal with n free electrons/m³ from Chapter 10. At $T = 0$ K will all the electrons have zero energy? Will they all be at rest? Why not? How many distinguishable states will the electrons in the metal have at $T = 0$ K? From Eq. (12.30), what is the entropy?

12.13   Free energy

A system in equilibrium with a heat reservoir at temperature T that could exist in two or more states will spontaneously transition to the state that represents the most probable state: the state of highest entropy. This represents a change in the way we normally think about systems with multiple chemical components or with more than one possible phase evolving to their lowest energy state. In the "real world," it is relatively uncommon for a system to be in its lowest energy state. Even if we use enthalpy to allow for the First Law equivalence of work and energy, systems are not usually found in their lowest enthalpy state.

We are actually quite familiar with this fact. We know that water as a solid is a lower energy state than as a liquid – ice requires the addition of 80 calories per gram to make the transformation to liquid water. Nevertheless, we know that thermal contact with a heat reservoir at a temperature above 0 °C, ice will freely absorb 80 calories per gram from the environment and melt to the higher energy liquid state. On the other hand, below 0 °C the water will transfer 80 calories per gram to the heat reservoir and transform to solid ice.

We need, of course, to consider both the system and heat reservoir. Between the two the energy is conserved, so the total energy is the same before and after. It's just a matter of comparing the number of states available to the system and to the reservoir and considering which state is the most statistically probable. A given quantity of heat dQ leaving or entering the heat reservoir changes the entropy of the reservoir by $S_r = dQ/T$. (dQ and, consequently, S_r are positive if heat enters the heat reservoir, negative if heat leaves the reservoir.) If we compare this with the change in entropy of the *system* and apply the Second Law condition that the entropy of the universe cannot decrease in any process, we can determine the equilibrium state of the system at any temperature. This will involve both the change of energy of the system (which determines the heat entering or leaving the reservoir), the temperature of the reservoir, and the change in entropy of the system itself as it changes phase or undergoes a chemical reaction.

These three considerations can be brought together in the concept of *free energy*. The *Helmholtz free energy* is defined by

$$F = E_{int} - TS \tag{12.33}$$

while the change in the F in a process at a fixed temperature (appropriate if the thermal reservoir is large compared to the system) is

$$\Delta F = \Delta E_{int} - T\Delta S. \tag{12.34}$$

In this equation ΔE_{int} represents the change in internal energy of system, which we have seen can be caused (at fixed temperature) by a change in phase (solid to liquid to gas, or vice versa) or a change in chemical bonding, and ΔS represents the change in entropy of the system. Both ΔE_{int} and ΔS can be calculated at a fixed temperature T for the specific system if the nature of the change in phase or bonding is specified. The significance of the combination is that if the system does no work, then $\Delta E_{int}/T = -Q_r/T = -\Delta S_r$ where Q_r is the heat transferred to the reservoir and ΔS_r is the change in entropy of the reservoir. The Helmholtz free energy compares the change in entropy of the system with the change in entropy of the thermal reservoir (both multiplied by the temperature T to give units of energy). Obviously the only permissible equilibrium state is one where the change in entropy of the reservoir plus the change in entropy of the system is positive.

$$\Delta S_{Tot} = \Delta S_r + \Delta S = -\Delta F/_T > 0. \tag{12.35}$$

Thus the system will be in equilibrium when it reaches a state of minimum Helmholtz free energy (maximum entropy).

Obviously, if the change in entropy of the system is negative (the system transitions to a more ordered state) and the change in system energy is positive (requiring energy to be transferred from the reservoir), the process will always decrease the total entropy and the process will never occur spontaneously. Conversely, a process that results in decreased system energy and also increases the entropy of the system will always be thermodynamically favored. If the energy of the system increases and the entropy increases (such as when ice melts) or if the energy decreases and the entropy decreases (as when steam condenses), the equilibrium state of the system depends on the details of the process and the temperature at which the system is held.

Question 12.8

Explain why ice melting increases both the energy and the entropy of the system (the volume of ice). How do these changes compare with the change in energy and entropy of the environment in thermal contact with the ice? Does this depend on the temperature of the environment?

In calculating the change in Helmholtz free energy in Eq. (12.34), we specified that no work was done by the system. This is appropriate if the volume is held constant because in that case $W = P\Delta V = 0$. However, a more common experimental condition is that the volume of a system is free to change but the pressure is constant, typically at a pressure near 1 atmosphere $= 1.013\times10^5$ Pa. In this case the appropriate quantity to consider is the *Gibbs free energy* (or the *Gibbs energy*)

$$G = E_{int} + PV - TS = H - TS \tag{12.36}$$

where $H = E_{int} + PV$ is the enthalpy that we introduced earlier in the chapter.[4] The change in the Gibbs energy of a system in equilibrium with a reservoir at constant T is given by

$$\Delta G = \Delta H - T\Delta S \tag{12.37}$$

where $-\Delta H = Q_r$ the net heat (change in system internal energy minus work done) transferred to the reservoir under the condition of constant pressure, increasing the entropy of the reservoir by $\Delta S_r = Q_r/_T$. A system under constant pressure will be in equilibrium when it reaches a state of minimum Gibbs free energy.

Tables of the Gibbs energy for chemical compounds, typically in units of kJ/mole or kcal/mole at $T = 25\ °C$ (= 298.15 K) and $P = 1$ bar (1 bar $= 1 \times 10^5$ Pa $= 0.987$

---

[4] See, for example, www.chem1.com/acad/webtext/thermeq/TE4.html for a discussion of Gibbs energy.

atmosphere), are published and can be consulted to compare the total Gibbs energy of chemical reactants with the total Gibbs energy of the products to determine the equilibrium state of the system. Since the change in internal energy (and the heat absorbed from or transferred to the thermal reservoir) can be found from the heat capacity of the material, the Gibbs energy can be found as a function of temperature by integrating over C_p (the heat capacity *at constant pressure*) to find the new enthalpy

$$H(T) = H(T_o) + \int_{T_o}^{T} C_p \, dT \,, \tag{12.38}$$

and the entropy at a new temperature can be found by integrating $\frac{dQ}{T} = C_V \frac{dT}{T}$

$$S(T) = S(T_o) + \int_{T_o}^{T} \frac{C_V}{T} \, dT \tag{12.39}$$

where C_V is the heat capacity *at constant volume*. If a material undergoes a phase transition at a critical temperature T_c in the temperature range between T_o and T, the latent heat of the transformation L_H needs to be added or subtracted from the enthalpy and an entropy change $\Delta S = {L_H}/{T_c}$ added or subtracted from the entropy, depending on whether the transition is endothermic (absorbing energy from the reservoir and increasing the system enthalpy and entropy) or exothermic (transferring energy to the reservoir and decreasing the system enthalpy and entropy).

The enthalpy, entropy, and Gibbs energies of materials at 298 K (25 °C) can be found from the National Institutes of Standards and Technology at http://kinetics.nist.gov/janaf/. The Gibbs energy is usually defined with respect to a zero defined as the Gibbs energy of elements in their most stable form at 25 °C and a pressure of 1 bar. This gives a different value than $G = H - TS$ in the website. However, the difference in Gibbs energy between two states, the critical parameter for any reaction, is not changed by adding a constant offset energy.

Question 12.9

Explain why the change in enthalpy in Eq. (12.38) uses the heat capacity at *constant pressure*, whereas the change in entropy in Eq. (12.39) uses the heat capacity at *constant volume*. (The definition of the change in enthalpy is given in Eq. (12.16) whereas the change in entropy is defined in Eq. (12.31).)

Example 12.4 Dissociation of dinitrogen tetroxide

Gaseous N_2O_4 can dissociate into two molecules of NO_2 by the following balanced reaction: $N_2O_4 \rightleftharpoons 2NO_2$. The enthalpy, Gibbs energy, entropy, and molar heat capacities of the two compounds at 298 K and 1 atmosphere are listed below. What is the

Continued

Example 12.4 (*cont.*)

stable equilibrium phase at 298 K? Assuming that the specific heat is temperature independent, is there a temperature at which the other phase is the favored?

Compound	H (kJ/mole)	G (kJ/mole)	S (J/K mole)	C_P (J/K mole)
NO_2 (gas)	33.10	51.26	240.04	36.97
N_2O_4 (gas)	9.08	97.79	304.38	77.26

Solution: Note that **one** mole of N_2O_4 dissociates into **two** moles of NO_2. The Gibbs energy of one mole of N_2O_4 is given by the table as 97.79 kJ, while the Gibbs energy of two moles of NO_2 is $2 \cdot (51.26) = 102.52$ kJ. The equilibrium state is the state at the lower Gibbs energy, which at 298 K is dinitrogen tetroxide.

The heat capacity of two moles of NO_2 is $2 \cdot 37.93 = 75.86$ J/K, slightly less than the heat capacity of N_2O_4, so as the temperature increases the enthalpy of the NO_2 will decrease with respect to the N_2O_4 making NO_2 more favorable. Using Eqs. (12.37) and (12.38), and assuming that the heat capacity is constant, we find

$$H(T) = H(298 \text{ K}) + C_p(T - 298)$$

$$S(T) = S(298 \text{ K}) + \int_{298}^{T} \frac{C_V}{T} \, dT = S(298 \text{ K}) + C_V \ln\left(\frac{T}{298.15}\right)$$

$$= S(298 \text{ K}) + (C_p - R) \ln\left(\frac{T}{298.15}\right)$$

using the ideal gas relation between the molar heat capacities at constant volume and at constant pressure $C_p = C_V + R$ from Section 12.6. Setting $G = H - TS$ equal for two moles of NO_2 and one mole of N_2O_4

$$9080 + 77.26(T - 298) - T\left(304.38 + 68.95 \ln\left(\frac{T}{298}\right)\right)$$

$$= 2(33100) + 2(36.97)(T - 298) - 2T\left(240.04 + 28.66 \ln\left(\frac{T}{298}\right)\right)$$

Solving graphically, we find that NO_2 will be the thermodynamically more stable state at temperatures above 326 K. We can also estimate the equilibrium temperature by recognizing that at equilibrium the change in Gibbs energy between the two phases is zero. This gives

$$\Delta G = \Delta H - T\Delta S = 0.$$

$$\Rightarrow T_{eq} = \frac{\Delta H}{\Delta S} = \frac{9080 - 2(33100)}{304.38 - 2(240.04)} = 325 \text{ K}.$$

We note two subtleties of this analysis. First, even given that a particular phase has the lower Gibbs energy and thus is thermodynamically favored at a particular temperature, this does not imply that the reaction rate to convert from the higher Gibbs energy state to the lower Gibbs energy state will be high enough that the system will actually convert to the thermodynamically favored state in a finite time. The classic case of this is solid carbon which exists at 298 K and one atmosphere in two states, graphite and diamond. The diamond form has a Gibbs energy which is 3 kJ/mole higher than graphite, implying that graphite is the thermodynamically stable state at 298 K. This is true, but the reaction rate for the conversion of diamond to graphite is so slow that for all practical purposes "diamonds are forever." The details of the reaction kinetics are such that the atoms in a diamond lattice would have to overcome an enormous energy barrier to convert to the graphite structure. This issue of very low reaction rates has been expressed as "if thermodynamics says 'No' the answer is 'No'; if thermodynamics says 'Yes' the answer is 'Maybe'."

Second, we should not think that, even if the product state is thermodynamically favored and the reaction rate is relatively rapid, the reactants will completely convert to the products. This result follows from the consideration that there are many more states where molecules of the reactants are mixed with molecules of the products than the number of states where the only molecules in the system are the reactant molecules, so entropy considerations should imply that reactants and products will coexist in the equilibrium state.

We can make this insight quantitative by noting that the number of states accessible to a system of a fixed number of molecules n is directly proportional to the volume available to the system since the molecules can arrange themselves in the added space in the same number of configurations per volume as they did in the previous space thus increasing the number of states of the system by the increase in volume. Thus the number of states Ω available to the molecules scales with the volume $\Omega = CV$ where C is a function of other variables (such as temperature) – not including the volume. This implies that the change in entropy when a system of n molecules is allowed to increase its volume from V_o to V_1 is

$$\Delta S = nk_B \ln V_1 - nk_B \ln V_o = n\,k_B \ln\left(\frac{V_1}{V_o}\right) = NR \ln\left(\frac{V_1}{V_o}\right) \tag{12.40}$$

where the last equality results from converting the number of particles n to the number of moles N. At a fixed temperature, the Ideal Gas Law tells us that the volume is inversely related to the pressure $V = {}^{NRT}\!/_P$ so the change in entropy for an ideal gas is

$$\Delta S = NR \ln\left(\frac{P_o}{P_1}\right). \tag{12.41}$$

For a mixture of gases, we can use the partial pressure of each gas as a measure of its volume. The change in the molar Gibbs energy as a function of the partial pressure of the gas is

$$\Delta G = \Delta H_{\text{conc}} - T\Delta S = \Delta H - RT \ln\left(\frac{P_o}{P_1}\right). \tag{12.42}$$

If we assume that the molecules are non-interacting so there is no change of enthalpy with concentration and recall that the standard reference pressure for the Gibbs energy G^o is $P_o = 1$ bar $(1$ bar $= 10^5$ N/m$^2 \simeq 1$ atmosphere), we find the contribution to the molar Gibbs energy from a gas at partial pressure P_1 measured in bars is

$$G_1 = G^o + RT \ln(P_1). \tag{12.43}$$

If we consider a reaction where N_A moles of reactant A and N_B moles of B react to form N_C moles of product C and N_D moles of product D, we can find the change in the Gibbs energy in the reaction by finding the difference in the standard Gibbs energies with the *mixing correction* due to the changing concentrations (partial pressures) of each species

$$\Delta G = N_C(G^o{}_C + RT \ln(P_C)) + N_D(G^o{}_D + RT \ln(P_D)) - N_A(G^o{}_A + RT \ln(P_A))$$

$$- N_B(G^o{}_B + RT \ln(P_B)) = \Delta G^o + RT \ln\left(\frac{P_C{}^{N_C} P_D{}^{N_D}}{P_A{}^{N_A} P_B{}^{N_B}}\right)$$

$$= \Delta G^o + RT \ln(Q)$$

$$\tag{12.44}$$

where we have introduced the change in the standard Gibbs energy ΔG^o and the *reaction quotient* $Q = \frac{P_C{}^{N_C} P_D{}^{N_D}}{P_A{}^{N_A} P_B{}^{N_B}}$. In equilibrium the Gibbs energy is at a minimum, so $\Delta G = 0$, and the reaction quotient in equilibrium is defined as K the *equilibrium constant*. Setting $\Delta G = 0$ and solving for Q in Eq. (12.44), we find

$$K = Q_{eq} = \left(\frac{P_C{}^{N_C} P_D{}^{N_D}}{P_A{}^{N_A} P_B{}^{N_B}}\right)_{eq} = e^{-\frac{\Delta G^o}{RT}}. \tag{12.45}$$

Example 12.5 Find the equilibrium concentration of N_2O_4 and NO_2 at equilibrium at 298 K.

Solution: From Example 12.4, the difference in the Gibbs energy of one mole of N_2O_4 at 298.15 K (25 °C) and the Gibbs energy of two moles of NO_2 is $\Delta G = 2(51.26) - (97.79) = 4.73$ kJ. The equilibrium constant is

$$K = e^{-\frac{\Delta G^o}{RT}} = e^{-\frac{4730}{8.314 \cdot 298}} = 0.148.$$

The ratio of the partial pressures (proportional to the number of moles) of NO_2 and N_2O_4 is given by $\frac{(P_{NO_2})^2}{(P_{N_2O_4})} = 0.148$. For every mole of N_2O_4 at equilibrium at 298 K there will be $\sqrt{0.148} = 0.385$ moles of NO_2.

This result can be generalized to non-ideal gases and to solutions by replacing the partial pressures in Eq. (12.44) by the *chemical activities* of the reactants and products. The chemical activity takes the intermolecular interactions into account. For relatively dilute solutions, the chemical activity is approximately the molar concentration of the reactants and products.

12.14   Heat transfer mechanisms

Heat is the energy that flows as a result of temperature difference between two objects and can be transferred via three mechanisms: *conduction*, *convection*, and *radiation*. Each of these mechanisms is discussed in some detail below.

12.14.1   Conduction

Conduction occurs when the two bodies at different temperatures are in direct contact. In this case, the higher average kinetic energy of molecules in the hotter body is transferred by contact to the cooler body by molecule-to-molecule collisions. In one dimension, the transfer of energy, measured in joules/s (watts), across a distance L is given by

$$P_{cond} = k_{th} A \frac{T_2 - T_1}{L} \Rightarrow P_{cond} = k_{th} A \frac{dT}{dx} \qquad (12.46)$$

where k_{th} is the *thermal conductivity*, A is the cross-sectional area of the conducting material, and the *thermal gradient* $\frac{dT}{dx}$ is the differential limit of the temperature difference over the length of the section as shown in Figure 12.14. The thermal conductivities of various materials are listed in Table 12.8.

In examining Table 12.8, we notice that, in general, metals are highly thermally conducting, whereas insulators are weak thermal conductors, as confirmed by everyday experience: a metal fork placed in a charcoal cooker will become quite hot, while the wooden handle will remain cool. In fact, thermal conductivity is strongly correlated with electrical conductivity. This suggests that the free electrons in metals contribute to the thermal conductivity: free electrons in the hotter part of the metal pick up kinetic energy from the lattice and travel freely to the cooler part of the metal. On the other hand, insulators, which lack free conduction electrons, conduct heat through the interaction of lattice vibrations. A quantum excitation of the lattice is referred to as a

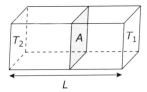

Figure 12.14 Geometry of linear heat transfer problem. A sample of length L and cross-sectional area A is held with one end at T_2 and the other end at T_1. The heat transferred per second across the sample is given by Eq. (12.46).

Table 12.8 Thermal conductivities of various materials

Material	Thermal Conductivity k_{th} (W/m K) at 300 K
Aluminum	235
Brass	109
Copper	401
Iron	67
Lead	35
Stainless steel	17
Diamond	2200
Silicon	149
Gallium arsenide	56
Sodium chloride	6.5
Limestone	1.3
Fused quartz (silica)	2.0
Window glass	1.0
Wood (white pine)	0.11
Plastic	0.5
Polyurethane foam	0.024
Nitrogen	0.025
Helium	0.15
Hydrogen	0.18

phonon. The efficiency of thermal transport through lattice vibrations depends on the phonon excitation spectrum and how easily phonons propagate through the lattice.

Crystalline diamond stands out in Table 12.8 because of its extremely high thermal conductivity despite being an insulator. Diamond is characterized by high hardness and small atomic mass resulting in high-frequency molecular vibrations. The resulting high-energy phonon spectrum and the weak phonon scattering contribute to the exceptional high thermal conductivity.

In three dimensions, thermal conductivity is described in vector calculus notation by

$$-\nabla \cdot (k_{th} \nabla T) = Q_{pd} \tag{12.47}$$

where Q_{pd} describes the power density in watts/m^3 associated with any heat sources in the material. Using the divergence theorem, we find

$$-\int \nabla \cdot (k_{th} \nabla T) \, dV = -\oint k_{th} \nabla T \, dA = \int Q_{pd} \, dV = P_T \tag{12.48}$$

which says that the total power (in watts), P_T produced within a volume is equal to the thermal conductivity times the temperature gradient, integrated over the area of the surface. When the problem depends only on the radial direction r, we find that $P_T = -\oint k_{th} \frac{\partial T}{\partial r} \, dA \rightarrow -k_{th} \frac{\partial T}{\partial r} A$ where A is the total area of the surface. The negative sign results because $\frac{\partial T}{\partial r}$ must be negative (temperature decreasing with radial distance) for power to be conducted out of the volume.

12.14.2   Convection

When heat is produced in a fluid, it may be transported either by molecule-to-molecule conduction or by the movement of macroscopic volumes of heated fluid, a process referred to as *convection*. Convection caused by the circulating flow of the fluid in macroscopic *convection cells* is usually a much more efficient heat transport mechanism than conduction. Despite their name, household radiators heat the surrounding area primarily through convection – air heated at the radiator rises due to the reduced density, flows across the room cooling as it does, then sinks and flows back across the floor to the radiator. Calculation of convective transport is quite complicated, requiring simultaneous calculation of thermal transport and fluid flow. Often an effective convective thermal transport coefficient h is used with heat conduction formulas to get an approximate result for convective thermal transport. The units of h are W/m^2 compared to $W/m\ K$, the units of k_{th}. The missing factor of distance reflects the fact that the moving fluid is always in direct contact with the heat source. The relation between the heat transferred and the distance L between the heat source and the heat sink is a complex fluid mechanics problem depending on the size of "convective cells" in the fluid. In some cases heat is transferred more efficiently when L is larger!

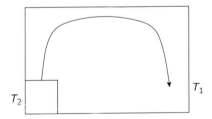

Figure 12.15 Convective heat transfer results from the movement of hot cells of heated fluid through the fluid from the high-temperature side T_2 to the low-temperature side T_1.

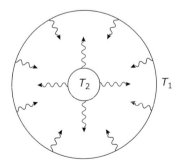

Figure 12.16 Heat transfer by radiation. The electromagnetic (e.g., infrared) radiation of the body is determined by its temperature T_2. If T_2 is higher than the temperature T_1 of the environment, there will be a net heat transfer by radiation from the body to the environment.

12.14.3   Radiation

The final thermal transport mechanism is radiation. Radiation is the electromagnetic radiation, primarily in the infrared region of the spectra, which is emitted from any body with $T > 0$. If the total heat emitted by a body and absorbed by the surroundings is greater than the heat emitted by the environment and absorbed by the body, there will be a net flow of heat from the body to the environment. Note that heat transfer by radiation does not require a physical medium between the heated body and the environment; radiation is the only heat transfer mechanism that will transport heat through a vacuum. In fact, radiation in the infrared, visible, and ultraviolet spectral range is the mechanism by which the Earth is heated by the Sun through 93 million miles of space.

The emission of electromagnetic radiation by a heated body will be considered in some detail in the next chapter. **The integrated power density emitted by a body of area A at temperature T** is given by the *Stefan–Boltzmann Law*

$$P_{rad} = \sigma \varepsilon A T^4 \tag{12.49}$$

where P_{rad} is the total radiated power in watts, $\sigma = 5.67 \times 10^{-8} \ \frac{W}{m^2 \cdot K^4}$ is the *Stefan–Boltzmann constant,* A is the surface area of the body, and T is the temperature of the body in kelvin. The *emissivity* ε is a property of the material of the surface of the radiating body. An object which has an emissivity of 1 is called a *blackbody radiator* and radiant heat transfer is often called blackbody radiation. The net heat flow from a body will depend on the total power radiated by the body, minus the power absorbed from the radiation of the environment. An object in thermal equilibrium with its environment (with the object temperature equal to the environment temperature) radiates exactly as much energy as it absorbs from the environment, a condition that requires that the *absorptivity* α of the object surface must be exactly equal to the emissivity of the surface ε.

Example 12.6 Find the net transfer of heat by radiation from a water tank with a surface area of 25 m^2 at night in the desert. Assume that the tank has a temperature of 90 °F from being in the Sun all day. Assume in addition, the desert floor has an effective temperature of 75 °F and the sky has an effective temperature of 40 °F (both measured assuming an emissivity of 1.0). The emissivity of the tank is 0.5.

In kelvin, the temperature of the tank, ground, and sky are respectively 305 K, 297 K, and 277 K. From the Stefan–Boltzmann Law, the emission of the tank is $P_{tank} = \sigma \varepsilon A T^4$ $= 5.67 \times 10^{-8} \cdot 0.5 \cdot 25 \cdot 305^4 = 6133$ W. If we assume the bottom half of the tank is exposed to 2π steradians of emission from the ground at 297 K, the power absorbed from the ground is $P_{ground} = \alpha_{tank} \sigma \varepsilon_{ground} A T^4 = 0.5 \cdot 5.67 \times 10^{-8} \cdot \frac{25}{2} \cdot 297^4 = 2757$ W. The power absorbed from the sky is $P_{sky} = \alpha_{tank} \sigma \varepsilon_{sky} A T^4 = 0.5 \cdot 5.67 \times 10^{-8} \cdot \frac{25}{2} \cdot 277^4 = 2086$ W.

The net emission from the tank is thus $6133 - 2757 - 2086 = 1290$ W.

Chapter 12 problems

1. How much heat does a human being radiate in a day if the outside temperature is 0 °C? How does this compare with the average energy from food intake in a day? (Assume that the person wears no clothes.) Can people lose weight by staying in a cold room?

2. If we have 30 people in a room, how long will it take to increase the room temperature by 2 °C? Estimate the dimensions of the room and make other necessary reasonable assumptions.

3. Where did the energy stored in oil and coal come from originally? Discuss the process.

4. From where does the energy in the hurricane come? How is it stored and released? Estimate the amount of energy stored in/released by Hurricane Sandy. How does this compare with the energy used by New York City in a day?

5. When you lose 10 kg, where does it go? How much weight do you lose by just breathing (no other function is performed) for a day (24 hours)?

6. Explain the following phenomena:
   a. the cloud formation when you open a beer bottle

   b. the fact that we can blow hot and cold air through our mouth

   c. the *desert effect*: one side of a mountain is covered with vegetation while the other has very little or no vegetation (i.e., the Rocky Mountains)

   d. the fact that it is colder on the top of a mountain than it is at the bottom, despite the top being closer to the Sun.

7. What would the equipartition theorem predict for the specific heat per mole for methane (chemical formula CH_4 – a carbon nucleus surrounded by four hydrogen atoms in a tetrahedral shape)? Compare your prediction with the room temperature specific heat (at constant volume) you find on-line.

8. Consider 2 moles of an **ideal diatomic** gas at 300 K and atmospheric pressure $P_o = 1 \times 10^5$ N/m². Assume the Ideal Gas Law holds ($PV = NRT$) with $N =$ the number of moles (2) and the universal gas constant $R = 8.314$ J/mole K.
   a. Using the equipartition theorem, what is the total internal energy, E_{int} (300 K)?

   b. What is the molar specific heat at constant volume, $c_V = \frac{1}{N} \frac{\partial E_{int}}{\partial T}$?

   c. If $Q = 3000$ J of energy is added to the gas as heat with the volume kept the same, what is the new temperature T? What is the new pressure? What is the total internal energy, E_{int} (T)?

   d. The gas is allowed to expand isothermally (heat is added to keep the temperature constant at T) until it is twice the volume. What is the total internal energy? How much work was done in this step? How much heat was added?

9. From what you know about spring–mass oscillation, can you make an argument why gas molecules with three or more atoms would have lower vibrational frequencies than molecules with two atoms?

10. Explain why in gases the specific heat at constant pressure c_p is higher than the specific heat at constant volume c_V. Would you expect there to be a large difference between c_p and c_V in solids? Why or why not?

11. In a metal, one might easily ask, why don't the free electrons in metals contribute $\frac{3}{2}k_BT$ per electron to the energy and $\frac{3}{2}k_B$ to the specific heat per electron (in addition to the degrees of freedom of the atoms in the lattice), thus raising the specific heat of metals above that of insulators. This question troubled physicists in the nineteenth century, and, as often is the case, the answer is found in quantum theory. If we imagine the electrons in a metal as filling up all states in the "Fermi sphere" with wave vector $k < k_f = \sqrt{2mE_f}/\hbar$, there are no states available for most of the electrons to transition to if additional heat is added to the metal. Only the electron states within about k_BT of the Fermi energy can adjust their energy in response to heat added to the system, because only those electrons have any appreciable density of states to transition to.

   Consider a metal with a Fermi energy $E_f = 5$ eV. Find the volume of k-space with $<k_f$. (Your answer will be units of $1/m^3$.) Convert this to a density per cubic meter of electron states, assuming there are two electron states (one spin up and one spin down) for each volume $(2\pi)^3$ of k-space. Compare this with the volume of k-space and number of states within k_BT of the Fermi energy at room temperature (300 K).

12. Consider a 3 cm column of mercury inside a glass tube 0.1mm in radius (without the bulb on the bottom). (a) How much would the mercury column move in mm/°C given that the mercury has a bulk expansion coefficient $\beta = 181 \times 10^{-6}/°C$ and neglecting the linear coefficient of expansion of the glass? b) How much would your answer change if you assume the linear coefficient of expansion of the glass is $\alpha = 3.2 \times 10^{-6}/°C$? (Remember that the radius of the column as well as its length will increase with temperature with the same α.)

13. An 0.5 kg sample of ice (specific heat $c = 0.5$ cal/(degree/g)) at $T = -10$ °C is heated by adding a constant power of 50 W.
   a. How long will it take to get the ice to 0 °C?

   b. How long would it take at a constant power of 50 W to melt the ice? (The *latent heat of fusion* of water is 80 cal/g.)

   c. How long would it take to heat the melted ice to 10 °C (specific heat of water $c = 1$ calorie/(degree/g))?

14. A person makes a drink by pouring 50 g of water at 20 °C over 50 g of ice initially at -5 °C. Assume the glass is well-insulated.

a. How much ice is left after the water and ice come to equilibrium at $0\ °C$?

b. How does you answer change if instead of water the liquid is 50 g of 60 proof whiskey (30% alcohol)?

15. One mole of an ideal monatomic gas with an equation of state $PV = RT$ begins at pressure $P = 1 \times 10^5\ N/m^2$ and temperature $T = 300\ K$ and is taken through the following cycle:

a. 2000 J of heat is added at constant volume

b. the gas is allowed to expand isothermally (at fixed temperature) until its pressure reaches its initial value

c. work is done at constant pressure to compress the gas and heat is removed until the gas returns to its original state ($P = 1 \times 10^5\ N/m^2$ and $T = 300\ K$).

How much does the temperature of the gas increase in the first step? How much work is done by the gas in the second step of the process? How much work is done and how much heat is removed in the third step? How much net work is done by the gas? Using the First Law of Thermodynamics, find the amount of heat added to the gas in the second step.

16. A 1.0 kg aluminum fixture is suspended in a vacuum chamber at an initial temperature $T = -50\ °C$. A 15 cm long copper rod 3 mm × 4 mm in cross-section connects the fixture to a large object fixed at $20\ °C$. How much power is transferred by conduction through the copper rod? If the aluminum has a surface area of $0.05\ m^2$, how much power would be transferred by radiation assuming an emissivity of 1.0? How long would the fixture take to warm to $15\ °C$?

17. In a thermally isolated system, 200 g of aluminum at $100\ °C$ is mixed with 50 g of water at $20\ °C$. What is the final temperature of the mixture? What is the entropy change of the aluminum? What is the entropy change of the water? What is the total entropy change of the system?

18. It takes 80 calories/g to melt ice at $0\ °C$ to water at $0\ °C$. One gram of ice has a volume of about $1.09\ cm^3$, and in the process of melting the volume decreases by about 9%. Does the liquid state have more entropy than the solid state? How much does the entropy change at atmospheric pressure ($P_{at} = 1.013 \times 10^5\ N/m^2$)? Assuming that the heat of fusion (80 calories/g) and melting point ($0\ °C = 273\ K$) don't change, how much would the entropy change be affected if the ambient pressure were doubled?

19. Consider the following processes: The temperatures of two identical gases are increased from the same initial temperature to the same final temperature using reversible processes in both cases. If for gas A, the process is carried out at constant volume, while for gas B it is carried out at constant pressure, in which process is the change in entropy the greatest? Justify your answer.

20. Calculate (estimate) the amount (mass) of sweat you generate when bicycling for 10 miles. Show your assumptions and calculations.

21. One mole of an ideal gas expands slowly and isothermally at temperature T until its volume is doubled. What is the change of entropy of this gas for this process?

22. 2 kg of water at $T_1 = 30\,^\circ$C is mixed with 4 kg of water at $T_2 = 90\,^\circ$C at constant pressure in a calorimeter of negligible heat capacitance. By how much has the entropy changed when the system reaches thermal equilibrium?

23. An eccentric inventor comes to you with a new useful material that he says has an emissivity that is high in the infrared and low in the visible, while its absorptivity is high in the visible and low in the infrared. Do you think this is possible? Why or why not?

24. An aluminum plate 30 cm in diameter and 1.0 cm thick is heated by an electric heating element with a power of 100 W.

 a. Assuming the top of the plate radiates to the sky with an effective temperature of 270 K and the bottom half of the plate radiates to the ground with an effective temperature of 300 K, what is the equilibrium temperature of the plate ignoring conduction cooling from the air? You can approximate the emissivity of the aluminum to be $\varepsilon = 0.2$. Do you think conductive cooling from the air would be important in this case?

 b. Assuming the water boils away quickly, how much is the temperature of the plate reduced instantaneously if 5 g of water at 20 $^\circ$C is poured on the plate?

13 Quantum statistics

As a young professor of electrical engineering, the first author returned for a visit to the institution where he had received his PhD in Physics. His former thesis advisor, a professor of physics, was on his way to be part of a PhD oral qualifying exam committee for students in their second year of graduate study in physics. "What should I ask this guy [the qualifying exam student]?" he asked. "Why don't you ask him how a p-n junction diode works?" your first author, who had just taught a course in physical electronics, suggested. His former advisor stopped in mid-stride and said "What, do you want him to faint dead away?"

It is our suspicion that not only physics graduate students but most engineers (including electrical engineers!) would have the same difficulty in describing the operation of a p-n junction, the most fundamental and ubiquitous device in all semiconductor electronics. So how *does* a p-n junction, the basis of virtually every electronic device that we use every day, work? And what does that have to do with the red-hot color of a burner on an electric stove or the radiation from the Sun? These are some of the questions we will address in this chapter where we look at statistical applications of quantum theory.

As we have seen in the last chapter, many of the fundamental results of thermal phenomena can be derived by physical arguments applied to the random behavior of molecules in a system in thermal equilibrium at a temperature T. The application of statistical methods to microscopic models of a system to derive thermal phenomena is known as *statistical mechanics*. In this chapter we will consider the quantum modifications of statistical mechanics and the implications for thermal effects.

To review, three of the fundamental results of quantum theory are listed below.

(1) Quantum theory **with its concept of systems with discrete allowed energy levels** replaces the classical ideas of "**degrees of freedom**" characterized by continuously increasing energy.

(2) **Quantum particles** (molecules, atoms, electrons, etc.) **are truly indistinguishable**. There is no way to label one electron to differentiate it from other electrons.

(3) Quantum particles are characterized by a quantized spin angular momentum. **Particles of non-integral spin are referred to as** *fermions*; **particles of integral spin are known as** *bosons*. **Fermions**, for example electrons, protons, and neutrons, which all **have a spin of 1/2, obey the Pauli Exclusion Principle** and **are restricted to having only one particle in each discrete quantum state**.

Table 13.1 Distribution of two distinguishable particles A and B in three quantum states

State 1	State 2	State 3
AB	–	–
–	AB	–
–	–	AB
A	B	–
B	A	–
A	–	B
B	–	A
–	A	B
–	B	A

Table 13.2 Distribution of two bosons into three states

State 1	State 2	State 3
AA	–	–
–	AA	–
–	–	AA
A	A	–
A	–	A
–	A	A

Bosons, such as photons, phonons (for which the number of particles is not conserved) and helium atoms (which have a fixed number), **do not obey the Exclusion Principle**.

In the last chapter we saw the effect of quantum energy levels in modifying the classical equipartition theorem result for the distribution of energy among different energy modes in a system. Quantum indistinguishability and the Exclusion Principle have important implications for the statistical mechanics of quantum particles.

For example, consider the distribution of two particles A and B, into three quantum states, State 1, State 2, and State 3. If we consider classical distinguishable particles the different system states are shown in Table 13.1.

There are nine distinct states of the system. **The statistics governing the behavior of classical distinguishable particles are called** *Maxwell–Boltzmann statistics*.

In contrast, consider the distribution of two quantum particles into three states. Since the quantum particles are indistinguishable, A = B. If we consider quantum bosons – particles without any Exclusion Principle restrictions – Table 13.2 gives the possible distinct states of the system.

There are six possible distinct states of the system. **The statistics that govern systems of quantum bosons are called** *Bose–Einstein statistics*. Bose–Einstein statistics determine the spectra dependence of *blackbody radiation* as explained in Section 13.2.

If we consider quantum fermions, particles that follow Pauli's Exclusion Principle, the first three states above are not possible because we have two particles in the same

Table 13.3 Distribution of two fermions into three states

State 1	State 2	State 3
A	A	–
A	–	A
–	A	A

Figure 13.1 A system at T_2 in thermal contact with a heat reservoir at T_1.

quantum state. The three possible states of the quantum fermion system are shown in Table 13.3.

The statistics that govern systems of quantum fermions are called *Fermi–Dirac statistics*. Fermi–Dirac statistics control the current–voltage characteristics of a diode as shown in Section 13.5.

13.1   Occupancy of states in equilibrium at temperature T

Consider a system kept in contact with a large "heat bath" at temperature T_1 as shown in Figure 13.1. The fundamental rule of statistical mechanics is that we count the number of states available to the system at a fixed energy; the equilibrium state will be the most probable state. The number of states available to a system increases exponentially with the temperature of the system. So if $T_2 > T_1$ there will be more states available to the system at T_2. However, there will be fewer states available to the heat bath at T_1, so the system as a whole will have fewer states available to it. The largest number of possible states for the system as a whole (the system at T_2 and the heat bath) will be when $T_2 = T_1$, which is the condition of thermal equilibrium.

13.1.1   Maxwell–Boltzmann statistics

The *occupancy* of a state s at energy E_s, $\bar{n}_s$, is defined as the most probable number of particles in state s at temperature T. **In Maxwell–Boltzmann statistics**, for classical particles, it can be shown that

$$\bar{n}_s = N \frac{e^{-E_s/k_B T}}{\sum_{all\ states} e^{-E_i/k_B T}} = N \frac{e^{-\beta E_s}}{\sum_{all\ states} e^{-\beta E_i}} \tag{13.1}$$

where N is the total number of particles, E_i is the energy of the ith level, and $\beta = 1/k_B T$.

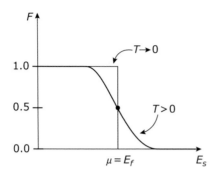

Figure 13.2 Fermi–Dirac occupancy of states as a function of energy for $T = 0$ and for positive temperature.

13.1.2　Fermi–Dirac statistics

In contrast, Fermi–Dirac statistics for quantum fermions leads to the following expression for the occupancy of a state

$$\bar{n}_s = \frac{1}{e^{(E_s-\mu)/k_BT} + 1} = \frac{1}{e^{\beta(E_s-\mu)} + 1}. \tag{13.2}$$

The constant μ in this equation is the **chemical potential** also referred to as the **Fermi energy, E_f. The chemical potential can be considered to be the energy required to add one particle to the system.** The value for the chemical potential (or Fermi energy) is determined from the condition that the sum over all states of the occupancy $\bar{n}_s$ must equal the total number of particles in the system N

$$\sum_s \bar{n}_s = N \Leftrightarrow \mu. \tag{13.3}$$

A plot of the occupancy as a function of the energy of the state for two different temperatures is shown in Figure 13.2. For $T \rightarrow 0$, $e^{(E_s-\mu)/k_BT}$ is either infinite (for $E_s > \mu$) or zero (for $E_s < \mu$). So at $T \rightarrow 0$ every state below the Fermi energy is filled with exactly one fermion particle of each spin, and every state above the Fermi energy is empty. As the temperature increases, some particles are thermally excited above the Fermi energy, leaving some states below the Fermi energy empty. **The area under the $\bar{n}(E)$ curve is equal to the total number of particles N in both cases.**

13.1.3　Bose–Einstein statistics

Quantum bosons have a state occupancy given by

$$\bar{n}_s = \frac{1}{e^{(E_s-\mu)/k_BT} - 1} = \frac{1}{e^{\beta(E_s-\mu)} - 1}. \tag{13.4}$$

Note that the only difference between bosons and fermions is the change of the sign in front of the 1 in the denominator and if $e^{\beta(E_s-\mu)} \gg 1$ both the Fermi–Dirac

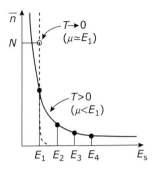

Figure 13.3 Bose–Einstein occupancy of states for a system with N particles and discrete states at E_n.

occupancy and the Bose–Einstein occupancy approach the classical Maxwell–Boltzmann result, $\bar{n}_s \propto e^{-E_s/k_B T}$.

The difference that the negative sign makes is enormous. At $T \to 0$, the occupancy diverges and all the bosons collect in the lowest energy state of the system. If the number of particles is fixed (as it would be for ^{4}He, for example), then the condition

$$\sum_s \bar{n}_s = N \tag{13.5}$$

where N is the total number of particles, fixes the value of the chemical potential μ. Figure 13.3 shows the distribution of bosons in a system with the allowed energy states at E_1, E_2, E_3, E_4, etc. As $T \to 0$ all N particles are collected in the lowest energy state E_1 and the chemical potential μ, representing the energy required to add a particle to the system, is set at $\mu = E_1$. As $T > 0$, the distribution of particles across the states is given by Eq. (13.4) with $\mu < E_1$ as determined by the particle conservation equation above.

For photons and phonons, where the number of particles is not conserved, the chemical potential is not a meaningful concept and the Bose–Einstein distribution equation is given by

$$\bar{n}_s = \frac{1}{e^{E_s/k_B T} - 1} = \frac{1}{e^{\beta E_s} - 1}. \tag{13.6}$$

13.2   Blackbody radiation

One of the most important applications of Bose–Einstein statistics is *blackbody radiation* – the electromagnetic radiation emitted by any body at $T > 0$ K. Consider a cavity in thermal equilibrium with the walls at temperature T (see Figure 13.4). Since the molecules of the enclosure are in motion, accelerating and decelerating, they are continually emitting and absorbing electromagnetic radiation. Assume for the time being that the enclosure is evacuated, so we don't have to worry about the absorption and emission from gases inside the cavity.

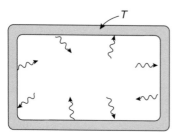

Figure 13.4 A cavity in thermal equilibrium with its walls at temperature T. The walls are emitting and absorbing electromagnetic radiation.

The photon population is in thermal equilibrium with the walls of the cavity at temperature T. The photon wave function is an electromagnetic plane wave:

$$\psi(r, t) = A \, e^{i(\vec{k}\cdot\vec{r} - \omega t)} \tag{13.7}$$

where ω is the frequency of the electromagnetic wave and $\vec{k}$ is the wave vector, with $|\vec{k}| = 2\pi/\lambda$. For propagation in free space, $\lambda = c/f = 2\pi c/\omega$ (where $c = 2.9979 \times 10^8$ m/s is the speed of light), from which we find the relation between ω and the magnitude of the wave vector, $k = \omega/c$. From quantum mechanics, we know that the energy of the photon state is given by

$$E_s = \hbar\omega = \hbar \, 2\pi \, c/\lambda = \hbar c k. \tag{13.8}$$

From the boson occupancy, we have the **photon occupancy**

$$\bar{n}_s = \frac{1}{e^{\beta E_s} - 1} = \frac{1}{e^{\beta \hbar \omega} - 1}. \tag{13.9}$$

To calculate the **energy density in the cavity** (J/m^3) at any given frequency, we need to (1) multiply the number of photons in the state $E_s = \hbar\omega$ by the energy of the state ($\hbar\omega$), and (2) sum up over all the states per unit volume (cubic meter) with frequency ω

$$\bar{u}(\omega, T)d\omega = \sum_{E_s} \hbar\omega \, \bar{n}_s(\omega, T) = \hbar\omega \, \bar{n}_s(\omega, T)D(\omega) \tag{13.10}$$

where $D(\omega)$ is the *density of states* for photons between ω and $\omega + d\omega$.

So how do we count the number of photon states in any interval $d\omega$? Using the same argument for counting electron states, we say that in one dimension, $\Delta x \, \Delta p_x = h \Rightarrow \Delta x \, \hbar \Delta k_x = h$. So in one dimension, a single photon state corresponds to a Δk_x interval of (2π) per unit length.

In three dimensions, a single photon state per unit volume corresponds to a $(2\pi)^3$ volume of "k-space." Each of these photon states can accommodate one photon of each of two polarizations (horizontal and vertical polarization, or left-hand and right-hand circular polarization). The density of photon states $D(\omega)$ thus depends on the k-space volume of states with frequencies between ω and $\omega + d\omega$ — divided by $(2\pi)^3$ and multiplied by 2 to account for the polarization degeneracy. The k-space volume

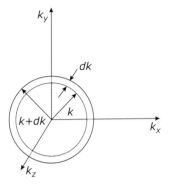

Figure 13.5 Volume of k-space between k and $k + dk$.

corresponding to frequencies between ω and $\omega + d\omega$ is shown in Figure 13.5, where $k = \omega/c$ and $dk = d\omega/c$.

The volume of the spherical shell in k-space is $4\pi k^2 \, dk$ and the number of photon states/m^3 in the volume is

$$\frac{2}{(2\pi)^3} \, 4\pi k^2 \, dk = \frac{2}{(2\pi)^3} \frac{4\pi\omega^2}{c^2} \frac{d\omega}{c} = D(\omega). \tag{13.11}$$

Multiplying $D(\omega)$ by the photon occupancy and the energy per photon gives us the **photon spectral energy density** in the enclosure

$$\bar{u}(\omega, T)d\omega = \hbar\omega \, \bar{n}_s(\omega, T)D(\omega) = \frac{\hbar\omega^3 \, d\omega}{\pi^2 \, c^3 \, (e^{\beta\hbar\omega} - 1)} \frac{\text{J}}{\text{m}^3}. \tag{13.12}$$

Figure 13.6 shows the photon spectral energy density, $\bar{u}(\omega, T)$, as a function of frequency for three different temperatures. **The low-frequency behavior** (when $e^{\hbar\omega/k_B T} \ll 1$) is dominated by the ω^3 dependence. **At high frequencies**, the exponential term takes over and drives the power density rapidly toward zero. The frequency of the maximum photon spectral energy density ω_{max} depends on the temperature, moving to higher frequencies at higher temperatures.

Question 13.1

Judging from Figure 13.6, what is the wavelength region that is the most important for transfer near room temperature (300 K)? What spectra k region does this correspond to? What about at $T = 600$ K and 900 K?

If we imagine an aperture of area A in the side of the cavity, as shown in Figure 13.7, the isotropic radiation inside the cavity will be emitted from the side of the cavity into the solid angle subtended by the aperture. As the angle θ increases the cross-sectional area of the cylinder decreases as $A \cos \theta$. The photon energy in the volume of a cylinder of length $c \, dt$ and cross-section $A \cos \theta$ is $\frac{\hbar\omega^3 A \cos\theta \, c \, dt \, d\omega}{\pi^2 \, c^3 \, (e^{\beta\hbar\omega} - 1)}$.

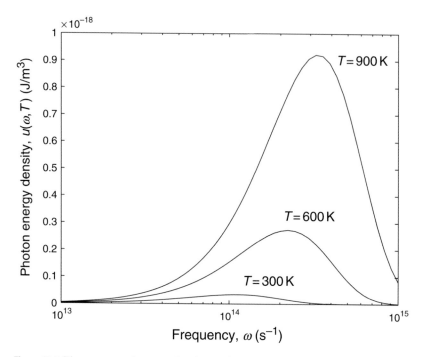

Figure 13.6 Photon spectral energy density vs. frequency at different temperatures from Equation (13.12).

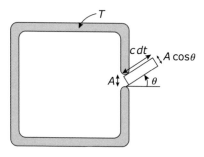

Figure 13.7 Thermal radiation emitted from an aperture in the cavity.

Dividing by dt, we find that the emitted power into a solid angle $d\Omega = \frac{dA}{R^2} = \sin\theta \, d\theta \, d\phi$ is

$$P_e(\omega, T, \theta)d\omega = A \, c \, \cos\theta \, \hbar\omega \, \frac{2}{(2\pi)^3} \frac{1}{(e^{\beta\hbar\omega} - 1)} \frac{\omega^2 \, d\omega}{c^3} \, d\Omega. \qquad (13.13)$$

Dividing by the area of the aperture A and integrating over all solid angles to find the total emitted power in the half-sphere of emission yields a factor of $\frac{1}{4}$ and the total power emitted through the aperture (per m^2 of aperture area) is

$$P_e(\omega, T)d\omega = \frac{1}{(e^{\beta\hbar\omega} - 1)} \frac{\hbar\,\omega^3 \, d\omega}{4\,\pi^2 c^2} \frac{W}{m^2}. \qquad (13.14)$$

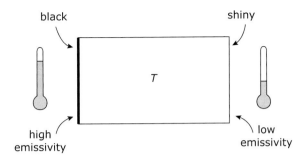

Figure 13.8 Higher apparent temperature from a high-emissivity surface than a low-emissivity surface.

This is the *blackbody spectral emission* formula. If one considers emission from an area of material rather than an aperture into a cavity, the properties of the material affect the emission of the surface. This is taken into account in the ***emissivity function*** $\epsilon(\omega, T)$ as below:

$$P_e(\omega, T)d\omega = \frac{\epsilon(\omega, T)}{(e^{\beta\hbar\omega} - 1)} \frac{\hbar\,\omega^3\,d\omega}{4\,\pi^2 c^2} \frac{W}{m^2} \tag{13.15}$$

where $0 < \epsilon(\omega,T) < 1$.

This formula (with appropriate values of the emissivity function for the materials involved) describes the emission of radiation from heated objects as disparate as the red-hot burner of your electric stove and the Sun, which can be modeled as blackbody emitters at temperatures 900 K and 5000 K respectively.

The values of the emissivity of real materials range from 0.98 (lampblack in the infrared spectral region) to 0.01 (a polished gold surface in the infrared spectral region). **A *blackbody* is an object with an emissivity of 1 for all frequencies.** A heated cavity with an aperture in the side such as we have been considering is an excellent approximation to a blackbody.

The effect of the differing emissivities can be observed in a demonstration as illustrated in Figure 13.8. The heat radiated from a high-emissivity side of the reservoir at temperature T, is much greater than heat radiated from the low-emissivity side – as reflected in the reading of the thermometers facing each side.

If we integrate the **power emission over all frequencies** ω, we derive the **Stefan–Boltzmann Law** we introduced in the last chapter

$$P_{rad} = \sigma \varepsilon A T^4 \tag{13.16}$$

with the emissivity ε here representing an average of $\epsilon(\omega,T)$ over all frequencies and σ is the Stefan–Boltzmann constant.

Question 13.2

What color are the heat sinks that are sometimes attached to semiconductor chips? Why do you think this is the case? Why are house radiators sometimes painted silver?

13.3   Consequence of thermodynamics for blackbody emission

The thermometer experiment above suggests that it might be possible to put a low-emissivity object in a cavity with high-emissivity walls and have the object arrive at a higher temperature than the walls, as shown in Figure 13.9. One might think that $T_2 > T_1$ because the object will emit less radiation than the cavity walls and thus gain heat from the cavity. But this result would violate the Second Law of Thermodynamics – if there is a temperature difference between the cavity and the object it should be possible to run a heat engine between T_2 and T_1 and get usable work out of a system at thermal equilibrium.

Since the world is not full of heat engines that are producing work for free from differing emissivity objects inside a cavity, we are forced to assume that $T_1 = T_2$ in the equilibrium reached if we let the cavity–object remain together for a long time. The answer to this paradox is that while the walls are emitting more radiation, they are also absorbing more radiation. Similarly, the low-emissivity object is emitting less radiation, but it is also absorbing less radiation. This result is because **the emissivity ϵ is equal to the *absorptivity* a, where a is the ratio of the power absorbed by the object to the power incident on the object (approximately equal to 1 minus the power reflectivity for opaque objects)**. This result, $\epsilon \equiv a$, is required by the Second Law of Thermodynamics.

In fact, the relation is even stronger than this. Consider the situation as before but with the enclosed object surrounded by a filter which is transparent to only one frequency. Clearly, if the emissivity of the object is different from the absorptivity at that frequency, the object would heat up or cool down with respect to the cavity. Since, as we have already discussed, this cannot happen due to thermodynamic fundamentals, the emissivity of the object (and the walls) must be exactly the same as the absorptivity at any frequency or wavelength. The same arguments hold for polarization, for temperature, for angle of emission/absorption, for a cavity with other objects or gases in it, or for any other variable; **the general Principle of *Detailed Balance*** can be written as

$$\epsilon(\omega, T) \equiv a(\omega, T). \qquad (13.17)$$

This principle is why infrared emission imaging of an environment in or near thermal equilibrium, for example a landscape with no heat sources on a rainy day with low

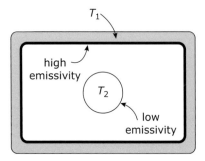

Figure 13.9 Low-emissivity object inside a high-emissivity cavity. At equilibrium, $T_1 = T_2$.

cloud cover, is completely featureless. Objects with a low emissivity reflect just enough more of the ambient radiation – because of their low absorptivity – to have the same apparent brightness as objects with high emissivity.

Question 13.3

How is the discussion in this section consistent with the situation shown in Figure 13.8? Shouldn't the two thermometers, one facing the high-emissivity side and one facing the low-emissivity side, be at the same temperature as we concluded for the situation in Figure 13.9? Does the result shown in Figure 13.8 violate the Second Law of Thermodynamics?

Example 13.1 Compare the amount of power emitted by a one-square meter 320 K blackbody in the 3–5 μm region of atmospheric transparency with the amount of power emitted in the 8–12 μm atmospheric window. Does this explain why night-vision systems are based on small band-gap semiconductor photodetector devices?

Solution: The power emitted in a particular wavelength range can be found by numerically integrating Eq. (13.15) from $\omega_1 = \frac{2\pi c}{\lambda_1}$ to $\omega_2 = \frac{2\pi c}{\lambda_2}$. A good approximation can be obtained from a simple trapezoidal rule integration:

$$\int_{x_1}^{x_n} f(x)\,dx \cong \left(\frac{1}{2}f(x_1) + f(x_2) + f(x_3) + \ldots + f(x_{n-1}) + \frac{1}{2}f(x_n) \right) \Delta x$$

where $f(x)$ is sampled at evenly spaced values of $x_1, x_2, x_3 \ldots x_{n-1}, x_n$ with spacing $\Delta x = x_i - x_{i-1}$. A MATLAB function m-file to do a trapezoidal integration of $P_e(\omega, T) = \frac{\epsilon(\omega, T)}{(e^{\beta \hbar \omega} - 1)} \frac{\hbar \omega^3 \, d\omega}{4 \pi^2 c^2}$ along with results for frequency ranges corresponding to 3 μm $< \lambda <$ 5 μm and 8 μm $< \lambda <$ 12 μm is shown below.

```
function P=Example13_1(lambda1, lambda2, T)
c=2.9979e8; hbar=1.054e-34; kB=1.381e-23;
beta=1/(kB*T);
w1=2*pi*c/lambda2; w2=2*pi*c/lambda1;
del_w=(w2-w1)/100;
w=(w1:del_w:w2);
N=length(w);
A=hbar/(4*pi^2*c^2);
B=exp(beta*hbar*w);
F=A*w.^3 ./ (B-1);      % emissivity = 1
P=( sum(F)-0.5*(F(1)+F(N)) )*del_w;
% the first and last points have weight=0.5; all others
  have weight=1
```

Continued

Example 13.1 (*cont.*)

```
>> Example13_1(3e-6, 5e-6, 320)
ans =
    11.6989
>> Example13_1(8e-6, 12e-6, 320)
ans =
    165.0480
```

At 320 K, a 1 m^2 blackbody emits 165.0 W in the 8–12 μm atmospheric window compared with 11.7 W in the 3–5 μm window. $Hg_xCd_{1-x}Te$ alloys with energy gaps $E_g \lesssim \hbar\omega = \hbar\, 2\pi c/12$ μm $= 0.10$ eV are used to produce sensitive night-vision systems that are said to be able to image a tank under trees with its engine turned off just from the body heat of its crew members!

13.4   Application of quantum statistics of fermions: doped semiconductors

The semiconductor revolution since the 1950s is based on the ability to *dope* semiconductors by adding certain impurities so that the semiconductor has an excess or deficiency of electrons. In this section, we will consider doped semiconductors using the concept of the quantum statistics of fermions (electrons), and how it leads to electrically interesting phenomena, such as a semiconductor junction diode.

In Chapter 11, we saw how in a periodic potential, such as a semiconductor lattice, electron quantum states are *Bloch functions*

$$\psi_{k,n}(x) = e^{ikx}\, u_{k,n}(x) \tag{13.18}$$

where the states are characterized by the continuous wave vector k (2π divided by the wavelength of the electron wave function, also equal to the *crystal momentum* divided by $\hbar$) and a band index n, which specifies the particular valence or conduction band state that the electron occupies. A volume of k-space of $\frac{1}{(2\pi)^3}$ corresponds to a single quantum state, which for fermions can be occupied by at most one electron of each spin.

A *band diagram*, a plot of the energy E vs. the wave vector (momentum), appropriate for a generic direct-gap semiconductor at temperature $T = 0$ K, is shown in Figure 13.10. For a pure semiconductor at $T = 0$ K, the valence band is completely filled with electrons and the conduction band is empty. The **Fermi level in the Fermi–Dirac equation represents the energy to add (or take away) a single electron particle from the system**. In the case where there is no available state at the Fermi energy, the Fermi level at $T = 0$ K is halfway between the highest filled and lowest empty state, in the middle of the energy gap. As we have seen, at $T = 0$ K the occupancy $\bar{n}_s$ of states below the Fermi energy is 1 and of the states above the Fermi energy is 0.

Doping of a semiconductor involves the addition of impurity atoms with a valence different from the underlying semiconductor lattice. For example, consider the doping of silicon $_{14}Si$, with an outer electron configuration of $3s^2 3p^2$, with atoms of phosphorus

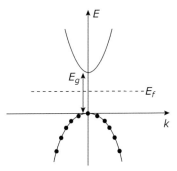

Figure 13.10 For a semiconductor at $T = 0$ the Fermi energy is in the middle of the energy gap.

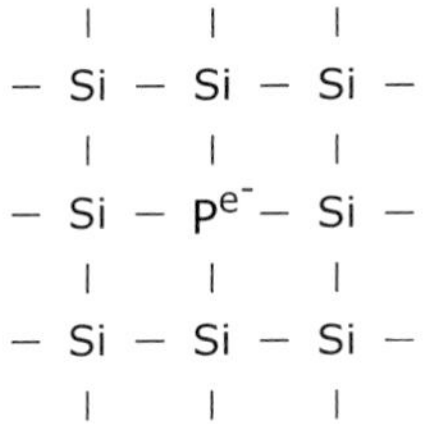

```
   |      |      |
 — Si  —  Si  —  Si  —
   |      |      |
 — Si  —  P^{e^-}  —  Si  —
   |      |      |
 — Si  —  Si  —  Si  —
   |      |      |
```

Figure 13.11 A silicon lattice doped with a phosphorus atom with one more electron than silicon.

$_{15}$P, with an outer electron configuration of $3s^2 3p^3$. The phosphorus atoms are incorporated into the silicon lattice, with a valence of four and an extra electron (compared to the silicon atoms) in an atomic p state of the phosphorus atom as shown in Figure 13.11. The energy of the extra electron is higher when it is confined in an atomic state on the phosphorus atom than if it transitions to one of the empty extended conduction band states of the lattice.

So at $T = 0$ K, the ground state of the electron is in an extended state bound to and orbiting around the phosphorus ion which has an extra positive charge due to 15 protons in the nucleus compared with the 14 protons in the silicon atoms. The electron is bound to the P$^+$ ion in a hydrogen-like *Rydberg state* (see Figure 13.12). The energy of an electron in the nth Rydberg state of hydrogen is given by

$$E_n = -\frac{m_o e^4}{8\epsilon_o^2 h^2}\frac{1}{n^2} = -\frac{13.6}{n^2}\,\text{eV} \tag{13.19}$$

and the Bohr radius of the $n = 1$ state is

$$a_o = \frac{h^2\,\epsilon_o}{\pi m_o e^2} = 0.0529\,\text{nm}. \tag{13.20}$$

For the extra phosphorus electron in silicon, these numbers need to be modified by replacing the permittivity of free space ϵ_o by the permittivity of silicon $\epsilon_{Si} = 11.8\,\epsilon_o$ and the electronic mass m_o by the *effective mass* of an electron in the conduction band of silicon $m^* = 0.26\,m_o$. With these modifications, we find that the Bohr radius of the impurity electron in silicon is 2.4 nm and the binding energy (with respect to the

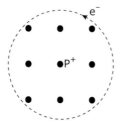

Figure 13.12 Phosphorus extra electron in an extended hydrogen-like orbit in silicon crystal.

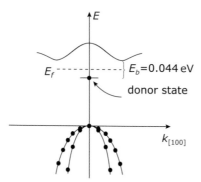

Figure 13.13 Phosphorus donor state with a binding energy of 0.044 eV below the conduction band in silicon.

conduction band) is $E_b = -0.025$ eV. The electron radius of 2.4 nm is equal to several lattice constants, which implies that the continuous media hydrogenic model is consistent. The actual bonding energy of an electron to a donor atom in silicon depends on the donor element. For phosphorus in silicon, for example, the donor binding energy is 0.044 eV, while arsenic has a donor binding energy of 0.049 eV.

The band diagram corresponding to the silicon doped with phosphorus at $T = 0$ K is shown in Figure 13.13. As we have seen in Chapter 11, the band gap of silicon is indirect with the minima in the conduction band located in six symmetrical pockets in the [100] direction. Since the phosphorus donor state is a localized state for which the crystal momentum k is not applicable, we will arbitrarily put the donor state at the zone center, 0.044 eV below the conduction band minima. The thermal excitation processes will provide any momentum necessary to put the excited electron into the conduction bands, so all that is important is the energy difference between the electron bound to the donor site and a free electron in the conduction band.

At $T = 0$, the highest filled state is at the donor level, 0.044 eV below the conduction band and the lowest empty state is the lower edge of the conduction band. The Fermi level at $T = 0$ is therefore halfway between the donor level and the conduction band edge $E_f = E_c - E_b/2 = E_c - 0.022$ eV. At $T > 0$ K, the conduction band states above E_f become occupied through thermal excitation and the donor states become unoccupied, with the Fermi level dropping to keep the number of electrons in the more numerous conduction band states and the number of vacancies in the donor level equal. **A semiconductor doped to have excess electrons in the conduction band is called *n-type*.**

This conceptual process, the extra electron on the phosphorus atom delocalizing to a hydrogen-like orbit around the charged phosphorus ion and then thermally ionizing to the conduction band, gives the phosphorus-doped silicon free-electron carriers in its conduction band. As the temperature increases further, vacancies begin to appear in the valence band as valence band electrons are thermally excited across the energy gap to the conduction band. **The vacancies in the valence band are referred to as *holes*,** and the position of the Fermi level is determined by charge balance: $n = N_d^+ + p$ where n is the number of electrons per m^3 in the conduction band, p is the number of holes per m^3 (vacancies) in the valence band, and N_d^+ is the number of ionized donor states per m^3. These quantities are found from the following relationship using the Fermi–Dirac equation for the occupancy of states $f(E) = \bar{n} = 1/\left(e^{(E-E_f)/k_B T} + 1\right)$:

$$n = \int_{E_g}^{\infty} f(E) N_c(E) \, dE \tag{13.21}$$

$$N_d^+ = N_d \left[1 - f(E_d)\right] \tag{13.22}$$

$$p = \int_{-\infty}^{0} [1 - f(E)] N_v(E) \, dE. \tag{13.23}$$

In these equations,

$$[1 - f(E)] = \left[1 - 1/\left(e^{(E-E_f)/k_B T} + 1\right)\right] = e^{(E-E_f)/k_B T} \Big/ \left(e^{(E-E_f)/k_B T} + 1\right)$$

represents the probability that a state at energy E is *unoccupied*, and $N_c(E)$ and $N_v(E)$ represent the density of states for the conduction band and valence band, respectively. The limits of integration are selected for the zero of energy to be at the top of the valence band; this gives the energy of the donor level $E_d = E_g - E_b$ where E_b is the donor binding energy . For three-dimensional isotropic bands, the density of states was found in Chapter 11 to be $N(E) = \frac{1}{2\pi^2 \hbar^3} (2m^*)^{3/2} (E)^{1/2}$ with E measured from the band edge. With the zero of energy measured from the valence band edge and the valence band energies negative, $N_c(E) = \frac{1}{2\pi^2 \hbar^3} (2m_c^*)^{3/2} (E - E_g)^{1/2}$ for the conduction band and $N_v(E) = \frac{1}{2\pi^2 \hbar^3} (2m_v^*)^{3/2} (-E)^{1/2}$ for the valence band. The *density of states masses* are $m_c^* = 1.1 \, m_o$ for the conduction band of silicon and $m_v^* = 0.56 \, m_o$ for the valence band of silicon.

Example 13.2 Calculate the density of electrons n, the density of holes p, and the density of ionized donor states N_d^+ in silicon at $T = 500$ K doped with $N_d = 1 \times 10^{23}$ m^{-3} phosphorus donors if the Fermi energy E_f is 0.7 eV above the top of the

Continued

Example 13.2 (cont.)

valence band. Is this the equilibrium value of E_f? How would you find the equilibrium value of E_f?

Solution: We can numerically integrate Eqs. (13.21)–(13.23) using the trapezoidal rule as in Example 13.1. Knowing that the Fermi function $f(E) = 1/\left(e^{(E-E_f)/k_BT} + 1\right)$ decreases exponentially for $(E - E_f) > k_BT$, we will integrate the conduction band from E_g to $E_g + 30\,k_BT$ and the valence band from $-30\,k_BT$ to 0. Since the density of states goes to zero at both the bottom of the conduction band edge and the top of the valence band, and the Fermi function goes to zero at $30\,k_BT$ above the conduction band edge since $e^{-30} = 9.3 \times 10^{-14}$ (and $[1 - f(d)]$ also goes to zero at $30\,k_BT$ below the valence band edge), we don't have to apply the trapezoidal rule weight of 1/2 to the points at the limits of integration since they are zero anyway. An m-file that calculates Eqs. (13.21)–(13.23) is below:

```
me=9.109e-31; hbar=1.0546e-34; eV2J=1.602e-19;
Mc=1.1*me; Mv=0.56*me; % conduction & valence band DOS
   masses
T=500;
kBT=1.381e-23*T;
Ef=0.7*eV2J;       % Fermi energy in Joules
Eg=1.1*eV2J;       % energy gap in Joules

Eb=0.044*eV2J;     % donor binding energy in Joules
Ed=Eg-Eb;          % donor energy wrt E=0 at top of valence band
Nd=3e23;           % density of donors
% unoccupied donor states = Nd*[1-f(Ed)]
ND_P=Nd*exp((Ed-Ef)/kBT)/(exp((Ed-Ef)/kBT)+1)

% conduction band energies from Eg to Eg+30*kB*T
EcA=[ Eg:(.001*kBT):(Eg+30*kBT)] ;
% density of states with E measured from edge of conduction band
NcE=(1/(2*pi^2))*(2*Mc/(hbar^2))^(3/2)*sqrt(EcA-Eg);
FcE=(1./(exp((EcA-Ef)./(kBT))+1));        % f(Ec)
n=sum(NcE .* FcE) * (.001*kBT)

% valence band energies from 0 to-30*kB*T
EvA=[ 0:(-.001*kBT):(-30*kBT)] ;
% density of states with E measured from edge of valence band at
Ev=0
NvE=(1/(2*pi^2))*(((2*Mv)/(hbar^2))^(3/2)).*sqrt(-EvA);
FvE=(exp((EvA-Ef)/(kBT))./(exp((EvA-Ef)/(kBT))+1));  %[ 1-f(Ev)]
p=sum(NvE .* FvE) * (.001*kBT)
```

Continued

Example 13.2 (*cont.*)

Running this m-file gives:

```
ND_P =
  2.9992e+023
n =
  5.8099e+021
p =
  2.0027e+018
```

At 500 K the donors are nearly completely ionized with $N_d^+ \simeq N_d$. Since $n = 5.8 \times 10^{21}$ m^{-3} $< (N_d^+ + p) = (3 \times 10^{23} + 2 \times 10^{18})$ m^{-3} the system is not in charge balance, and $E_f = 0.7$ eV is not the correct equilibrium value of E_f. The Fermi energy should be higher to give more electrons in the conduction band. The correct value of E_f could be found by gradually increasing E_f until $n \geq N_d^+ + p$, or by using a numerical root-finding tool such as MATLAB `fzero`.

Applying the condition of charge balance to silicon, with $E_g = 1.1$ eV containing $N_d = 3 \times 10^{23}$ phosphorus donors per m^3, we find the temperature dependence of the Fermi level and the electron, hole, and ionized impurity densities as a function of temperature, as shown in Figures 13.14 and 13.15.

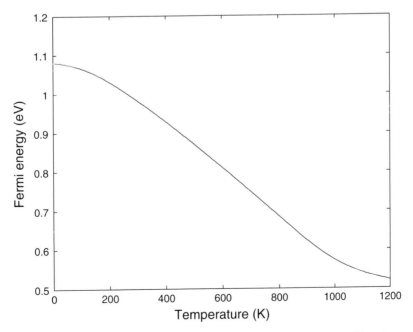

Figure 13.14 Fermi energy for phosphorus-doped silicon with $N_d = 3 \times 10^{23}$ m^{-3} as a function of temperature.

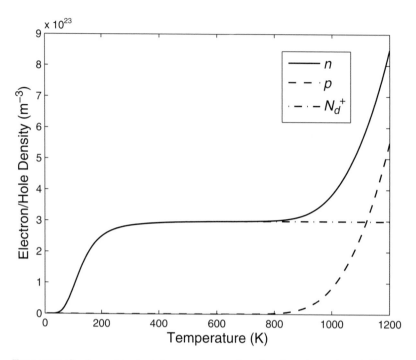

Figure 13.15 Carrier and ionized donor concentrations for phosphorus-doped silicon as a function of temperature.

We note that the Fermi level begins halfway between the donor level energy and conduction band edge and drops with temperature to near mid-gap. The number of electrons increases rapidly with increasing temperature as the donor states ionize, stays constant and equal to the doping density from about 200 K to 800 K, and then rapidly increases above 800 K as electrons from the valence band begin to thermally excite over the band gap to the conduction band. For the region between 200 K and 800 K, the number of electrons is essentially fixed and equal to the number of dopant donor atoms.

Similarly, if silicon is doped with an element with a valence less than silicon, the missing bonding electron provides a state close to the valence band edge that electrons from the silicon valence band can be thermally promoted into. Such dopant elements are called *acceptors* and a **semiconductor with an excess of vacancies in the valence band ("holes") due to acceptor doping is called *p-type*.** For example, aluminum $(3s^2 3p^1)$ and boron $(2s^2 2p^1)$ are acceptors in silicon. The band structure for p-type acceptor-doped silicon is show in Figure 13.16. For boron-doped silicon, the acceptor level is 0.045 electron volts (eV) above the valence band. At $T = 0$ the Fermi level will be halfway between the highest filled state (the top of the valence band) and lowest empty state (the acceptor level at $E_a = 0.045$ eV), and as the temperature increases the Fermi level moves farther from the valence band edge into the lower half of the band gap (see Figure 13.16).

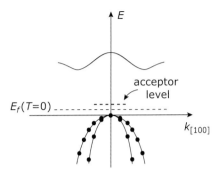

Figure 13.16 Acceptor level for boron-doped silicon. At $T = 0$ the Fermi energy at $T = 0$ is half-way between the acceptor level and valence band edge.

13.5   A p-n junction diode

The most elementary semiconductor device, ubiquitous in chip design, is the *p-n junction diode*. A diode is a device that conducts current in one direction only, and **a p-n junction diode, as its name indicates, is a diode made by joining together a p-type and an n-type semiconductor**. In Figure 13.17, we plot the energy of the bottom of the conduction band E_c and the top of the valence band E_v vs. position (at room temperature) in a p-type and n-type silicon brought next to each other, but not in contact. The p-type material has its Fermi level E_{fp} in the bottom half of the band and, at room temperature, has many vacancies (holes) in the valence band ($p \simeq N_a$ the number of acceptors), and just a few electrons in the conduction band. The n-type material, with its Fermi level E_{fn} in the top half of the band, has many electrons in the conduction band ($n \simeq N_d$ the number of donors), and only a few holes in the valence band. With the bands aligned, the Fermi energy in the top half of the band in the n-type material E_{fn} does not line up with the Fermi level in the bottom half of the band in the p-type material E_{fp}. Since the Fermi level represents the energy at which an electron is added to (or removed from) the material, energy can clearly be gained by removing an electron from the n-type material at energy E_{fn} and adding it to the p-type material at the lower energy E_{fp}.

If the two materials are brought together to form a *p-n junction*, electrons will flow spontaneously from the n-type material to fill the vacancies (holes) in the p-type material, leaving unbalanced negative acceptors N_a^-. Every electron that flows into the p-type material adds a negative charge to the p-type side and leaves an unbalanced positive ionized donor N_d^+ on the n-side. Since it requires more energy to add an electron to a negatively charged material, the electron energies on the p-side, including E_c, E_v, and E_{fp}, are all increased with respect to the n-side. This **process goes on until $E_{fp} = E_{fn}$**, as shown in Figure 13.18.

In the ***depletion region*** near the interface between the two materials, there are very few free electrons or holes – the region is depleted of free carries. The positively ionized donors on the n-side and negatively ionized acceptors on the p-side create a charge

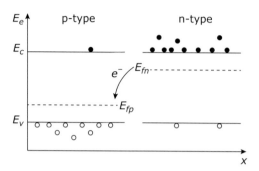

Figure 13.17 Band energies and Fermi energy for p-type and n-type semiconductor samples before charge transfer occurs.

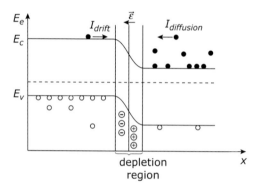

Figure 13.18 Band energies and Fermi energy for semiconductor p-n junction after charge transfer occurs. Electrons from the n-side have filled the holes on the p-side within the depletion region and the resultant electric field has created a band energy offset.

dipole supporting an electric field $\vec{\mathcal{E}}$ from the n-side toward the p-side. This electric field is shown by the tilted energy bands, which illustrate the electric force driving the electrons back toward the n-side and the holes back toward the p-side. (Holes, vacancies in the valence band, are at their lowest energy when they are at the top of the valence band – thus leading to the mnemonic "holes bubble up" to indicate that holes will tend to move up a section of sloped electron energy bands, while electrons tend to move down the sloped electron energy bands.)

If one of the few electrons in the p-type material (a *minority carrier*) happens to wander into the depletion region, it will be swept to the n-side by the electric field. Similarly, if a minority carrier hole on the n-side enters the depletion region it will be accelerated to the p-side. **The combination of these two currents is called the *drift current.*** On the other hand, the *majority carriers* – electrons on the n-side and holes on the p-side – are numerous and have a thermal distribution given by the Fermi–Dirac distribution. A few of the electrons will have enough energy to overcome the barrier and diffuse to the p-side. Similarly a few of the holes on the p-side will have enough energy

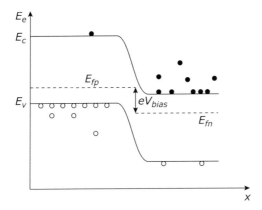

Figure 13.19 Energy bands in p-n junction under reverse bias with the band offset increased by eV_{bias}.

to overcome their barrier and diffuse to the p-side. **The combination of these two majority carrier currents is called the** *diffusion current*. Clearly, **at equilibrium the drift current and diffusion current must be exactly equal and cancel each other out so there is no net charge flow in the junction.**

Now consider an external bias voltage being applied to the junction. Assume first that a negative bias $V = -V_{bias}$ is applied to the p-side with respect to the n-side. Adding a negative bias will increase the electron band energies on the p-side by an energy $\Delta E = (-e)(-V_{bias}) = eV_{bias}$. The bands under a negative bias are shown in Figure 13.19. Since an external voltage is being applied, the junction is not in equilibrium and the Fermi levels no longer align, but instead

$$E_{fn} = E_{pn} - eV_{bias}. \tag{13.24}$$

The depletion region is widened and the depletion region electric field is increased, but since the **drift current** is controlled by the number of minority carriers (electrons on the p-side and holes on the n-side) which is fixed by the doping levels, it **isn't changed very much by changing** V_{bias}. On the other hand, the **diffusion current**, which depends on the number of carriers in thermal equilibrium with energy greater than the barrier, **is decreased exponentially**

$$I_{diffusion} \propto \bar{n} = \frac{1}{e^{[(E_o - eV_{bias})/k_B T]} + 1} \cong e^{-[(E_o - eV_{bias})/k_B T]} \rightarrow 0. \tag{13.25}$$

In reverse bias, only the drift current will contribute significantly to the current in the junction. The total current due to only the drift currents is the *reverse leakage current* I_o.

On the other hand, if **a positive voltage is applied to the p-side with respect to the n-side**, the barrier between the p-side and n-side is decreased by $(-e)V_{bias}$, as shown in Figure 13.20. **In this case the diffusion current increases exponentially, while the drift current remains about the same,**

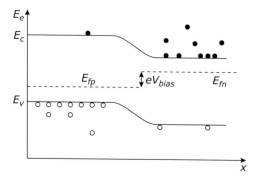

Figure 13.20 Energy bands in p-n junction under forward bias with the band offset decreased by eV_{bias}.

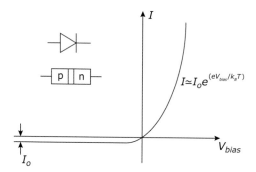

Figure 13.21 Current–voltage characteristics of p-n junction along with circuit symbol and junction geometry.

$$I_{diffusion} \propto e^{-(E_o - eV_{bias})/k_B T} \simeq I_o\, e^{+(eV_{bias})/k_B T}.$$ (13.26)

The current–voltage characteristics of the junction can be written

$$I = I_o \left(e^{+(eV_{bias})/k_B T} - 1 \right)$$ (13.27)

which is plotted in Figure 13.21. For typical diodes, I_o is on the order of nanoamps or picoamps, but approximately doubles for each 10 °C increase in temperature. **Essentially the p-n junction acts as a valve that permits current flow in one direction, but not in the other**. The electric circuit symbol for a diode and its realization in a p-n junction are sketched in Figure 13.21 along with the I–V characteristics.

Question 13.4

What is the current in a junction if $V_{bias} = 0$? Can you make an argument why the prefactor in the expression in Eq. (13.26) for the diffusion current should be equal to the drift current I_o?

Question 13.5

From our discussion of the source of the drift current, can you explain why I_o increases with temperature? At a constant forward bias voltage increasing the temperature increases I_o but decreases $e^{+(eV_{bias})/k_BT}$. Which effect do you think is more important?

13.6 Photodiode detectors and photocells

The built-in electric field in a p-n diode can be used as an optical detector or for energy production in solar cells. Consider light with $\hbar\omega > E_g$ incident on a p-n junction. Assume further that the depletion region is wide enough that a light photon is likely to be absorbed in the depletion region. As shown in Figure 13.22, a photon absorbed in the depletion region will create an electron and a hole that will be driven toward opposite sides of the junction. The effect is to produce an electron/hole current, due to the flux of incident photons, that flows in the opposite direction from the usual diode current. This can be modeled by a current source, which depends on the optical flux in parallel with the junction diode, as shown in Figure 13.23.

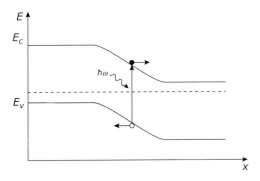

Figure 13.22 p-n junction under optical illumination used as a photodetector or photocell.

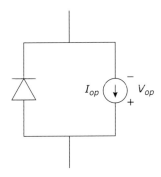

Figure 13.23 Circuit model for an illuminated photocell junction.

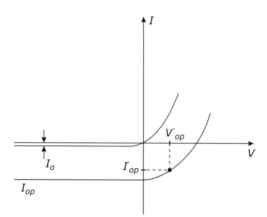

Figure 13.24 Current–voltage characteristics of optically illuminated p-n junction I_{op} compared with the dark junction. The operating point of the junction used as a solar cell with current I_{op} extracted from the junction is shown.

The voltage–current characteristics of the illuminated junction are shown in Figure 13.24, where the current is the sum of the diode current and the optically induced current. **For use as a photodiode detector**, the junction can be run in reverse bias, where the difference between the reverse leakage current I_o and the optically induced current I_{op} is a measure of the optical flux on the device.

For solar cell applications the junction acts like a voltage/current source, producing a current I_{op} into a load at a voltage V_{op}. As the V_{op} increases, the current is shorted through the diode current. The difference between I_{op} and the diode current is the current delivered to the load. The operating point is chosen at (I_{op}', V_{op}'), so that the power delivered to the load, $P_L = I_{op}' \cdot V_{op}'$ (the area of the rectangle outlined in the figure) is maximum.

Chapter 13 problems

1. Consider a semiconductor with the band structure shown doped with $N_d = 1 \times 10^{22}$ m^{-3} donors. The energy gap is 1.1 eV and the donor state is $E_d = 0.044$ eV below the conduction band edge. The energy of the conduction band can be represented by $E = E_g + \frac{\hbar^2 k^2}{2m_c*}$ and the energy of the valence band by $E = -\frac{\hbar^2 k^2}{2m_v*}$ with $m_c* = 1.1 \, m_o$ and $m_v* = 0.56 \, m_o$. (These parameters, although not the simplified E vs. k diagram, closely represent silicon with phosphorus donors.)

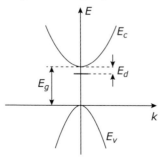

The total number of occupied states in the conduction band is given by $n = \int_{E_g}^{\infty} D_c(E) f(E) dE$ and the total number of unoccupied states in the conduction band (holes) is given by: $p = \int_{-\infty}^{0} D_v(E)[1 - f(E)] dE$. The number of donor states that are ionized is: $N_d^+ = N_d[1 - f(E_g - E_d)]$.

a. Using the equation of charge balance $n = N_d^+ + p$ and numerically integrating over the bands, find and plot the position of the Fermi level (E_f) as a function of temperature from $T = 5$ K to $T = 1200$ K.

b. On a single graph, plot the number of electrons n, the number of holes p, and the number of ionized donors N_d^+ as a function of temperature from $T = 5$ K to $T = 1200$ K. Where do most of the electrons come from at temperatures below 800 K? Above 1000 K? What happens at temperatures below 200 K?

c. Plot the percentage of donors that are ionized N_d^+/N_d as a function of temperature from $T = 5$ K to $T = 1200$ K.

2. Based on the quantum statistics of bosons (integral spin particles), we have derived the formula for the blackbody emission per unit area at temperature T at frequencies between ω and $\omega + d\omega$:

$$P_e(\omega, T) d\omega = \varepsilon(\omega, T) \frac{\hbar}{4\pi^2 c^2} \frac{\omega^3 d\omega}{e^{\beta\hbar\omega} - 1} \; (W/m^2)$$

where $\varepsilon(\omega, T)$ is the object's emissivity. The blackbody spectral emission function is an example of a *distribution function* that describes how a quantity (emitted power in this case) is distributed as a function of a continuous variable ω. Because the quantity depends on the size of the differential interval, the characteristics of a distribution function can depend on what variable it is expressed in terms of. On the other hand, the integral of the distribution function across any interval should not depend on the units used.

a. Plot $P_e(\omega, T)$ as a function of ω from $\omega_1 = 2\pi c/10000$ nm to $\omega_2 = 2\pi c/100$ nm for an object with $\varepsilon(\lambda, T) = 1$ ("blackbody") at $T = 5000$ K. Compare linear and semilog plots. Which is more useful?

b. What is the frequency of maximum energy emission? What wavelength does this frequency correspond to?

c. Use numerical integration to determine how much total power is emitted between $\omega = 2\pi c/450$ nm and $\omega = 2\pi c/600$ nm from a 1 cm^2 area of this object.

3. For Problem 2:

a. Find an expression for $P_e(\lambda, T) d\lambda$, the power emitted per unit area between λ and $\lambda + d\lambda$ by an object at temperature T with emissivity $\varepsilon(\lambda, T)$.

b. Plot $P_e(\lambda, T)$ as a function of λ for 10 nm $< \lambda <$ 10000 nm for an object with $\varepsilon(\lambda, T) = 1$ ("blackbody") at $T = 5000$ K. Use a semilog plot.

c. What is the wavelength of maximum energy emission?

d. Use numerical integration to determine how much total power is emitted between $\lambda = 450$ nm and $\lambda = 600$ nm from a 1 cm^2 area of this object.

4. For Problem 2:

    a. Find an expression for $N_e(\lambda, T)d\lambda$, the number of photons emitted per unit area between λ and $\lambda + d\lambda$ by an object at temperature T with emissivity $\varepsilon(\lambda, T)$.

    b. Plot $I(\lambda, T)$ as a function of λ for 10 nm $< \lambda < 10000$ nm for an object with $\varepsilon(\lambda, T) = 1$ ("blackbody") at $T = 5000$ K. Use a semilog plot.

    c. What is the wavelength at which the maximum number of photons is emitted?

5. For Problem 2:

    a. Find an expression for $N_e(\omega, T)d\omega$, the number of photons emitted per unit area between ω and $\omega + d\omega$ by an object at temperature T with emissivity $\varepsilon(\omega, T)$.

    b. Plot $N_e(\omega, T)$ as a function of ω from $\omega_1 = 2\pi c/10000$ nm to $\omega_2 = 2\pi c/100$ nm for an object with $\varepsilon(\omega, T) = 1$ ("blackbody") at $T = 5000$ K. Use a semilog plot.

    c. At what frequency is the maximum number of photons emitted? What wavelength does this frequency correspond to?

    d. It is sometimes stated that the human eye, with a maximum sensitivity at around 550 nm, has evolved to be most sensitive at the wavelength of the maximum output of the Sun (modeled as a blackbody of $T = 5000$ K). Based on the results of this exercise, would you have any misgivings about this statement?

6. Choose three applications of p-n junction diodes and discuss the science behind each.

7. Find the probability that an energy level of energy E is NOT occupied, if the probability of the same energy level being occupied is

$$P(E) = \frac{1}{e^{(E-E_f)/k_B T} + 1}.$$

8. Find the probability that a state at the Fermi energy $E = E_f$ will be occupied for any $T > 0$.

9. Find the Fermi energy of aluminum if you know that it has three conduction electrons per atom, mass density 2.70 g/cm^3 and molar mass 27.0 g/mol.

10. The resistivity of a metal increases with temperature, as the number of collisions for electrons increases. Discuss what happens to the resistivity of semiconductors as temperature increases.

11. Plot the probability of a state with energy E, with $E - E_f = 1.10$ eV, being occupied as a function of the absolute temperature T using Fermi–Dirac and classical Maxwell–Boltzmann statistics. For which temperatures do the theories show better than 5% agreement?

12. If the band gap of a KCl crystal is 7.6 eV, to which wavelengths is this crystal opaque?

13. A metal has a work function $\Phi = 1.5$ eV as shown below.

    b. Suppose this metal was heated to 1500 K. What is the probability that an electron state 1.5 eV above the Fermi energy would be occupied?

    c. Describe in words the steps that you might take to estimate the number of electrons that would be *thermionically emitted* from the metal if it were heated to 1500 K.

14 Maxwell's equations and electromagnetism

On laboratory triple-beam balances used in chemistry or biology laboratories, there is often a metallic tab on the end of the moving beam which passes through a magnet to damp out the rocking motion of the balance arm. This is described as *eddy current damping*. What exactly is going on here and how does it help reduce the time for the scale balance to come to rest? This is one of the questions that we will address in this chapter, where we consider the theoretical foundations and practical applications of electromagnetic theory.

It is one of the great triumphs of nineteenth-century physics that all electromagnetic phenomena can be described by the four equations that James Clerk Maxwell developed in the 1860s. With some corrections for relativistic and quantum effects, Maxwell's equations are still sufficient to explain phenomena from electric generators and motors to radio waves and optics. In this chapter we will introduce the concept of a *vector field* description of the electric and magnetic fields, develop the *vector calculus* mathematics that will allow us to express Maxwell's equations as local differential relations between the electric and magnetic fields and their source charges and current. We will then show how Maxwell's equations and electromagnetic material properties contained in the *constitutive relations* can describe low-frequency electromagnetic effects as well as how propagating electromagnetic waves interact with materials. Since our purpose is conceptual understanding rather than computational capability, we will restrict our three-dimensional analysis primarily to Cartesian x–y–z coordinates.

14.1 Vector fields

Consider a room half-filled with water. Assume that water is flowing in one door and flowing out of another door, as shown in Figure 14.1. The velocity of the water in the room can be described by a two-dimensional *vector field*. At each point in the room the velocity of the water will be different in direction and magnitude. So we can associate with each point (x, y) a vector that defines the direction and magnitude of the velocity of the water flow at that point

$$\vec{v}\,(x, y) = v_x(x, y)\hat{x} + v_y(x, y)\hat{y}. \tag{14.1}$$

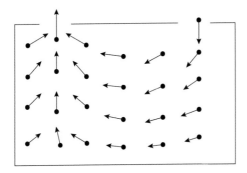

Figure 14.1 Two-dimensional vector field of water flow in one door and out of another door in a room.

Every point has a different value of the x-velocity of the water v_x and the y-velocity of the water. The direction of the flow is given by the ratio of v_y to v_x and the speed of the flow is given by our usual Pythagorean theorem result: $|v| = \sqrt{v_x{}^2 + v_y{}^2}$. So we can describe the flow equally well by a continuously changing vector velocity with a distinct magnitude and direction at each point $\vec{v}\,(x, y)$ or as the scalar value of the components $v_x(x, y)$ and $v_y(x, y)$ at each (x, y) point.

The electric and magnetic fields are similarly defined at every point in three-dimensional space as vector fields, $\vec{E}\,(x, y, z)$ and $\vec{H}\,(x, y, z)$, remembering that this is a shorthand for six scalar component fields: $E_x(x, y, z)$, $E_y(x, y, z)$, $E_z(x, y, z)$, $H_x(x, y, z)$, $H_y(x, y, z)$, and $H_z(x, y, z)$. At any point (x, y, z) we have:

$$\vec{E} = E_x(x, y, z)\hat{x} + E_y(x, y, z)\hat{y} + E_z(x, y, z)\hat{z} \tag{14.2}$$

$$\vec{H} = H_x(x, y, z)\hat{x} + H_y(x, y, z)\hat{y} + H_z(x, y, z)\hat{z}. \tag{14.3}$$

Note that in this chapter we will use the symbol E to represent the electric field and not the energy. If we make a reference to energy, we will make the distinction clear.

Question 14.1

Can you name two other examples of vector fields and two examples of scalar fields?

14.2   Vector operators

Since the electric and magnetic fields are continuous in space, we can talk about spatial derivatives of the fields, or of the field components. For example, the derivative $\frac{\partial E_x}{\partial y}$ is a well-defined quantity. However, it will be useful to have a shortcut to refer to the variation in space of all three components of the electric and magnetic fields. The vector differential operator, *del*, is defined as

$$\vec{\nabla} = \hat{x}\frac{\partial}{\partial x} + \hat{y}\frac{\partial}{\partial y} + \hat{z}\frac{\partial}{\partial z} \tag{14.4}$$

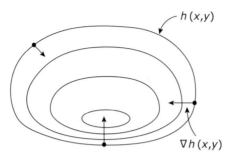

Figure 14.2 Topographic contours of height $h(x, y)$ with arrows indicating the gradient $\nabla h(x, y)$.

where the vector properties are determined by the unit vectors, and the partial derivatives act on the scalar fields selected out by the vector operation to which del is applied.

14.2.1 Gradient

The simplest vector calculus operation is the *gradient* in which del is applied to a single scalar field $\phi(x, y, z)$. The unit vectors are simply multiplied by ϕ giving

$$\vec{\nabla} \phi = \hat{x}\frac{\partial \phi}{\partial x} + \hat{y}\frac{\partial \phi}{\partial y} + \hat{z}\frac{\partial \phi}{\partial z}. \tag{14.5}$$

For example, assume that the scalar field ϕ represents the elevation at the point (x, y), $\phi = h(x, y)$. The lines that connect points of equal elevation are the contours on a topographic map. The gradient of h at any point is a vector that is directed directly up the slope at that point; the magnitude of the gradient is proportional to the steepness of the slope. A topographic representation of a hill with one steep side is shown in Figure 14.2 and the vector gradient of the elevation is shown at three points. Since h is increasing toward the top of the hill, the partial derivatives are positive in the direction toward the top of the hill. *The gradient operates on a scalar field and produces a vector field.*

14.2.2 Divergence

The *divergence* operation is the *dot product* of the del operator with a *vector field*. For example, if we have a vector field $\vec{U}(x, y, z) = \hat{x}U_x(x, y, z) + \hat{y}U_y(x, y, z) + \hat{z}U_z(x, y, z)$ the divergence of $\vec{U}$ is

$$\vec{\nabla} \cdot \vec{U} = \left(\hat{x}\frac{\partial}{\partial x} + \hat{y}\frac{\partial}{\partial y} + \hat{z}\frac{\partial}{\partial z}\right) \cdot (\hat{x}U_x + \hat{y}U_y + \hat{z}U_z) = \frac{\partial U_x}{\partial x} + \frac{\partial U_y}{\partial y} + \frac{\partial U_z}{\partial z} \tag{14.6}$$

using the rules of the vector dot product in Cartesian coordinates $\hat{x} \cdot \hat{x} = \hat{y} \cdot \hat{y} = \hat{z} \cdot \hat{z} = 1$ while the dot product of any two different unit vectors is zero, $\hat{x} \cdot \hat{y} = 0$ for example. Since the dot product between two vectors gives a scalar result, the dot product between del and a vector field yields a scalar field in the divergence: *the divergence of a vector field is a scalar field.*

Question 14.2

Find the divergence of the vector field $\vec{U}(x, y, z) = x^2\hat{x} + \cos z\hat{y} + (1 - y)\hat{z}$.

14.2.3 Curl

The *curl* operator is the *cross product* of the *del* operator with a *vector field*. Using the determinant form for the cross product in Cartesian coordinates, we find

$$\vec{\nabla} \times \vec{U} = \left(\hat{x}\frac{\partial}{\partial x} + \hat{y}\frac{\partial}{\partial y} + \hat{z}\frac{\partial}{\partial z}\right) \times (\hat{x}U_x + \hat{y}U_y + \hat{z}U_z) = \begin{vmatrix} \hat{x} & \hat{y} & \hat{z} \\ \frac{\partial}{\partial x} & \frac{\partial}{\partial y} & \frac{\partial}{\partial z} \\ U_x & U_y & U_z \end{vmatrix}$$

$$= \hat{x}\left(\frac{\partial U_z}{\partial y} - \frac{\partial U_y}{\partial z}\right) - \hat{y}\left(\frac{\partial U_z}{\partial x} - \frac{\partial U_x}{\partial z}\right) + \hat{z}\left(\frac{\partial U_y}{\partial x} - \frac{\partial U_x}{\partial y}\right). \tag{14.7}$$

Question 14.3

Find the curl of the vector field $\vec{U}(x, y, z) = x^2\hat{x} + \cos z\hat{y} + (1 - y)\hat{z}$.

14.3 Gauss's theorem

To arrive at a physical understanding of the divergence and curl operations, we need to consider two mathematical theorems that relate to integrals of vector operators. Consider first *Gauss's theorem* that connects the *volume integral* of the divergence of a vector field $\vec{U}$ with the *surface integral* of that vector field over the surface that encloses the volume:

$$\int_V (\vec{\nabla} \cdot \vec{U})dV = \oint_S \vec{U} \cdot d\vec{S}. \tag{14.8}$$

The left-hand side is the integral over the volume V of the scalar field $\vec{\nabla} \cdot \vec{U}(x, y, z)$. Gauss's theorem equates the three-dimensional volume integral to the two-dimensional surface integral – over the entire surface that confines the volume as the closed loop on the integral symbol reminds us – of the dot product of vector field with an infinitesimal *surface element* $d\vec{S}$. The magnitude of $d\vec{S}$ is the infinitesimal area of the surface at a particular point (in m²) and the direction of $d\vec{S}$ is the normal to the surface at that point which points *out* of the volume, as shown in Figure 14.3.

Figure 14.3 The surface element $d\vec{S}$ has a magnitude of an infinitesimal surface area dS and a direction normal to the surface.

As an example, consider the *charge continuity equation*, written in differential form,

$$\frac{\partial \rho}{\partial t} + \vec{\nabla} \cdot \vec{J} = 0 \tag{14.9}$$

where ρ is the *charge density* (C/m^3) and $\vec{J}$ is the *current density* ($A/m^2 = C/s\ m^2$). In differential form, the continuity equation is a local relation between the time derivative of the charge density at a point and the divergence (spatial derivative) at the same point. Applying Gauss's theorem can illuminate the physical implications of continuity. Integrating the continuity equation over a volume of space and then applying Gauss's theorem yields

$$\int_V \left(\frac{\partial \rho}{\partial t}\right) dV = -\int_V \vec{\nabla} \cdot \vec{J}\, dV = -\oint_S \vec{J} \cdot d\vec{S}. \tag{14.10}$$

The integral of the charge density over the volume is just the enclosed charge $Q_{enclosed}$. With that replacement the charge continuity equation can be written as

$$\frac{\partial}{\partial t} Q_{enclosed} = -\oint_S \vec{J} \cdot d\vec{S}. \tag{14.11}$$

So the time rate of change in the total charge within the volume is equal to the negative of the integral of $\vec{J} \cdot d\vec{S}$ – the current directed outward through the surface element $d\vec{S}$ – over the entire enclosing surface. In this form, it is easy to understand that if there is net current perpendicular to the surface leaving the surface, the surface integral is positive and, because of the negative sign, the net enclosed charge will decrease. If there is net current flowing perpendicular to the surface into the volume, the enclosed charge will increase. If the volume is in a region where $\vec{J}$ is uniform, on the other hand, the integral of $\vec{J} \cdot d\vec{S}$ over the enclosing surface will be equal to zero because the integral of $\vec{J} \cdot d\vec{S}$ on the side where the current is out of the surface will be exactly canceled by the integral of $\vec{J} \cdot d\vec{S}$ over the other half of the surface where the current is into the surface. Thus, in the case of uniform current density, there will be no change in the enclosed charge.

This example leads us to a physical interpretation of the divergence of a vector field at any point. Consider a small volume δV containing the point and let δV shrink to zero around the point. Then the divergence of a vector field $\vec{U}(x, y, z)$ at that point is given by:

$$\vec{\nabla} \cdot \vec{U} = \lim_{\delta V \to 0} \frac{1}{\delta V} \oint_S \vec{U} \cdot d\vec{S} \tag{14.12}$$

where S is the surface surrounding the infinitesimal volume δV. If there is a net flux of the vector field $\vec{U}$ out of the volume $\vec{\nabla} \cdot \vec{U}$ is positive. If there is a net flux into the volume, $\vec{\nabla} \cdot \vec{U}$ is negative, and if there is no net flux through the surface S, the divergence is zero. So, $(\vec{\nabla} \cdot \vec{U})$ *gives us the net flux of the vector field* $\vec{U}$ *into or out of a volume* δV.

Question 14.4

Is temperature $T(x, y, z)$ a vector field? What about the temperature gradient $\vec{\nabla} T(x, y, z)$? If the gradient of the temperature integrated over a closed surface ($\oint_S \vec{\nabla} T \cdot d\vec{S}$) is non-zero, what can you say about what might be inside that surface? Using Gauss's theorem can you write an equation involving the divergence of the gradient of temperature $\vec{\nabla} \cdot \vec{\nabla} T(x, y, z)$?

14.4 Stokes' theorem

Stokes' theorem relates the integral of the curl of a vector field over a surface to the line integral of the dot product of the vector field and an element of length on the line that is the border of the surface

$$\int_S (\vec{\nabla} \times \vec{U}) \cdot d\vec{S} = \oint_\ell \vec{U} \cdot d\vec{\ell}. \tag{14.13}$$

Since there are two directions that the line integral could be taken, clockwise or counter-clockwise, the definition of the curl is that the direction of integration is established by the right-hand rule: the direction of $d\vec{\ell}$ is given by the direction that the fingers of your right hand curl if your thumb is in the direction of $\vec{U}$ through the surface. In Figure 14.4,

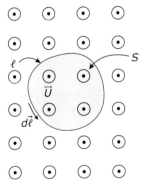

Figure 14.4 Relation of the surface S and the bounding line ℓ. The direction of $d\vec{\ell}$ is given by the right-hand rule.

Figure 14.5 Infinitesimal surface area dS with normal $\hat{s}$ and bounding line ℓ. The direction of $d\vec{\ell}$ is defined by the right-hand rule with respect to $\hat{s}$.

with the vector field directed out through the surface, $d\vec{\ell}$ is in the counter-clockwise direction as given by the direction of the fingers of the right hand if the thumb is directed out of the page.

Stokes' theorem can be used to create a definition of the curl of a vector field at a particular point. If we consider an infinitesimal surface element dS bordered by a closed line ℓ with a unit vector $\hat{s}$ directed normal to the surface (see Figure 14.5), then the $\hat{s}$ component of the curl of the vector field $\vec{U}(x, y, z)$ at any point is given by

$$(\vec{\nabla} \times \vec{U})_{\hat{s}} = \lim_{dS \to 0} \frac{\hat{s}}{dS} \oint_{\ell} \vec{U} \cdot d\vec{\ell} \tag{14.14}$$

where it is understood that the direction of the line integral is defined by the right-hand rule with respect to the direction of $\hat{s}$.

Given this definition, we understand that *the curl at any point is non-zero if the closed loop integral of the vector field is not zero, i.e. if the vector field has a component that is circulating around the point, or is larger on one side of the point than on the other side.*

14.5 Vector calculus identities

Below are three vector identities, which we will use later in the chapter. These identities can be verified by expansion of the dels in Cartesian coordinates where $\vec{U}$ is any vector field and ϕ is any scalar field.

$$\vec{\nabla} \cdot (\vec{\nabla} \times \vec{U}) = 0, \tag{14.15}$$

$$\vec{\nabla} \times \vec{\nabla} \phi = 0, \tag{14.16}$$

$$\vec{\nabla} \times (\vec{\nabla} \times \vec{U}) = \vec{\nabla}(\vec{\nabla} \cdot \vec{U}) - \vec{\nabla}^2 \vec{U}. \tag{14.17}$$

The *vector Laplacian* $\vec{\nabla}^2 \vec{U}$ introduced in the last identity in Cartesian coordinates is given by

$$\vec{\nabla}^2 \vec{U} = \hat{x}\nabla^2 U_x + \hat{y}\nabla^2 U_y + \hat{z}\nabla^2 U_z \tag{14.18}$$

where $\nabla^2 \phi = \frac{\partial^2 \phi}{\partial x^2} + \frac{\partial^2 \phi}{\partial y^2} + \frac{\partial^2 \phi}{\partial z^2}$ is the usual scalar Laplacian. (In other coordinate systems, Eq. (14.16) can be considered a definition of $\vec{\nabla}^2 \vec{U}$.)

Question 14.5

Show by writing out the component equations that $\vec{\nabla} \times \vec{\nabla} \phi = 0$.

14.6 Maxwell's equations

With these preliminaries we can write Maxwell's equations in differential (vector calculus) form as

$$\vec{\nabla} \cdot \vec{D} = \rho \tag{14.19}$$

$$\vec{\nabla} \times \vec{E} = -\frac{\partial \vec{B}}{\partial t} \tag{14.20}$$

$$\vec{\nabla} \cdot \vec{B} = 0 \tag{14.21}$$

$$\vec{\nabla} \times \vec{H} = \vec{J} + \frac{\partial \vec{D}}{\partial t}. \tag{14.22}$$

These equations are *local equations*: the divergence of $\vec{D}$ at point $\vec{r}_1$, for example, is related only to the free charge density at $\vec{r}_1$, etc. The equations contain two source quantities that we saw in the continuity equation, the scalar charge density ρ in coulombs/m^3 and the vector current density $\vec{J}$ in amperes/m^2, and four vector fields, the *electric field* $\vec{E}$ in newtons/coulomb or volts/m, the *electric displacement* $\vec{D}$ in coulombs/m^2, the *magnetic field* $\vec{H}$ in amps/m, and the *magnetic flux density* $\vec{B}$ (also called *magnetic induction field*) in teslas or volt s/m^2. The relations between $\vec{D}$ and $\vec{E}$, between $\vec{H}$ and $\vec{B}$, and between $\vec{J}$ and $\vec{E}$ depend on the properties of the materials that the fields are in. The electromagnetic properties of materials are completely determined by the three *constitutive relations*

$$\vec{D} = \varepsilon \vec{E} \tag{14.23}$$

$$\vec{B} = \mu \vec{H} \tag{14.24}$$

$$\vec{J} = \sigma \vec{E} \tag{14.25}$$

where the *permittivity* ε, in units of farads/m ($=$ coulombs/volt m), describes the effect of a *dielectric material* on the external $\vec{D}$ field that yields the $\vec{E}$ field inside the material; the *permeability* μ, in units of henries/m ($=$ kg m/coulombs^2) describes the effect of a *magnetic material* on the external $\vec{H}$ field that yields the $\vec{B}$ field inside the material; and the *conductivity* σ in siemens/m ($=$ 1/ohm m) which describes the charge current density $\vec{J}$ that is caused by a given electric field $\vec{E}$ in a *conducting material*. In free space (vacuum) $\epsilon = \varepsilon_o = 8.8543 \times 10^{-12}$ F/m, $\mu = \mu_o = 4\pi \times 10^{-7}$ H/m, and $\sigma = 0$, since there are no charges in complete vacuum.

The last major equation of electromagnetics describes the forces that electromagnetic fields exert on objects. The electromagnetic force on an object with charge q moving with velocity $\vec{v}$ is described by the *Lorentz force*

$$\vec{F} = q\vec{E} + q\vec{v} \times \vec{B}. \tag{14.26}$$

The Lorentz force is a combination of the familiar Coulomb's Law force between two charges – opposite charges attract, like charges repel – (first term) and the magnetic force on a moving charge, perpendicular to the particle's velocity and to the magnetic flux density (second term). All electromagnetic forces derive from the Lorentz force equation.

To the first approximation, macroscopic objects are most commonly electrically neutral. The electromagnetic force on large objects typically arises either (1) because the electric field induces charge separation and thus creates a net *electrostatic* attraction or repulsion, or (2) because only one type of charge – usually the negative electrons in a bulk metal or metal wire – is moving, while the charge carriers of the other sign – typically the positive ion cores – are immobile. In this second case, the force on the object can be expressed in terms of the current carried by the mobile carriers.

Question 14.6

Which term from the Lorentz force equation Eq. (14.26) is responsible for the force described in (1) in the paragraph above? For the force described in (2) above?

Consider a wire of length ℓ carrying a current I in a region of magnetic flux density $\vec{B}$ as shown in Figure 14.6. The differential form of the Lorentz force on an infinitesimal amount of charge dq in an infinitesimal length $d\vec{l}$ is given by

$$d\vec{F} = dq\vec{E} + dq\vec{v} \times \vec{B}. \tag{14.27}$$

We can neglect the first term proportional to $\vec{E}$ because the force due to the electric field on the moving charge carriers (electrons) is canceled out by the opposite force on the equal density of background (positive) lattice charge. The second term, however, does not cancel out because only the electrons are moving. Since the direction of the velocity $\vec{v}$ is parallel to the direction of the wire $d\vec{\ell}$ we can write $\vec{v} = d\vec{\ell}/dt$. The time differential dt is a scalar and can be moved out of the cross product, yielding

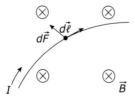

Figure 14.6 Element of current $d\vec{\ell}$. The direction of the force on $d\vec{\ell}$ is perpendicular to $\vec{B}$ and to $d\vec{\ell}$, in the plane of the page and normal to $d\vec{\ell}$ in this example.

$$dF = \frac{dq}{dt} d\vec{\ell} \times \vec{B} = I d\vec{\ell} \times \vec{B} \tag{14.28}$$

where we have used the definition of current $I = \frac{dq}{dt}$. The total force on a current-carrying wire can thus be found by integrating $I d\vec{\ell} \times \vec{B}$ over the entire length of the wire in the magnetic field.

Note that although the force is exerted on the moving charge – usually the electrons – the negative charge on the electrons makes the force in the opposite direction to $\vec{v}_e \times \vec{B}$ where $\vec{v}_e$ is the electron velocity, i.e., it is in the direction of $\vec{I} \times \vec{B}$ where $\vec{I}$ is the conventional current given by the equivalent direction of *positive charge movement*. Thus $d\vec{\ell}$ in Eq. (14.28) is in the direction of I as shown in Figure 14.6.

Example 14.1 Find the force on a semicircular segment of wire of radius 0.35 m carrying 0.25 A of current in a magnetic field $B = 0.5$ tesla as shown.

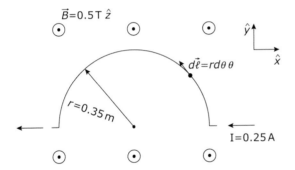

Solution: The force on any segment of arc is given by $d\vec{F} = I \, d\vec{\ell} \times \vec{B} = I \, r \, d\theta \, \hat{\theta} \times \vec{B}$. Expressing the $\hat{\theta}$ unit vector in terms of Cartesian coordinates $\hat{\theta} = -\hat{x} \sin \theta + \hat{y} \cos \theta$, we find $d\vec{F} = IBr \, d\theta(-\hat{x} \sin \theta + \hat{y} \cos \theta) \times \hat{z} = IBr \, d\theta(\hat{y} \sin \theta + \hat{x} \cos \theta)$. Integrating from $\theta = 0$ to π we find:

$$\vec{F} = IBr \int_0^\pi (\hat{y} \sin \theta + \hat{x} \cos \theta) d\theta = IBr[-\hat{y} \cos \theta + \hat{x} \sin \theta]_0^\pi$$

$$= IBr \, \hat{y}(1 + 1) = 0.25 \cdot 0.5 \cdot 0.35 \cdot 2\hat{y} = 0.0875\hat{y} \text{ N.}$$

The complete differential form of the Lorentz force, useful to find the force on an object containing a net infinitesimal charge dq and a current element of infinitesimal length $I d\vec{\ell}$ is

$$d\vec{F} = dq\vec{E} + I d\vec{\ell} \times \vec{B}. \tag{14.29}$$

Question 14.7

A 1 mm long section of wire parallel to the x-axis has a positive charge of 1.1 μC and a negative charge of 1.0 μC. The negative charge is moving in the $+\hat{x}$ direction at a velocity of 1 mm/s while the positive charge is not moving. What is the dq of this wire section? What is the current I? What is the size and direction of $d\,\vec{l}$? If both $\vec{E}$ and $\vec{B}$ are in the $+\hat{z}$ direction, what direction is the electric field force on the wire segment? What direction is the magnetic field force on the wire segment?

14.7 Electromagnetic potentials

We know from our vector identities that $\vec{\nabla} \cdot (\vec{\nabla} \times \vec{V}) = 0$ where $\vec{V}$ is any vector field. This implies that any vector field with a zero divergence can be expressed as the curl of another vector field. Since Maxwell's equations tell us that the divergence of the magnetic induction is zero, $\vec{\nabla} \cdot \vec{B} = 0$, this implies that the magnetic induction $\vec{B}$ can be described as the curl of another vector field $\vec{A}$

$$\vec{B} = \vec{\nabla} \times \vec{A}. \tag{14.30}$$

$\vec{A}$ is the *magnetic vector potential* in units of tesla·m. Since we also know that the curl of the gradient of any scalar field is zero, the vector potential is not unique. Any potential that is created by adding the gradient of a scalar field to $\vec{A}$ will also yield the same physical $\vec{B}$ field:

$$\vec{\nabla} \times \vec{A}' = \vec{\nabla} \times (\vec{A} + \vec{\nabla \psi}) = \vec{\nabla} \times \vec{A} = \vec{B}. \tag{14.31}$$

The transformation of adding the gradient of a scalar field to the vector potential to create another physically identical vector potential is called a **gage transformation**.

Rewriting Maxwell's equation $\vec{\nabla} \times \vec{E} = -\frac{\partial \vec{B}}{\partial t}$, as $\vec{\nabla} \times \vec{E} + \frac{\partial \vec{B}}{\partial t} = 0$, substituting $\vec{B} = \vec{\nabla} \times \vec{A}$ and interchanging the order of time and spatial differentiation, we find

$$\vec{\nabla} \times \vec{E} + \frac{\partial}{\partial t}(\vec{\nabla} \times \vec{A}) = \vec{\nabla} \times \vec{E} + \left(\vec{\nabla} \times \frac{\partial \vec{A}}{\partial t}\right) = \vec{\nabla} \times \left(\vec{E} + \frac{\partial \vec{A}}{\partial t}\right) = 0. \tag{14.32}$$

The vector identity $\vec{\nabla} \times \vec{\nabla} \phi = 0$ and the fact that the curl equals zero suggests that the quantity $\left(\vec{E} + \frac{\partial \vec{A}}{\partial t}\right)$ can be expressed as the gradient of a scalar field. By convention, the negative of the *electric potential*, measured in volts, is defined as

$$-\vec{\nabla}\varphi = \vec{E} + \frac{\partial \vec{A}}{\partial t}. \tag{14.33}$$

Since the gradient of a constant is zero, we can add any constant to the electric potential $\varphi' = \varphi + \varphi_o$ without changing any physical effect. This corresponds to the fact that the zero of potential (the electrical *ground* of the system) is arbitrary.

At low frequencies where we can ignore the time dependence of $\vec{A}$, $\frac{\partial \vec{A}}{\partial t} = 0$, and we have the well-known *electrostatic* relationship:

$$\vec{E} = -\vec{\nabla}\varphi \tag{14.34}$$

where φ is the familiar electric potential or voltage (as in a *12-volt battery*, for example) of any point with respect to a ground point where $\varphi = 0$. This equation allows us to define the units of electric field to be volts/m.

14.8 Maxwell's equations in integral form

Maxwell's equations as we have written them are strictly *local* relationships. The vector derivatives of the magnetic and electric fields at any point are related to the sources (charges and currents) and the time derivatives of the fields *at the same point*. We can, however, apply Gauss's and Stokes' theorems to Maxwell's equations in differential form and express them in *integral forms*. We note that this transformation to integral form will result in *non-local* relationships – the fields at one point will be related to sources and fields elsewhere. On the other hand, the integral relations involve the electric and magnetic fields themselves, not the spatial derivatives of the fields as in the differential form. The integral forms are particularly useful in situations of high symmetry where the integrals can be evaluated easily to find the fields.

In nearly all cases except for a handful of special symmetries, however, the electric and magnetic vector fields need to be calculated by solving the vector differential equations that result from the differential form of Maxwell's equations with known *boundary conditions* – given values of the fields (or the field derivatives) on the boundary of the volume where the solution is to be obtained. Exact solutions of Maxwell's equations involve complex programming or software packages, and full three-dimensional solutions are significantly limited in the size of the volume that can be addressed.

Because of their historical interest and because of the relation between fields and sources given by the high symmetry cases, we will review the integral forms of Maxwell's equations and some of the simple solutions below.

14.8.1 Coulomb's Law: $\vec{\nabla} \cdot \vec{D} = \rho$

Taking the integral of the first Maxwell equation over a volume $\mathcal{V}$ inside a surface S, we find

$$\int_{\mathcal{V}} \vec{\nabla} \cdot \vec{D} \, d\mathcal{V} = \int_{\mathcal{V}} \rho \, d\mathcal{V}. \tag{14.35}$$

Applying Gauss's theorem, and recognizing the volume integral of the charge density is just the charge enclosed inside the surface of integration, we find

$$\oint_S \vec{D} \cdot d\vec{S} = Q_{enclosed}. \tag{14.36}$$

where $d\vec{S}$ has the dimension of a small element of the surface with a direction normal to the surface at the given point.

As the simplest example, consider a charge Q in vacuum with the surface S a sphere centered on Q. By symmetry, the electric displacement field $\vec{D}$ is radially directed (parallel to $d\vec{S}$ everywhere) and, since there is no preferred direction, can only depend on the distance from the charge to the sphere. The integral equation reduces to

$$4\pi r^2 D = Q_{enclosed}. \tag{14.37}$$

With the relationship between $\vec{D}$ and $\vec{E}$ given in vacuum by $\vec{D} = \epsilon_o \vec{E}$, we find

$$\vec{E} = \frac{1}{4\pi\epsilon_o} \frac{Q_{enclosed}}{r^2} \hat{r}. \tag{14.38}$$

The static force on a charge q at any point is given by the Lorentz force

$$\vec{F} = \frac{1}{4\pi\epsilon_o} \frac{q\, Q_{enclosed}}{r^2} \hat{r}$$

which we recognize as Coulomb's Law for the force between two charges, q and $Q_{enclosed}$, separated by a distance r. Since fields and forces are vectors, the total electric field or force due to a collection of charges is just the vector sum of the fields or forces from the individual charges, calculated as if the other charges were not present. This is the principle of *superposition*.

Example 14.2   Electric dipole

Find the net electric field in the x–y plane created by two equal charges $+q$ and $-q$ separated by a distance d.

Solution: Putting the origin of the coordinate system halfway between the two charges, the electric field at any point $\vec{r} = x\hat{x} + y\hat{y}$ can be found by adding the electric field of the $+q$ charge at a relative position $\vec{r}_1 = x\hat{x} + \left(y - \frac{d}{2}\right)\hat{y}$ to the electric field of the $-q$ charge at a relative position $\vec{r}_2 = x\hat{x} + \left(y + \frac{d}{2}\right)\hat{y}$. Using Eq. (14.38), we find

$$\vec{E} = \vec{E}_1 + \vec{E}_2 = \frac{1}{4\pi\epsilon_o} \frac{q}{r_1{}^3} \vec{r}_1 - \frac{1}{4\pi\epsilon_o} \frac{q}{r_2{}^3} \vec{r}_2$$

where we have used the definition of the unit vector $\hat{r} = \frac{\vec{r}}{r}$ where $r = \sqrt{x^2 + y^2}$ is the magnitude of the vector. The MATLAB code to create a "quiver plot" of the electric field direction and magnitude is below:

```
d=.8;
[x,y] =meshgrid(-2.1:.2:2.1);
y1=y-(d/2)*ones(size(y));
y2=y+(d/2)*ones(size(y));
magr1=sqrt(x.^2 +y1.^2);
magr2=sqrt(x.^2 + y2.^2);
Ex=x./ magr1.^3 - x./ magr2.^3;
Ey=y1./ magr1.^3 - y2./ magr2.^3;
quiver(x,y,Ex,Ey,.5)
axis([-1,1,-1, 1])
axis square
```

Continued

Example 14.2 (*cont.*)

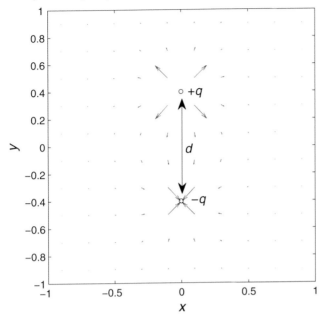

Another high-symmetry geometry is a parallel-plate capacitor: two infinite parallel plates with equal and opposite surface charge density (coulombs/m^2) ρ_s and $-\rho_s$ on the two plates as shown in Figure 14.7. First we will assume that the two plates are in vacuum. Later we will assume that the space between the plates is filled with a polarizable medium – one which undergoes some degree of charge separation when exposed to an electric field.

If the plates are in vacuum, it is clear that the electric fields are normal to the plates, directed away from the positive plate and toward the negative plate in the direction that a positive test charge would be accelerated. If the charge densities on the two plates are equal in magnitude and opposite in sign, there is no field outside where the fields from the individual plates cancel (one plate pushing and one plate pulling the test charge). Between the plates, however the fields of the plates reinforce each other, the positive plate pushing and the negative plate pulling the test charge in the same direction.

For the surface integration S in this case we can use a "Gaussian pillbox" – a cylinder with ends parallel to the plates and sides perpendicular to the plates as shown in Figure 14.7. The area of one cylindrical end is A. Again we have $\oint_S \vec{D} \cdot d\vec{S} = Q_{enclosed}$. The integral can be broken up into an integral over the top surface of the "pillbox," an integral over the bottom surface, and an integral over the cylindrical curved surface. The integral over the top surface is zero because, due to the cancelation of the fields from the positive and negative plates, there are no fields outside the plates. The integral over the curved surface is also zero because, at each point on the surface, the normal to the

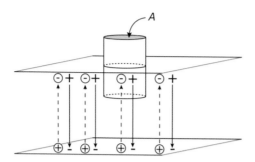

Figure 14.7 Electric field in a parallel-plate capacitor with positive charge on the top plate and negative charge on the bottom plate. The electric field is in the direction that a positive test charge would move (down). If the capacitor is filled with a dielectric material the positive bound charges are induced at the bottom of the dielectric and negative bound charges at the top. The bound charges create an induced electric field that opposes (and reduces) the electric field from the charges on the plates.

surface is parallel to the plates while the fields are perpendicular to the plates. Thus $\vec{D} \perp d\vec{S}$ and $\vec{D} \cdot d\vec{S} = 0$. The only contribution to the integral is from the bottom surface – assuming the field is uniform (as it must be from symmetry!) – the integral is $\oint_S \vec{D} \cdot d\vec{S} = \vec{D} \cdot \hat{n} A = DA$. Since the field is perpendicular to the bottom surface the dot product with the normal unit vector just gives $\cos\theta = \cos(0) = 1$. The enclosed charge is just the surface charge density times the cross-section area of the cylinder, $Q_{enclosed} = \rho_s A$. Putting this together we find,

$$\oint_S \vec{D} \cdot d\vec{S} = Q_{enclosed} \Rightarrow DA = \rho_s A \Rightarrow D = \rho_s. \tag{14.39}$$

If the plates are in free space, the dielectric constant $\varepsilon = \epsilon_o$ and so, from the constitutive relation $\vec{D} = \varepsilon \vec{E}$, the electric field inside two parallel plates in free space is $\vec{E} = \rho_s/\epsilon_o(-\hat{z})$. If the plates are filled with a polarizable medium, then the electric field between the plates causes charge separation in the internal medium leading to extra "bound charges" on the surface of the material next to the plates. This is shown in Figure 14.7. Note that the bound charges are in the opposite orientation to the charge on the plates – the positive surface charge on the plates attracts a negative surface charge layer in the medium – reducing the effective electric field in the material.

This is formally accomplished by defining an *electric polarization* field $\vec{P}$ as the dipole moment density per unit volume, where the dipole moment is the induced polarization charge (equal positive and negative) times the separation between them, $\vec{p} = q\vec{d}$, where $\vec{d}$ points from the *negative to the positive charge*. For an induced charge as shown in Figure 14.7, $\vec{P}$ is proportional to the electric field $\vec{E}$ and in the same direction as $\vec{E}$, thus $\vec{P} = \epsilon_o \chi_e \vec{E}$. Here χ_e is the *dielectric susceptibility*, a dimensionless material parameter which characterizes the electric polarization of the medium in response to an electric field. The factor of ϵ_o, the permittivity of free space, is a constant which maintains the unit convention.

In terms of the electric field and the polarization field, the displacement field is defined as:

$$\vec{D} = \epsilon_o \vec{E} + \vec{P} = \epsilon_o \vec{E} + \epsilon_o \chi_e \vec{E} = (\epsilon_o + \epsilon_o \chi_e)\vec{E}. \tag{14.40}$$

By comparing with the constitutive relation above, we see that the dielectric permittivity of the material is $\varepsilon = (\epsilon_o + \epsilon_o \chi_e)$. The dielectric permittivity thus depends on the susceptibility, which measures the dielectric polarization of the medium in response to an electric field. The susceptibility depends on the properties of the material. In general, the susceptibility χ_e – and thus the dielectric permittivity ε – are frequency dependent, and in anisotropic materials both are 3×3 tensor quantities. In the case of an isotropic material, the dielectric permittivity reduces to a constant, and the electric field inside the material filling the parallel plates is $\vec{E} = \rho_s/\varepsilon(-\hat{z})$. In most cases $\varepsilon > 1$, and the internal electric field is decreased by the material polarization as suggested in the figure. But, as we will see later, in some cases there are frequency ranges where $\varepsilon < 1$ or even $\varepsilon < 0$ in which case the fields inside the material can be larger than and out-of-phase with the driving field. The dielectric and optical properties of materials are contained in the details of the dielectric permittivity ε and in the material conductivity σ.

Question 14.8

How could Gauss's Law be applied to just one infinite sheet of positive charge? How would the result differ from the parallel-plate capacitor? What if the charge was negative?

14.8.2 No magnetic monopoles: $\vec{\nabla} \cdot \vec{B} = 0$

Applying Gauss's Law to the divergence of the magnetic induction $\vec{B}$, we find that

$$\int_V \vec{\nabla} \cdot \vec{B}\, dV = \oint_S \vec{B} \cdot d\vec{S} = 0 \tag{14.41}$$

which tells us that lines of the $\vec{B}$ field can neither originate nor terminate inside any volume. This means that there are no magnetic charges analogous to electric charges which can be enclosed inside the surface. Magnetic fields originate from currents or from dipolar sources that create continuous lines of magnetic induction $\vec{B}$ that enter and leave a volume in exactly equal fluxes. The vanishing divergence of $\vec{B}$ tells us that *there are no magnetic monopoles.*

14.8.3 Faraday's Law: $\vec{\nabla} \times \vec{E} = -\frac{\partial \vec{B}}{\partial t}$

If we integrate the curl $\vec{E}$ in Maxwell's equation over a surface S and apply Stokes' theorem, we find

$$\int_S (\vec{\nabla} \times \vec{E}) \cdot d\vec{S} = \oint_\ell \vec{E} \cdot d\vec{\ell} = -\int_S \frac{\partial \vec{B}}{\partial t} \cdot d\vec{S}. \qquad (14.42)$$

Interchanging the order of the spatial integration and time derivative, we find the integral form of *Faraday's Law*:

$$\oint_\ell \vec{E} \cdot d\vec{\ell} = -\frac{\partial}{\partial t} \int_S \vec{B} \cdot d\vec{S} = -\frac{\partial}{\partial t} \Phi_B \qquad (14.43)$$

where we have defined the *magnetic flux* through the surface S as $\Phi_B = \int_S \vec{B} \cdot d\vec{S}$, ℓ as the line which borders the surface S, and $d\vec{\ell}$ as an element of length along ℓ. Note that $\vec{E} \cdot d\vec{\ell}$ has the units of electric potential – volts – and the closed loop integral around the line ℓ is the total electric potential difference induced around the loop, the *electromotive force $\mathcal{E}MF$* around the loop.

When there is no significant time-varying magnetic flux through a surface, for example at low frequencies, we find $\oint_\ell \vec{E} \cdot d\vec{\ell} = 0$ which tells us the $\mathcal{E}MF$, or the sum of the voltage drops, around a loop is zero. This result is known as *Kirchhoff's Voltage Law* and is a mainstay of electric circuit theory. It is important to note that this is a *low-frequency approximation* and is violated, for example, when a wire loop antenna picks up a radio-frequency voltage from an electromagnetic wave.

As an example of Faraday's Law, consider the situation illustrated in Figure 14.8. A rail laid across a U-shaped wire with a separation of 0.5 m in a magnetic induction field of 1.5 tesla is pulled to the right at a constant velocity of 4 m/s. If we consider the area defined by the U-shaped wire and the rail as our surface S, we see that as the rail moves to the right the magnetic flux through the surface is increasing because the area of S is increasing. Therefore an $\mathcal{E}MF$ should be induced in the loop formed by the wire and the rail. The size of the $\mathcal{E}MF$ can be calculated by Faraday's Law

$$\oint_\ell \vec{E} \cdot d\vec{\ell} = \mathcal{E}MF = -\frac{\partial}{\partial t} \int_S \vec{B} \cdot d\vec{S}. \qquad (14.44)$$

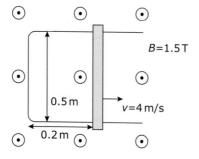

Figure 14.8 U-shaped wire with an electrically connected cross-bar being pulled to the right with velocity $\vec{v}$ in a magnetic field $\vec{B}$ directed out of the paper.

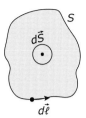

Figure 14.9 The direction of $d\vec{S}$ (in same direction as $\vec{B}$) and $d\vec{\ell}$ (counter-clockwise in same sense as the $\mathcal{EMF}$) in the example illustrated in Figure 14.8.

The magnetic induction field $\vec{B}$ is constant in this problem, but the surface is increasing in area as the rail is pulled to the right. The magnetic flux is the field strength, 1.5 T, times the area of the surface $\Phi_B = 1.5 \text{ T} \cdot 0.5 \text{ m} \cdot (0.2 + vt)$. The $\mathcal{EMF}$ can then be calculated from the time derivative of the flux:

$$\mathcal{EMF} = -\frac{\partial}{\partial t}\Phi_B = -\frac{\partial}{\partial t}[1.5 \text{ T} \cdot 0.5 \text{ m} \cdot (0.2 + vt)] = -1.5 \text{ T} \cdot 0.5 \text{ m} \cdot v$$
$$= -1.5 \cdot 0.5 \cdot 4 = -3 \text{ V}.$$

The direction of the $\mathcal{EMF}$ can be deduced from careful consideration of the right-hand rule. Since the flux through the loop (the direction of $\vec{B}$) is up, the positive direction of the loop integral $\vec{E} \cdot d\vec{\ell}$ is counter-clockwise by the right-hand rule as shown in Figure 14.9. The negative sign then implies that the $\mathcal{EMF}$ is in the clockwise direction (would cause a current in the clockwise direction). Later we will introduce **Lenz's Law**, a physical way to determine the direction of the induced $\mathcal{EMF}$.

The induced $\mathcal{EMF}$ in this case can also be described as the local Lorentz force on the electrons in the moving rail $\vec{F} = q\vec{v} \times \vec{B}$. The force on the electron can be compared with the force due to an induced electric field $\vec{F} = q\vec{E}_{in}$. Comparing these two equations, it is clear that the magnetic force on an electron in the moving rail is equivalent to the force of an electric field $\vec{E}_{in} = \vec{v} \times \vec{B} = vB$, since the velocity is perpendicular to the field. Since $\vec{E} = -\vec{\nabla}\phi$, where ϕ is the electric potential (voltage), the $\mathcal{EMF}$ induced in the moving rail is found by the line integral over the moving wire as before $\mathcal{EMF} = -\int_0^{0.5 \text{ m}} \vec{E}_{in} \cdot d\vec{\ell} = -vB(0.5 \text{ m}) = -(4)(1.5)(0.5) = -3 \text{ V}$. We note that application of Faraday's Law is consistent in this case with consideration of the Lorentz force, but Faraday's Law would also yield the correct result if the magnetic field were changing in time even if the wire was not moving.

Example 14.3 Find the current induced in a 0.5 m radius circular loop of copper wire in the x–y plane induced by a time-varying magnetic field $\vec{H} = H_o \cos \omega t \hat{z}$ where the magnetic field strength $H_o = 0.6$ A/m and the frequency $\omega = 1200 \ \pi$ rad/s ($f = 600$ Hz). The copper wire has a diameter of 1 mm and the conductivity of copper is 6.0×10^7 S/m.

Continued

Example 14.3 (*cont.*)

Solution: We apply the integral form of Faraday's Law

$$\oint_{\ell} \vec{E} \cdot d\vec{\ell} = -\frac{\partial}{\partial t} \int_{S} \vec{B} \cdot d\vec{S} .$$

Recognizing that, by symmetry, the electric field must be the same everywhere around the circular loop and using the constitutive relation $\vec{B} = \mu \vec{H}$ with $\mu = \mu_o = 4\pi \times 10^{-7}$ H/m in free space we find

$$2\pi(0.5)E = -\frac{\partial}{\partial t} B\pi(0.5)^2 = -\mu_o \pi(0.5)^2 \frac{\partial}{\partial t} H_o \cos \omega t.$$

Since we have assumed that $d\vec{S}$ is in the same direction as $\vec{B}$, parallel to the z-axis, the direction of $d\vec{\ell}$ must curl around the z-axis in the direction given by the right-hand rule: in a counter-clockwise direction when viewed from the positive z-axis. The magnitude of $\vec{E}$ everywhere around the loop is then

$$E = -\mu_o \frac{0.5}{2} \omega H_o(-\sin \omega t) = 4\pi \times 10^{-7} \frac{0.5}{2} 1200\pi(0.6 \sin \omega t) = 0.71 \sin \omega t \text{ mV}.$$

The current is found from the constitutive relation

$$I = JA = \sigma E(\pi r^2) = (6.0 \times 10^7)(0.71 \times 10^{-3} \sin \omega t)\left(\pi \left(\frac{1 \times 10^{-3}}{2}\right)^2\right)$$

$$= 0.0335 \sin \omega t \text{ A}.$$

Careful circuit layout to prevent large loops is important to avoid currents induced by stray magnetic fields. Radio antennae are able to pick up much smaller fields because the frequency is a factor of 10^3–10^6 higher.

Question 14.9

How large would the correction to Kirchhoff's Voltage Law be in volts if a magnetic field of $B(t) = 1.0 \cos (2\pi ft)$ µT at a frequency $f = 60$ Hz were perpendicular to a 1 cm × 2 cm circuit loop? (1 tesla = 10 kilogauss is an extremely large magnetic field, comparable to the largest iron core magnets.)

14.8.4   Ampere's and Maxwell's Laws: $\vec{\nabla} \times \vec{H} = \vec{J} + \frac{\partial \vec{D}}{\partial t}$

If we integrate the curl $\vec{H}$ equation over a surface S, we find:

$$\int_{S} (\vec{\nabla} \times \vec{H}) \cdot d\vec{S} = \int_{S} \vec{J} \cdot d\vec{S} + \int_{S} \frac{\partial \vec{D}}{\partial t} \cdot d\vec{S}. \tag{14.45}$$

Since $\vec{J}$ is the *current density* in amps/m^2, we can immediately identify the surface integral of $\vec{J} \cdot d\vec{S}$ as the total current (in amperes) flowing normally across the surface, and we denote it as I_S. The second term on the right-hand side must also have units of amperes. This term is referred to as the *displacement current* through the surface. We will discuss the displacement current in detail later, and just note at this time that it depends on the time derivative of the fields and can be neglected for steady-state DC or low-frequency fields.

Applying Stokes' theorem to convert the surface integral of curl $\vec{H}$ into a line integral around the boundary of the surface ℓ, we find the integral form of the equation:

$$\oint_\ell \vec{H} \cdot d\vec{\ell} = I_S + \int_S \frac{\partial \vec{D}}{\partial t} \cdot d\vec{S}. \tag{14.46}$$

If we ignore the displacement current term initially, we find *Ampere's Law* relating the magnetic field, integrated around a closed line, to the source current – in amperes or coulombs/s – crossing over the surface that the line encloses

$$\oint_\ell \vec{H} \cdot d\vec{\ell} = I_S. \tag{14.47}$$

This can be considered a low-frequency or DC approximation of the full Maxwell's equation. In cases of high symmetry, this integral form can be used to find the magnetic field around a path in terms of the source current(s). For example, consider a thin wire carrying a current I as shown in Figure 14.10. We know that the magnetic field must curl around the current and, by symmetry, the $\vec{H}$ field must be constant on a circular path with the wire at the center and directed tangentially to the circle. Since $\vec{H}$ is parallel to $d\vec{\ell}$ everywhere around the circle, Ampere's Law reduces to

$$\oint_\ell \vec{H} \cdot d\vec{\ell} = H2\pi r = I_S \Rightarrow H = \frac{I_S}{2\pi r} \tag{14.48}$$

directed tangentially around the wire in the direction given by the right-hand rule: the $\vec{H}$ field curls around the wire in the direction that the fingers of your right hand curl if your

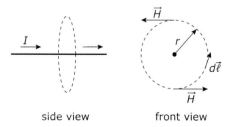

side view   front view

Figure 14.10 The magnetic field $\vec{H}$ curls around a current carrying wire the direction given by the right-hand rule with respect to the direction of the current.

thumb is in the direction of the current. Looking from the end that the current is flowing toward, this gives us

$$\vec{H} = \frac{I_S}{2\pi r}\hat{\theta} \tag{14.49}$$

with $\hat{\theta}$ being a unit vector in the conventional direction of increasing angle (i.e., counter-clockwise) as shown in Figure 14.10.

Example 14.4 Consider the configuration of two infinite parallel wires separated by 1 m, each with a current of $I = 2$ A directed in opposite directions. Find the direction and magnitude of the magnetic field (a) at a point directly between the two wires, and (b) at a point equidistant from the wires and 0.5 m above the plane defined by the two wires.

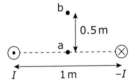

Hint: Since the magnetic field $\vec{H}$ is a *vector* quantity, the magnetic fields from two source currents should add like vectors.

Solution: Using the right-hand rule, the field from the wire on the left is $H_1 = \frac{I}{2\pi r_1}$ directed perpendicular to the vector from the wire to point a or point b in a counter-clockwise direction. Similarly the field from the wire on the right is $H_2 = \frac{I}{2\pi r_2}$ directed perpendicular from the wire to the point in a clockwise direction. At point a, both fields are vertically up and the total field is $\vec{H} = \frac{2}{2\pi(0.5)}\hat{y} + \frac{2}{2\pi(0.5)}\hat{y} = \frac{4}{\pi}\hat{y}$. At point b, a distance of $\sqrt{2}/2$ m from each wire at an angle of 45°, the field $\vec{H}_1 = \frac{2}{\pi\sqrt{2}}(-\cos 45° \,\hat{x} + \cos 45° \,\hat{y})$ and $\vec{H}_2 = \frac{2}{\pi\sqrt{2}}(+\cos 45° \,\hat{x} + \cos 45° \,\hat{y})$. The total field at b is thus $\vec{H} = \frac{2}{\pi\sqrt{2}}\left(1/\sqrt{2} + 1/\sqrt{2}\right)\hat{y} = \frac{2}{\pi}\hat{y}$.

A second case of high symmetry is the magnetic field caused by an infinite sheet of current as shown in Figure 14.11. The symmetry of the problem implies that the magnetic field doesn't depend on lateral position; the $\vec{H}$ field must be horizontal and oppositely directed above and below the sheet as given by the right-hand rule. Since the current that crosses the surface doesn't change depending on how far away from the sheet the surface is (path a or path b), the strength of the magnetic field doesn't depend on the distance from the current sheet. If I/w is the current/unit width, by applying Ampere's Law the magnetic field above and below the sheet is $\vec{H} = \pm\frac{1}{2}\frac{I}{w}\hat{x}$, where $\hat{x}$ is the horizontal direction perpendicular to the current.

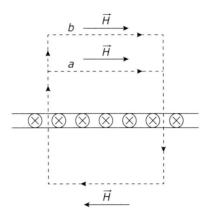

Figure 14.11 Magnetic field created by an infinite sheet of current. By symmetry the field is horizontal and is independent of distance from the plate since the same current is enclosed by path a and path b.

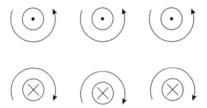

Figure 14.12 The magnetic field created by solenoid coil is the sum of the field from each coil with current going into or out of page.

A final high-symmetry application of Ampere's Law is the field inside an infinitely long *solenoid*, a coil of wire wrapped in a cylindrical shape around an open core. A small section of a solenoid is shown enlarged in Figure 14.12 with the magnetic fields curling around the wire. Since the figure is a cross-section of a single wire, the current is the same in each loop coming out of the page and going in. As is clear, the fields between the loops will cancel out, whereas the fields in the bore of the solenoid will add. By symmetry, the field inside the coil is horizontal and uniform across the cross-section of the solenoid. Outside the coil, the field is zero from the cancelation of the fields of the wire segments going in and coming out. The field can be determined by applying Maxwell's equation around the path shown in Figure 14.13. The only segment of the closed-loop integral that contributes is the segment down the center of the solenoid – there is no field outside and on the perpendicular segments $\vec{H} \cdot d\vec{\ell} = 0$ because $\vec{H} \perp d\vec{\ell}$. The total current through the loop is NI where N is the number of wire segments inside the loop and I is the current in the wire. Ampere's Law gives:

$$\oint_{\ell} \vec{H} \cdot d\vec{\ell} = Hw = NI \Rightarrow H = I\frac{N}{w} \tag{14.50}$$

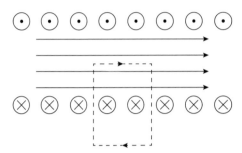

Figure 14.13 Calculation of magnetic field inside solenoid by application of Ampere's Law around square path shown.

where $\frac{N}{w}$ is the number of turns per unit length (turns/m) of the solenoid. The direction of the field is to the right in the figure as can be deduced by applying the right-hand rule to any of the solenoid segments: with the fingers of your right hand curled in the direction of the current, the field points in the direction of your thumb – toward the right in the example shown in Figure 14.13.

In cases where it isn't possible to evaluate $\oint_{\ell} \vec{H} \cdot d\vec{\ell}$ by symmetry arguments, it is necessary to apply the *Biot–Savart Law*, an empirical relationship consistent with the curl-H Maxwell's equation,

$$d\vec{H} = \frac{I}{4\pi} \frac{d\vec{\ell} \times \hat{r}}{r^2} = \frac{I}{4\pi} \frac{d\vec{\ell} \times \vec{r}}{r^3} \tag{14.51}$$

where $I d\vec{\ell}$ is an element of current and $\vec{r}$ is a vector from the current element to the point at which the field is calculated. The total $\vec{H}$ field is found by integrating $d\vec{H}$ from the entire wire.

Example 14.5 Use the Biot–Savart Law to find expressions for (a) the magnetic field a distance h from an infinitely long wire, and (b) the field along the x-axis of a circular loop of wire in the y–z plane centered on the x-axis with diameter h.

Solution: In the geometry of (a), we have $d\vec{\ell} = dx\hat{x}$ (positive in the direction of the current) while the vector from $d\vec{\ell}$ to the point at $-h\hat{y}$, where the field is to be evaluated is $\vec{r} = x\hat{x} - h\hat{y}$ and its magnitude $r = \sqrt{x^2 + h^2}$. Using the definition of the curl we have

$$d\vec{\ell} \times \vec{r} = \begin{vmatrix} \hat{x} & \hat{y} & \hat{z} \\ dx & 0 & 0 \\ x & -h & 0 \end{vmatrix} = -hdx\hat{z}.$$

To integrate over the wire from $x = -\infty$ to $+\infty$, we will set the limits of integration as $-L$ to $+L$ and then let $L \to \infty$

Continued

Example 14.5 (*cont.*)

$$\vec{H} = \frac{I}{4\pi}\hat{z}\int_{-\infty}^{\infty}\frac{-hdx}{(\sqrt{x^2+h^2})^3} = \frac{-Ih}{4\pi}\hat{z}\lim_{L\to\infty}\left(\frac{L}{h^2\sqrt{L^2+h^2}} - \frac{-L}{h^2\sqrt{L^2+h^2}}\right) = \frac{-I}{4\pi h}\hat{z}(2) = -\frac{I}{2\pi h}\hat{z}$$

which is the same result we got from applying the integral form of Maxwell's equation.

In the geometry of (b), we will use the arc length formula to relate $d\ell$ to a differential element of angle around the loop $d\ell = h\,d\phi$ and then integrate $d\phi$ from 0 to 2π around the loop. The vector length element is given by $d\,\vec{\ell} = hd\phi\sin\phi\hat{y} + hd\phi\cos\phi\hat{z}$ where ϕ is the angle around the loop starting at 0 at the position at $-h\hat{y}$ and increasing in the clockwise direction of the current I when viewed from the right. In the same system $\vec{r} = x\hat{x} + h\cos\phi\hat{y} - h\sin\phi\hat{z}$, and the magnitude $r = \sqrt{x^2+h^2}$ for all angles ϕ.

$$d\vec{\ell}\times\vec{r} = \begin{vmatrix} \hat{x} & \hat{y} & \hat{z} \\ 0 & hd\phi\sin\phi & hd\phi\cos\phi \\ x & h\cos\phi & -h\sin\phi \end{vmatrix}$$

$$= -\hat{x}h^2(\sin^2\phi + \cos^2\phi)d\phi + \hat{y}hx\cos\phi d\phi - \hat{z}hx\sin\phi d\phi.$$

The integrals over ϕ of the $\cos\phi$ and $\sin\phi$ terms from 0 to 2π are zero, while the trigonometric identity $\sin^2\phi + \cos^2\phi = 1$ simplifies the $\hat{x}$ term giving just 2π for the integral. The field is then

$$\vec{H} = -\hat{x}\frac{I}{4\pi}\int_{0}^{2\pi}\frac{h^2 d\phi}{(\sqrt{x^2+h^2})^3} = -\hat{x}\frac{I}{2}\frac{h^2}{(\sqrt{x^2+h^2})^3}.$$

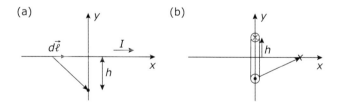

(a)   (b)

14.8.5 **Displacement current**

The necessity of the displacement current term in the curl $\vec{H}$ equation at finite frequency can be demonstrated by considering the situation in Figure 14.14. Alternating current is flowing in a wire which contains a parallel-plate capacitor. Alternating current flows across the capacitor because the two plates of the capacitor change polarity and charge and discharge in unison, drawing current into the wire on the right side of the capacitor

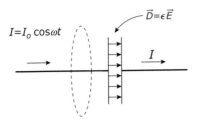

Figure 14.14 A wire carrying AC current through a capacitor shown with an Ampere's Law surface crossing the current-carrying wire.

in-phase with the current on the left side of the capacitor. Such an alternating current clearly gives an alternating magnetic $\vec{H}$ field

$$\vec{H} = \frac{I_o \cos \omega t}{2\pi r} \hat{\theta} \tag{14.52}$$

as can be shown by an application of Ampere's Law across the surface shown in Figure 14.14. However, what happens if, without changing the position of the bordering loop, the surface is deformed to go down the center of the capacitor between the two plates? Since there is no current in the gap between the capacitor plates, now the surface has no current through it, and Ampere's Law would therefore imply no magnetic field $\vec{H} = 0!$

The ambiguity in the result for these two surfaces is one indication of the necessity to generalize Ampere's Law through the inclusion of the displacement current. The addition of the term proportional to $\partial \vec{D}/\partial t$, to the $\vec{\nabla} \times \vec{H}$ equation (or to the $\oint_{\ell} \vec{H} \cdot d\vec{\ell}$ equation) is sometimes referred to as *Maxwell's Law*. In the case we have been discussing, while there is no physical current through the gap between the two plates of the capacitor, there is an electric displacement field between the positive and negative charged plates which we calculated from our discussion of the $\vec{\nabla} \cdot \vec{D}$ equation as $D = \rho_s$, where ρ_s is the surface charge density on one of the capacitor plates. Thus we find the displacement current

$$\int_S \frac{\partial \vec{D}}{\partial t} \cdot d\vec{S} = \frac{\partial}{\partial t} \int_S \vec{D} \cdot d\vec{S} = \frac{\partial}{\partial t} \int_S \rho_s \, dS = \frac{\partial}{\partial t}\rho_s \, A = \frac{\partial}{\partial t}Q_s = I \tag{14.53}$$

where we have used the fact that the integral over the capacitor cross-section of the surface charge density is equal to the total surface charge, and the time rate of change of the total surface charge is the current flowing into the capacitor. Thus the displacement current is exactly equal to the current in the wire and the magnetic field $\vec{H}$ is the same whether the surface passes through the wire or through the electric displacement field between the capacitor plates!

While the consistency of Ampere's Law for the magnetic field around a wire containing a parallel-plate capacitor is satisfying, the largest import of the addition of Maxwell's Law to Ampere's Law is what it says about the electric and magnetic fields in the absence of currents or other sources. In free space, the current density $\vec{J} = 0$ and the two curl equations become

$$\vec{\nabla} \times \vec{E} = -\frac{\partial \vec{B}}{\partial t} = -\mu_o \frac{\partial \vec{H}}{\partial t} \tag{14.54}$$

$$\vec{\nabla} \times \vec{H} = \frac{\partial \vec{D}}{\partial t} = \epsilon_o \frac{\partial \vec{E}}{\partial t} \tag{14.55}$$

thus indicating the presence of an electric field supported by the time-varying magnetic field, and the presence of a magnetic field supported by the time-varying electric field, even in the absence of any local sources. This creates the possibility of self-propagating electromagnetic waves in free space, thus making possible electromagnetic wave propagation from radio (and lower) frequencies to optical (and higher) frequencies. The effects of materials on the optical properties – or the properties of an electromagnetic wave at any frequency – are the result of $\varepsilon(\omega)$, $\mu(\omega)$, and $\sigma(\omega)$ as specified in the constitutive relations. We will examine the phenomena of propagating electromagnetic waves in more detail in the next chapter.

Question 14.10

The displacement current can be expressed as $I_D = \frac{\partial}{\partial t} \int_S \vec{D} \cdot d\vec{S}$. Find the magnitude of the displacement current for an electric field $\vec{E} = 1 \cos(2\pi ft) \text{V/m}$ at a frequency $f = 60$ Hz normal to a 1 cm×2 cm surface. Assume that the dielectric medium is free space with $\epsilon_o = 8.85 \times 10^{-12}$ F/m. Does this give you a hint as to why, although the connection between currents and magnetic field was known by Jean-Baptiste Biot and Félix Savart in 1820 and Michael Faraday discovered that currents could be induced by a magnetic field in 1831, the displacement current term in the Ampere–Maxwell Law wasn't conceived by James Clerk Maxwell until 1861?

14.9 Applications of Faraday's Law

14.9.1 Eddy currents

We will conclude this chapter by examining three more applications of Faraday's Law, the phenomenon of *eddy currents,* the operation of an *electric generator,* and voltage-level conversion with a transformer. First, we will look at the generation of electrical currents due to changing magnetic flux through a conductor, and use the currents to damp oscillatory motion in a scientific balance scale.

Consider a 1 cm thick aluminum cylindrical disk, 20 centimeters in diameter, in a region of space with a uniform, time-varying magnetic induction field $\vec{B} = (0.5 - 0.04t)\hat{z}$ tesla as shown in Figure 14.15. Applying the integral form of Faraday's Law, we find

$$\oint \vec{E} \cdot d\vec{\ell} = -\frac{\partial}{\partial t} \int \vec{B} \cdot d\vec{S}.$$

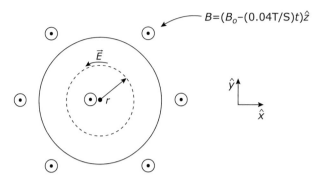

Figure 14.15 A conducting disk in a region of time-varying magnetic induction field. By symmetry, the induced electric field is in the tangential direction and depends on the distance from the center of the disk r.

By symmetry, the induced electric field is tangentially directed around the disk and can depend only on the radial distance from the center of the disk, and not on the angle. At a distance r from the center of the disk, we have

$$E(2\pi r) = -\frac{d}{dt}(0.5 - 0.04t)\pi r^2 \Rightarrow E = 0.02r\frac{V}{m}.$$

To find the direction of the electric field, we will apply *Lenz's Law*, which states that *the direction of the induced electric field around the path is in the direction that would create currents that would* oppose the change *in magnetic flux through the surface defined by the path*. Lenz's Law follows from a consideration of energy conservation: if the direction of $\vec{E}$ could create currents to reinforce the change in flux, there is the potential that the process could be self-reinforcing and continue even in the absence of the magnetic flux. In the case we are considering, since the magnetic induction field is decreasing, the electric field will be in the direction that would create a current which would *increase* the magnetic induction field. From our discussion of Ampere's Law, we know that a *counter-clockwise* loop of current will create a $\vec{B}$ field that is upwardly directed (opposing the decrease in flux). Thus the direction of the electric field is counter-clockwise.

To find the magnitude of the eddy currents that are induced by the changing magnetic induction field, we need to use the constitutive relation $\vec{J} = \sigma\vec{E}$ where σ is the *conductivity* of aluminum. From standard tables, the *resistivity* of aluminum is $\rho = 2.8 \times 10^{-8}$ $\Omega \cdot$m. The conductivity is the reciprocal of the resistivity: $\sigma = 1/\rho = 1/2.8 \times 10^{-8} = 3.57 \times 10^7$ Ω^{-1} m^{-1} = 3.57×10^7 S/m where S/m is the symbol for *siemens per meter*, the SI unit for conductivity. The magnitude of the current density is thus $J = \sigma E = 3.57 \times 10^7 \cdot 0.02\ r = 7.14 \times 10^5\ r$ (A/m^2). The current in the outer 1 mm of the disk, for example, is thus approximately:

$$I = JA = \left(7.14 \times 10^5 \cdot 1 \times 10^{-3}\ \frac{A}{m^2}\right)(0.1\ m)(0.01\ m) = 0.714\ \text{ampere}.$$

This is a significant current that will persist as long as the magnetic induction field continues to change at the same rate.

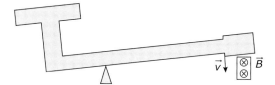

Figure 14.16 A conducting metal piece on the lever arm of a scientific balance moves in and out of a magnet with magnetic induction field $\vec{B}$.

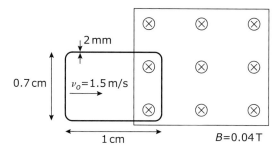

Figure 14.17 A coil of wire moving into an area of magnetic induction field $\vec{B}$.

Due to Lenz's Law, the eddy currents oppose the change in flux. Thus eddy currents can be used to damp out motion, such as on the lever arm of a triple-beam balance scale. As shown in Figure 14.16, a conductive piece on the end of the moving arm moves in and out of a magnet as the arm oscillates.

With a moving conductor, we can consider the problem from the point of view of the Lorentz force: (1) the velocity of the coil creates a force on the electrons in the moving wire which produces a current I_o, and (2) once the current is flowing there is a Lorentz force $\vec{F} = I_o \, \vec{L} \times \vec{B}$ on the wire which slows down and stops the motion. Equivalently, we can find the $\mathcal{EMF}$ and hence the current from Faraday's Law, and then apply the Lorentz force to calculate the slowing/stopping force. We will use the second technique applied to a single coil of copper wire moving horizontally into a region of magnetic field $B = 0.04$ T at an initial velocity of 1.5 m/s as in Figure 14.17. We will assume the conductor coil is a 2 mm diameter copper wire rectangle 1 cm × 0.7 cm in size.

Faraday's Law in integral form gives us

$$\oint_{\ell} \vec{E} \cdot d\vec{\ell} = \mathcal{EMF} = -\frac{\partial}{\partial t} \int_{S} \vec{B} \cdot d\vec{S}.$$

In this problem the magnetic field is not changing in time, only the area of the loop inside the magnetic field area is changing as given by $A_{loop} = A_o + 0.007 \, v_o t$ where $v_o = 1.5$ m/s is the initial velocity of the loop and t is the time with the initial time $t = 0$ selected when the loop has an area A_o in the magnetic field. The $\mathcal{EMF}$ around the loop is then

$$\mathcal{EMF} = -B\frac{\partial}{\partial t}\int_S dS = -0.04\frac{\partial}{\partial t}(A_o + 0.007v_o t) = -(0.04)(0.007)(1.5) = -0.42 \text{ mV}.$$

Since the flux through the loop is increasing, using Lenz's Law we can determine the direction of the $\mathcal{EMF}$ as the direction which acts to decrease the flux. Using the right-hand rule we find a counter-clockwise current gives a magnetic field which opposes the DC magnetic field.

Using the conductivity of copper as $\sigma = 1/\rho = 6.0 \times 10^7 \text{s/m}$ we find the *resistance* of the coil as $R = \frac{1}{\sigma}\frac{L}{A}$ where L is the length of the wire in the loop and A is the cross-section area of the wire. From this we find

$$R = \frac{1}{6.0 \times 10^7}\frac{2(0.01 + 0.007)}{\pi(1 \times 10^{-3})^2} = 2.85 \times 10^{-4}\ \Omega.$$

From Ohm's Law the instantaneous current in the loop is $I = \mathcal{EMF}/R = 4.2\times10^{-4}/2.85\times10^{-4} = 1.47$ A.

The stopping force on the wire is given by the Lorentz force

$$F = I\vec{\ell} \times \vec{B} = (1.47)(0.007)(0.04) = 4.11 \times 10^{-4} \text{ N}.$$

Using the right-hand rule on a current going up (counter-clockwise) in the end segment of the loop, we find the force is to the left, opposing the motion of the coil. The mass of the loop is quite small; using the mass density of copper $\rho_{mass} = 8.9 \times 10^3 \text{ kg/m}^3$ we find a total mass for the loop as

$$m = \rho_{mass}\ LA = (8.9 \times 10^3)2(0.01 + 0.007)\pi(1 \times 10^{-3})^2 = 9.5 \times 10^{-4} \text{ kg}.$$

Newton's Law gives us the acceleration $a = F/m = 4.11\times10^{-4}\text{ N}/9.5\times10^{-4}\text{ kg} = 0.43 \text{ m/s}^2$. If we assumed that the force were constant, the stopping time is $t = v_o/a = 1.5/0.43 = 3.46$ s. In fact, the force is not constant – it decreases as the velocity of the coil decreases – but the deceleration torque due to Faraday's and Lenz's Laws contributes significantly to the damping of the motion in a triple-beam balance scale.

14.9.2   Electric generators

An electric generator is another important application of Faraday's Law. Consider a loop of wire being rotated at a circular frequency ω in a fixed magnetic field of 0.04 T, as shown in Figure 14.18. In this case both the magnetic field and the coil area are fixed in time; what is changing in time is the angle θ between the field $\vec{B}$ and the normal vector to the coil area $d\vec{S}$, as shown in Figure 14.18(b). The magnetic flux through the coil is thus

$$\int_S \vec{B} \cdot d\vec{S} = BA \cos \theta = BA \cos \omega t.$$

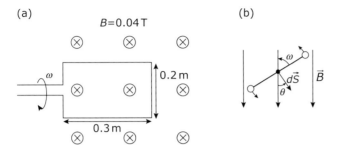

Figure 14.18 (a) A coil of wire is rotated around its central axis at a frequency ω in a region of constant $\vec{B}$ field. (b) End view of the rotating coil. As the coil rotates the angle between the normal to the surface enclosed by the wire and the magnetic induction field is continually changing by $\theta = \cos \omega t$.

The EMF is given by the time derivative of the flux

$$\text{EMF} = -\frac{\partial}{\partial t}(B A \, \cos \, \omega t) = B A \, \omega \sin \, \omega t.$$

Assuming that the wire loop is 0.3 m $\times$ 0.2 m and rotating at 500 rpm, we find the $\mathcal{E}MF$ induced in the loop to be:

$$\text{EMF} = -\frac{\partial}{\partial t}(B A \, \cos \, \omega t) = B A \, \omega \sin \, \omega t = (0.04)(0.2 \times 0.3)\left(\frac{2\pi 500}{60}\right) \sin \, \omega t$$

$$= 0.126 \, \sin \, \omega t \text{ V}.$$

If the single loop is replaced by a coil consisting of 1000 loops in series, the $\mathcal{E}MF$ generated by the rotating coil creates a net voltage $V(t) = 126 \sin \omega t$ volts.

14.9.3 Transformers

A transformer is an electrical component that uses Faraday's Law and Ampere's Law to convert an alternating current signal from one voltage level to another voltage level. The ability to easily transform signals to any voltage level desired is a major advantage of alternating current and the main reason why George Westinghouse's alternating current distribution systems won out in the commercial competition with Thomas Edison's direct current system in the late nineteenth century.

A transformer consists of an input primary coil of wire (solenoid) with N_1 turns that creates a magnetic field by Ampere's Law and a secondary coil with N_2 turns, as shown in Figure 14.19. The primary coil is wrapped around a high-permeability soft magnetic core that converts the magnetic field H to a large magnetic flux density B by the constitutive relation $\vec{B} = \mu \vec{H}$. The soft magnet confines the magnetic flux and transfers the magnetic induction field B to the secondary coil with N_2 turns. (At electrical circuit frequencies, μ for a soft magnetic material such as iron can be

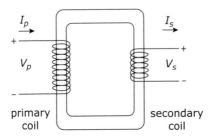

Figure 14.19 A transformer linking the current in the primary coil I_p with N_p turns to the current in the secondary coil I_s with N_s turns by channeling the magnetic induction flux through a soft magnet loop that connects the two coils.

hundreds of times larger than the permittivity of free space μ_o.) The time-varying flux in the secondary coil then induces a voltage in the secondary coil by Faraday's Law. If the number of turns in the secondary coil is greater than in the primary coil, the voltage in the secondary is "stepped-up" to a higher voltage. If the number of coils in the secondary is less than in the primary, the voltage is "stepped-down" to a lower voltage. The ability to easily transmit electrical power through high-voltage/low-current transmission lines which have lower "I-squared R" ohmic losses and then step the voltage down to safe levels for household use is the basis of modern electric power grids.

We will consider an ideal loss less transformer, with N_p primary turns and N_s secondary turns driven by a primary current $I_p = I_o\, e^{-i\omega t}$ with the secondary circuit connected to an open switch so that $I_s = 0$. As we have seen in our discussion of Ampere's Law, a long solenoid produces a magnetic field $B = I(N/w)$ through its center. Taking into account the permeability μ of the core material, the primary coil creates a magnetic flux density

$$B = \mu(N/w)_p I_o e^{-i\omega t}. \tag{14.56}$$

The time-varying B creates a voltage across the coil by Faraday's Law:

$$V_p = N_p \text{ EMF} = -N_p \frac{\partial}{\partial t}\int_{S_p} \vec{B} \cdot d\vec{S} = -N_p \frac{\partial}{\partial t}\mu(N/w)_p I_o e^{-i\omega t} A_p$$

$$= i\omega N_p \mu A_p (N/w)_p I_o e^{-i\omega t}. \tag{14.57}$$

Electrical engineers will recognize this as the generalization of Ohm's Law to reactive elements, $V = ZI$ with the *impedance* $Z = i\omega L$ and the *inductance* given by $L = N_p \mu A_p (N/w)_p$.

We note that the factor of i in the equation relating V_p to $I_p = I_o\, e^{-i\omega t}$ implies that the voltage is 90° out-of-phase with the current. It is easy to see that when the voltage and current are 90° out-of-phase, during one half of the cycle we have the current and voltage both positive or both negative, the charges are moving in the same direction as the electric field, and work is being done on the charges by the electric field. During the other half of the cycle, however, the current and voltage are in opposite directions and negative work is done on the charges by the electric field. The total work done over one cycle is zero and *no net energy is gained or lost when the voltage and current are 90° out-of-phase.* In electrical engineering notation the average complex power across the inductor $S_p = \frac{1}{2}VI^* = P + iQ = \frac{1}{2}i\omega N_p\mu A_p(N/w)_p|I_o|^2$ is entirely imaginary, which implies that $P = 0$: the power is all *reactive*, with current and voltage 90° out-of-phase, and the average power dissipated is zero.

In a well-designed transformer the B field is confined within the soft magnetic transformer core and the flux is nearly completely transferred to the core of the secondary coil, where it also induces a voltage by Faraday's Law,

$$V_s = N_s \text{ EMF} = -N_s \frac{\partial}{\partial t} \int_{S_s} \vec{B} \cdot d\vec{S} = -N_s \frac{\partial}{\partial t} \mu(N/L)_p I_o e^{-i\omega t} A_s$$

$$= i\omega N_s \mu A_s (N/L)_p I_o e^{-i\omega t}. \tag{14.58}$$

If the area of the magnetic core is the same on the primary and secondary coils, the voltage ratio is thus given by

$$\frac{V_s}{V_p} = \frac{N_s}{N_p}. \tag{14.59}$$

If the switch in the secondary circuit is closed so current I_s flows to a load in the secondary circuit, we can see that Ampere's Law applied to the secondary circuit will cause a magnetic flux that will be transferred by the magnetic core back to the primary circuit. This will change the effective inductance of the coil in the primary circuit by the *mutual inductance.* It can be shown that the voltage ratio is still given by the ratio of the number of turns as given in Eq. (14.59).

The flux transferred back to the primary circuit by the current in the secondary circuit will also shift the phase of the current in the primary circuit so that the current I_p is no longer 90° out-of-phase with V_p and power is dissipated in the primary circuit. Conservation of energy tells us that the power dissipated – the real part of the complex power – in the primary circuit must equal the power delivered to the load in the secondary circuit.

Chapter 14 problems

1. Consider two solid spheres of radius R. One is a conductor and the other an insulator. Each has a total charge Q on it. Find the electric field inside and outside of the

   a. charged solid conducting sphere; and

   b. uniformly charged solid spherical insulator.

2. A loop of material on a pendulum swings down into a magnetic field (directed into the page) slows down and stops. Using Maxwell's equations, constitutive relations, and the Lorentz force, explain what is happening here. Indicate the direction of currents, fields, and forces.

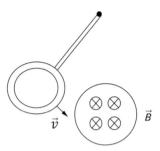

3. What will happen if a strong cylindrical magnet is dropped in a vertical hollow Cu tube?

   a. Explain the phenomenon qualitatively. If you were to drop a non-magnetic material i.e., a piece of Pb which one will fall faster?

   b. What if the tube was made out of Al?

   c. What if the tube was made out of wood?

   d. How can you slow down or speed up the falling magnet inside the metallic tube?

4. Eddy current damping: A square copper loop of wire 20 cm × 30 cm of radius 1 mm is moving horizontally at velocity 15 cm/s into a region of uniform magnetic field of 0.4 T.

   a. Find the EMF induced in the copper wire when it first moves into the magnetic field.

   b. If the conductivity of copper is $60 \times 10^6 \ \Omega^{-1} \ \mathrm{m}^{-1}$, find the resistance of the loop $R = \frac{1}{\sigma} \frac{\ell}{A}$ (ℓ = length of wire, A = cross-section area).

   c. What is the current $I = $ EMF/R in the copper loop when the loop first moves into the magnetic field?

   d. Calculate the Lorentz force on the bottom wire of the copper loop as it moves into the magnet. What direction is the Lorentz force on the side wires of the loop?

   e. If the mass density of copper is $8.9 \times 10^3 \ \mathrm{kg/m^3}$, find the deceleration of the copper loop due to the Lorentz force when it first enters the region of the magnetic field. Estimate how long it would take the loop to come to rest under the eddy current damping.

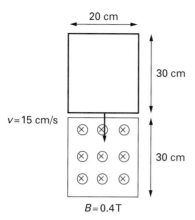

5. Consider a horizontal circular loop of 1 mm-diameter aluminum wire of radius 7 cm placed in a vertical pulsed magnetic field which is ramped up with $B(t) = 0.2 + 3t^2$ tesla where t is the time in seconds.

   a. Find the induced voltage in the aluminum wire as a function of time.

   b. If the resistivity of aluminum is $\rho = 2.75 \times 10^{-8}$ Ω m, find the current in the wire as a function of time.

6. Calculate the E-field due to two oppositely charged infinite **conducting sheets** each of surface charge density (charge/unit area) ρ_s and $-\rho_s$, respectively, both in the space between the sheets and outside the sheets, if they are a distance d apart.

7. Calculate the E-field due to two oppositely charged infinite **insulating sheets** each of surface charge density (charge/unit area) ρ_s and $-\rho_s$, respectively, both in the space between the sheets and outside the sheets if they are a distance d apart.

8. What path will a charged particle follow if it enters a uniform magnetic field at a random angle with the magnetic field? Examine and discuss all possible cases.

9. Discuss the phenomenon of Aurora Borealis in terms of Maxwell's equations and the Lorentz force.

10. Calculate the E-field due to a charged conducting thin circular disk of radius R at any point along the axis going through its center and perpendicular to the disk of radius R. Will the field be the same or different if the disk of radius R is made of insulating material, but has the same charge density?

11. In Example 14.1, calculate the force (magnitude and direction) on the semicircular wire, if the magnetic field is $B = 0.5$ T $\hat{y}$.

12. It can be shown that in a charged parallel-plate capacitor the energy density (energy/unit volume) stored in the electric field is given by $u = \frac{1}{2}\varepsilon E^2$, ε is the dielectric constant of the material between the plates and E is the strength of the electric field between the plates. Discuss what happens to the stored energy when

a dielectric is inserted between the plates of a charged capacitor, which previously had air between its plates. Does the energy change? If so, where does "unaccounted" energy go?

13. Discuss how a microwave oven heats up food.

14. Consider two parallel coils of the same diameter, with their centers lying on an axis that is perpendicular to them, having the same number of turns and carrying current in the same direction. These are known as Helmholtz coils. Calculate the magnetic field at the midpoint between their centers.

15. Consider a metallic strip carrying a current in a direction perpendicular to a constant magnetic field. Discuss what happens inside the conducting strip as a result of the magnetic field (Hall Effect).

16. Discuss what happens when you bring two parallel wires close to each other with currents in the same and in opposite directions. What are the forces on the wires?

15 Electromagnetic waves

What makes some materials transparent to electromagnetic waves and other materials opaque to the same waves? A *screen room* is an enclosure completely surrounded by conducting (usually copper) screen where low-noise experiments can be isolated from external electromagnetic interference. If a radio is playing inside a screen room when the copper screen door is shut, the radio immediately goes silent. How thick should the screen be to have the desired effect? Does this effect mean that there is no way to electromagnetically detect an explosive charge concealed inside an aluminum soda can?

This is one of the problems that we will address in this chapter and the next where we look at solutions to Maxwell's equations in the form of traveling waves and examine the propagation properties of these waves in various materials. It is a remarkable success of electromagnetic theory that the same four Maxwell's equations can describe phenomena from near DC through radio-frequency up to optical properties. While we will not attempt to describe the mathematics around the extremely non-linear quantum regime described by the "photonics" field, we will at least point out in Chapter 16 where classical Maxwellian theory has limitations and where we would have to expand our assumptions to explain photonic phenomena.

15.1 Maxwell's equations and electromagnetic waves in free space

We will begin our treatment with Maxwell's equations in differential form, as introduced in the last chapter. In fact, the derivation of electromagnetic waves is one of the strongest motivations for studying the differential form of Maxwell's equation. We recall that the differential Maxwell's equations are *local* equations, connecting the spatial and time derivatives of the electric and magnetic fields at the same point

$$\vec{\nabla} \cdot \vec{D} = \rho \qquad \vec{\nabla} \times \vec{E} = -\frac{\partial \vec{B}}{\partial t}$$

$$\vec{\nabla} \cdot \vec{B} = 0 \qquad \vec{\nabla} \times \vec{H} = \vec{J} + \frac{\partial \vec{D}}{\partial t}. \tag{15.1}$$

As we pointed out in the last chapter, the relations between $\vec{D}$ and $\vec{E}$, between $\vec{H}$ and $\vec{B}$, and between $\vec{J}$ and $\vec{E}$ depend on the properties of the materials that the fields are in. The electromagnetic properties of materials are completely determined by the three *consti-tutive relations*:

$$\vec{D} = \varepsilon \vec{E}$$

$$\vec{B} = \mu \vec{H}$$

$$\vec{J} = \sigma \vec{E} \tag{15.2}$$

where the *permittivity* ε, the *permeability* μ, and the *conductivity* σ completely describe the electromagnetic properties of materials. Along with the *Lorentz force* $\vec{F} = q\vec{E} + q\vec{v} \times \vec{B}$, these equations and materials functions are sufficient to describe all of classical electro-magnetic, radio-frequency, and optical phenomena.

We will begin by considering electromagnetic waves in free space (vacuum), where the permittivity and permeability are *scalar quantities*, $\varepsilon = \epsilon_o = 8.8543 \times 10^{-12}$ F/m, $\mu = \mu_o = 4\pi \times 10^{-7}$ H/m, with $\rho = 0$ and $\sigma = 0$, since there are no charges or currents in total vacuum. First take the curl of the $\vec{\nabla} \times \vec{E}$ in the Faraday's Law equation

$$\vec{\nabla} \times (\vec{\nabla} \times \vec{E}) = -\vec{\nabla} \times \frac{\partial \vec{B}}{\partial t}. \tag{15.3}$$

We can interchange the order of space and time differentiation on the right-hand side, substitute $\vec{B} = \mu_o \vec{H}$ and insert the $\vec{\nabla} \times \vec{H}$ Maxwell's Law equation to get

$$\vec{\nabla} \times \left(\vec{\nabla} \times \vec{E}\right) = -\frac{\partial}{\partial t}\left(\vec{\nabla} \times \mu_o \vec{H}\right) = -\mu_o \frac{\partial}{\partial t}\left(\frac{\partial \vec{D}}{\partial t}\right) = -\mu_o \frac{\partial^2 \vec{D}}{\partial t^2} = -\mu_o \epsilon_o \frac{\partial^2 \vec{E}}{\partial t^2} \tag{15.4}$$

where we have used $\vec{D} = \epsilon_o \vec{E}$ and used the fact that the current density $\vec{J}$ is zero in the vacuum. Now, we apply the vector identity $\vec{\nabla} \times (\vec{\nabla} \times \vec{V}) = \vec{\nabla}(\vec{\nabla} \cdot \vec{V}) - \vec{\nabla}^2 \vec{V}$, where $\vec{\nabla}^2 \vec{V}$ is the *vector Laplacian*, to the left-hand side:

$$\vec{\nabla}(\vec{\nabla} \cdot \vec{E}) - \vec{\nabla}^2 \vec{E} = -\mu_o \epsilon_o \frac{\partial^2 \vec{E}}{\partial t^2}. \tag{15.5}$$

We note that there is no charge density in free space, so $\rho = 0$. The first Maxwell's equation thus gives $\vec{\nabla} \cdot \vec{D} = 0$ and therefore $\vec{\nabla} \cdot \vec{E} = \frac{1}{\epsilon_o} \vec{\nabla} \cdot \vec{D} = 0$. This then gives us the vector equation for the electric field in free space

$$\vec{\nabla}^2 \vec{E} = \mu_o \epsilon_o \frac{\partial^2 \vec{E}}{\partial t^2}. \tag{15.6}$$

We can recognize this as the three-dimensional *wave equation* with a solution in Cartesian coordinates of a three-dimensional plane wave

$$\vec{E}(\vec{r}, t) = \vec{E}_o e^{i(\vec{k} \cdot \vec{r} - \omega t)} \tag{15.7}$$

with $\vec{r}$ being the position vector $\vec{r} = x\hat{x} + y\hat{y} + z\hat{z}$ and $\vec{k} = k_x\hat{x} + k_y\hat{y} + k_z\hat{z}$ the *wave vector* with a magnitude $|\vec{k}| = 2\pi/\lambda$, where λ is the wavelength of the plane wave. The direction of $\vec{k}$ is the direction of propagation of the wave.

Question 15.1

Substitute the plane-wave solution into the wave equation and show that the expression given for $\vec{E}(\vec{r}, t)$ is a solution of the wave equation Eq. (15.6) if $(\vec{k} \cdot \vec{k}) = k^2 = \mu_o \epsilon_o \omega^2$.

The *dispersion relation* $k^2 = \mu_o\epsilon_o\omega^2$ implies that $\frac{\omega}{k} = \lambda f = v_{ph} = \frac{1}{\sqrt{\mu_o\epsilon_o}}$, which tells us that the wave velocity (the *phase velocity* of the wave) is given by $v_{ph} = \frac{1}{\sqrt{\mu_o\epsilon_o}} = \frac{1}{\sqrt{(4\pi \times 10^{-7})(8.854 \times 10^{-12})}} = 2.9979 \times 10^8$ m/s, which we recognize as the speed of light in a vacuum. The ability to derive the exact speed of light from the DC electric and magnetic constants ϵ_o and μ_o was a triumph of nineteenth-century physics and provided proof that light is an electromagnetic phenomenon. Maxwell's equations also predicted the existence of other electromagnetic waves beyond the visible spectrum before anyone knew that such waves existed.

We can find the geometry of the traveling wave by substituting the plane-wave solution $\vec{E}(\vec{r}, t) = \vec{E}_o e^{i(\vec{k}\cdot\vec{r} - \omega t)}$ back into Maxwell's equations. First, note that the $\vec{\nabla} \cdot \vec{D} = \rho$ equation gives us in free space $\vec{\nabla} \cdot \epsilon_o \vec{E} = 0$, which implies that:

$$0 = \vec{\nabla} \cdot \vec{E} = \frac{\partial}{\partial x}E_{ox}e^{i(k_x x + k_y y + k_z z - \omega t)} + \frac{\partial}{\partial y}E_{oy}e^{i(k_x x + k_y y + k_z z - \omega t)}$$

$$+ \frac{\partial}{\partial z}E_{oz}e^{i(k_x x + k_y y + k_z z - \omega t)}. \tag{15.8}$$

The derivatives operate only on the exponentials, not on the components of $\vec{E}_o$, and pull out i times the x, y, or z-component of the wave vector $\vec{k}$ from each term

$$\vec{\nabla} \cdot \vec{E} = ik_x E_{ox}e^{i(k_x x + k_y y + k_z z - \omega t)} + ik_y E_{oy}e^{i(k_x x + k_y y + k_z z - \omega t)} + ik_z E_{oz}e^{i(k_x x + k_y y + k_z z - \omega t)}$$

$$= (ik_x E_{ox} + ik_y E_{oy} + ik_z E_{oz})e^{i(\vec{k}\cdot\vec{r} - \omega t)} = i\vec{k} \cdot \vec{E}_o e^{i(\vec{k}\cdot\vec{r} - \omega t)} = 0. \tag{15.9}$$

Since the dot product of $\vec{k}$ and $\vec{E}_o$ is zero, this implies that $\vec{E}_o$ is perpendicular to the wave propagation direction $\vec{k}$. The electromagnetic wave is a *transverse* wave – the electric field vector of the wave $\vec{E}_o$ is perpendicular to the direction of propagation $\vec{k}$. Substituting the same form for the propagating wave into the *curl* $\vec{E}$ equation and recognizing that the *del* operator, $\vec{\nabla}$, operating on the plane-wave exponential form just brings out a factor of $i\vec{k}$, we find that

$$\vec{\nabla} \times \vec{E} = -\frac{\partial \vec{B}}{\partial t} \Rightarrow$$

$$i\vec{k} \times \vec{E}_o e^{i(\vec{k}\cdot\vec{r} - \omega t)} = \mu_o i\omega \vec{H}_o e^{i(\vec{k}\cdot\vec{r} - \omega t)} \tag{15.10}$$

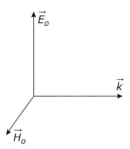

Figure 15.1 The electric field, magnetic field, and propagation vector for a plane wave in free space are mutually perpendicular vectors with $\vec{H}_o = \frac{1}{\mu_o \omega} \vec{k} \times \vec{E}_o$.

from which we find $\vec{H}_o = \frac{1}{\mu_o \omega} \vec{k} \times \vec{E}_o$, i.e., the magnetic field of the wave, $\vec{H}_o$, is perpendicular to $\vec{E}_o$ and to the direction of propagation $\vec{k}$, as shown in Figure 15.1. Substituting $|\vec{k}| = \sqrt{\mu_o \epsilon_o}\,\omega$, we find $|\vec{H}_o| = \sqrt{\frac{\epsilon_o}{\mu_o}}|\vec{E}_o|$.

The instantaneous power carried by the wave (W/m^2) is defined by the *Poynting vector* $\vec{S} = \vec{E} \times \vec{H}^*$, where $\vec{H}^*$ is the complex conjugate of $\vec{H}$. Averaging over the sinusoidal period of the complex exponential, we find the time average power transfer of the wave is given by

$$\langle \vec{S} \rangle = \frac{1}{2}\vec{E}_o \times \vec{H}_o^*. \tag{15.11}$$

The direction of the Poynting vector is perpendicular to $\vec{E}_o$ and $\vec{H}_o$ in the direction of the wave propagation vector $\vec{k}$. The magnitude of the time average value of S above, $|\langle \vec{S} \rangle|$, is also known as the intensity of the wave (power/unit area).

15.2 EM wave propagation in non-conducting materials

When we are considering light or other electromagnetic waves propagating inside materials, we can't automatically set the charge density ρ and the current density $\vec{J}$ in Maxwell's equations to zero. However, if the conductivity of the material is negligible and the charge is all bound polarization charge included in the permittivity ε, we can follow the same procedure, taking the curl of the curl $\vec{E}$ equation to get

$$\vec{\nabla} \times (\vec{\nabla} \times \vec{E}) = \vec{\nabla}(\vec{\nabla} \cdot \vec{E}) - \vec{\nabla}^2 \vec{E} = -\frac{\partial}{\partial t}(\vec{\nabla} \times \mu \vec{H}). \tag{15.12}$$

In general ε and μ are *tensors*, reflecting the anisotropic electric and magnetic properties of the material the electromagnetic wave is propagating through. Assuming non-conductive media and that ε and μ are not changing in space or time, we can use the curl $\vec{H}$ equation with $\vec{J} = 0$ on the right-hand side to get

$$\vec{\nabla}(\vec{\nabla} \cdot \vec{E}) - \vec{\nabla}^2 \vec{E} = -\mu\varepsilon\frac{\partial^2 \vec{E}}{\partial t^2}. \tag{15.13}$$

If the material is anisotropic, then ε is not proportional to the identity matrix and $\vec{E}$ is not parallel to $\vec{D}$. So the lack of any free charge[5] gives us $\vec{\nabla} \cdot \vec{D} = \rho = 0$ but, in general, $\vec{\nabla} \cdot \vec{E} \neq 0$ and both terms on the left-hand side must be retained. Many materials, such as crystal quartz, calcite, and graphite, have anisotropic dielectric materials, and many magnetic materials have anisotropic susceptibilities at microwave/rf frequencies and below. In general in these materials, solution of the above equation implies anisotropic propagation with $\vec{E}_o$ and $\vec{H}_o$ no longer perpendicular to $\vec{k}$, and $\vec{k}$ depends on the propagation direction. But in isotropic dielectric materials at infrared/optical frequencies and higher (and even in anisotropic materials when the electric field is along a high-symmetry direction), the divergence of $\vec{E}_o$ is zero and the wave equation reduces to

$$\vec{\nabla}^2 \vec{E} = \mu\varepsilon \frac{\partial^2 \vec{E}}{\partial t^2} \tag{15.14}$$

with a transverse plane-wave solution $\vec{E}(\vec{r},t) = \vec{E}_o e^{i(\vec{k}\cdot\vec{r}-\omega t)}$ and a dispersion relation $k^2 = \mu\varepsilon\omega^2$. This gives us a propagation phase velocity $v_{ph} = \omega/k = 1/\sqrt{\mu\varepsilon}$. Pulling out $1/\sqrt{\mu_o\varepsilon_o} = c$, we find a propagation velocity $v_{ph} = c/\sqrt{\mu_r\varepsilon_r}$, where $\mu_r = \mu/\mu_o$ and $\varepsilon_r = \varepsilon/\varepsilon_o$ are the *relative permeability* and *relative permittivity*. At optical frequencies where $\mu_r \simeq 1$, $\sqrt{\varepsilon_r} = n$ is the *index of refraction,* which describes how the propagation velocity of light through a material is slowed down, compared with propagation through the vacuum $v_{ph} = c/n$.

Materials with loss mechanisms at the frequency of the electromagnetic radiation have permittivity or permeability with imaginary parts, as we saw in our study of harmonic oscillators. Thus, for example, $\varepsilon = \varepsilon' + i\varepsilon''$ and $\mu = \mu' + i\mu''$. This makes the wave vector k a complex quantity

$$k = \omega\sqrt{\mu\varepsilon} = k' + ik'' \tag{15.15}$$

with the square root of the complex quantity taken by the usual rules $\sqrt{z} = \sqrt{ri^\theta} = \sqrt{r}e^{i\theta/2}$, where r is the modulus of the complex quantity and θ is the phase (in radians).

As we have seen in Chapter 7, a complex wave number k leads to a damped wave with amplitude that decreases as the wave propagates through the material

$$\vec{E}(\vec{r},t) = \vec{E}_o e^{-k''r}e^{i(\vec{k}'\cdot\vec{r}-\omega t)}. \tag{15.16}$$

---

[5] Free charge here (ρ) refers to the excess charge of the material, not to the electrons in metals that can move freely when a potential difference along the conductor is applied and whose charge is balanced out by the uniform positive lattice charges.

All non-transparent materials have a complex wave vector arising from complex permittivity ε or permeability μ.

Example 15.1 Find the wavelength, propagation velocity, and $1/e$ damping distance of a wave with a free-space wavelength of 550 nm in a material with permeability $\mu = \mu_o$ and permittivity $\varepsilon = \epsilon_o(2 + 0.05i)$.

Solution: This wavelength corresponds to visible light in the green spectral region. The frequency ω is given in terms of the free-space wavelength by $\omega = \frac{2\pi c}{\lambda_o} = \frac{2\pi 2.9979 \times 10^8}{550 \times 10^{-9}} = 3.42 \times 10^{15}$ radians/s. The wave vector is given by $k = \omega\sqrt{\mu\varepsilon} = \frac{2\pi c}{\lambda_o}\sqrt{\mu_o\epsilon_o(2 + 0.05i)} = \frac{2\pi}{\lambda_o}\sqrt{(2 + 0.05i)}$ where the last equality follows from $\sqrt{\mu_o\epsilon_o} = 1/c$. This gives $k = 1.62 \times 10^7 + i\,2.02 \times 10^5$ m^{-1} and the propagating electric field in this material is $\vec{E}(\vec{r},t) = \vec{E}_o e^{-2.02\times 10^5\,r} e^{i(1.62\times 10^7 r - 3.42\times 10^{15}\,t)}$. The real field is thus $\mathcal{R}e\{\vec{E}(\vec{r},t)\} = \vec{E}_o e^{-2.02\times 10^5\,r}\cos\left(1.62 \times 10^7 r - 3.42 \times 10^{15}\,t\right)$, which we recognize as a propagating wave with wavelength $\lambda = \frac{2\pi}{1.62\times 10^7} = 389$ nm and amplitude that decreases with r as $\vec{E}_o e^{-2.02\times 10^5\,r}$. The propagation velocity is

$$v_{ph} = \omega/\mathcal{R}e\{k\} = c\Big/\mathcal{R}e\{\sqrt{\varepsilon/\epsilon_o}\} = 2.9979\times 10^8/1.414 = 2.12 \times 10^8 \text{ m/s}.$$ The field amplitude

will decrease to E_o/e in a distance $r = 1/2.02\times 10^5 = 4.92 \times 10^{-6}$ m $= 4.92$ microns. The amplitude of the electric field as a function of propagation distance in this material is shown in Figure 15.2.

Note on notation: The propagation velocity of an electromagnetic wave in non-lossy (transparent) material is given by $v_{ph} = \omega/k = \frac{1}{\sqrt{\mu\varepsilon}} = \frac{c}{\sqrt{\mu_r\varepsilon_r}}$, where $c = 1/\sqrt{\epsilon_o\mu_o} = 2.9979 \times 10^8$ m/s is the speed of light in vacuum and we have defined the *relative permittivity* $\varepsilon_r = \varepsilon/\epsilon_o$ and *relative permeability* $\mu_r = \mu/\mu_o$. At optical frequencies, the quantity $\sqrt{\mu_r\varepsilon_r}$ is defined as the *index of refraction*, denoted by the symbol n, and indicates how much the speed of light is reduced in a material $v_{ph} = c/n$. At optical frequencies the relative permeability is almost always very close to 1 and $n \simeq \sqrt{\varepsilon_r}$. In lossy materials, where the permittivity is complex, the index of refraction has an imaginary part. We can define a *complex index of refraction* $\tilde{n} = \sqrt{\mu_r\varepsilon_r} = n + i\kappa$, where n is the (real) index of refractions and κ is the *extinction coefficient*.[6] In terms of $\tilde{n}$, the wave vector is given by $k = \frac{\omega}{c}\tilde{n}$ and the electric field of a propagating wave is given by $\vec{E}(z,t) = \vec{E}_o e^{i(\tilde{n}(\omega/c)z - \omega t)} = \vec{E}_o e^{-\kappa z} e^{i(\frac{n\omega z}{c} - \omega t)}$.

---

[6] The imaginary part of the complex index of refraction is often denoted by k, $\tilde{n} = n + ik$, an unfortunate historical usage which gives the same symbol to the (dimensionless) imaginary part of the index as to the wave vector k (units of inverse distance) as in e^{ikx}.

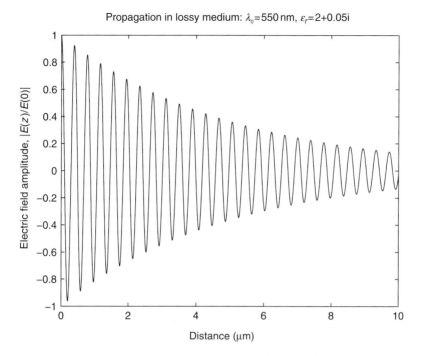

Propagation in lossy medium: $\lambda_o = 550$ nm, $\varepsilon_r = 2+0.05i$

Figure 15.2 Electric field amplitude of propagating optical wave in lossy medium with parameters as in Example 15.1.

Example 15.2 Calcite ($CaCO_3$) is an insulating anisotropic trigonal-rhombohedral crystal with a conductivity $\sigma \cong 0$ and a permittivity tensor at a wavelength of 500 nm given by $\varepsilon = \epsilon_o \begin{pmatrix} 2.75 & 0 & 0 \\ 0 & 2.75 & 0 \\ 0 & 0 & 2.21 \end{pmatrix}$ where the z-axis is along the uniaxial crystal symmetry direction. For propagation in an arbitrary direction the $\vec{E}$ field is not parallel to $\vec{D}$, and we cannot assume that the $\vec{\nabla}(\vec{\nabla} \cdot \vec{E})$ term in the wave equation vanishes even if there is no free charge density so that $\vec{\nabla} \cdot \vec{D} = 0$. If the propagation vector $\vec{k}$ is parallel to or perpendicular to the *optical axis* in the z-direction, however, the problem simplifies, and we can apply the analysis above.

First, for propagation parallel to the optic axis with $\vec{E}(\vec{r},t) = \vec{E}_o e^{i(\vec{k}\cdot\vec{r}-\omega t)}$ and $\vec{k} = k\hat{z}$, Maxwell's equation $\vec{\nabla} \times \vec{H} = \frac{\partial \vec{D}}{\partial t}$ implies $i\vec{k} \times \vec{H}_o = -i\omega\varepsilon\vec{E}_o$. Expanding in components, we find

$$\begin{vmatrix} \hat{x} & \hat{y} & \hat{z} \\ 0 & 0 & k \\ H_{ox} & H_{oy} & H_{oz} \end{vmatrix} = \begin{pmatrix} -kH_{oy} \\ kH_{ox} \\ 0 \end{pmatrix} = -\omega\epsilon_o \begin{pmatrix} 2.75\,E_{ox} \\ 2.75\,E_{oy} \\ 2.21\,E_{oz} \end{pmatrix}$$

which implies that $E_{oz} = 0$ and $\vec{E}$ is in the x–y plane with $\vec{D} = \epsilon_o 2.75\vec{E}$. Thus, $\vec{\nabla} \cdot \vec{E} = \epsilon_o 2.75 \vec{\nabla} \cdot \vec{D} = 0$ because there is no free charge density in the crystal. Taking

Continued

Example 15.2 (*cont.*)

$\vec{\nabla} \times (\vec{\nabla} \times \vec{E}) = -\frac{\partial(\vec{\nabla} \times \vec{B})}{\partial t} = -\mu \frac{\partial^2 \vec{D}}{\partial t^2}$, applying our vector identity $\vec{\nabla} \times (\vec{\nabla} \times \vec{E}) = \vec{\nabla}(\vec{\nabla} \cdot \vec{E}) - \vec{\nabla}^2 \vec{E}$, and setting $\vec{\nabla} \cdot \vec{E} = 0$, we find

$$\vec{\nabla}^2 \vec{E} = -k^2 \vec{E} = \mu_o \epsilon_o \, 2.75 \frac{\partial^2 \vec{E}}{\partial t^2} = -\mu_o \epsilon_o \, 2.75 \, \omega^2 \vec{E}$$

where we have used the fact that at optical frequencies the permeability is a constant $\mu = \mu_o$. The solution is $\vec{E}(\vec{r}, t) = (E_{ox}\hat{x} + E_{oy}\hat{y})e^{i(kz-\omega t)}$ with $k = \frac{\omega}{c}\sqrt{2.75}$. The wave propagates with an index $n = n_o = \sqrt{2.75} = 1.66$ for all polarizations in the x–y plane, where n_o is the *ordinary index of refraction*.

If the propagation is perpendicular to the optic axis, for example in the $\hat{x}$-direction, then $\vec{k} = k\hat{x}$ and expanding the curl H equation yields

$$\begin{vmatrix} \hat{x} & \hat{y} & \hat{z} \\ k & 0 & 0 \\ H_{ox} & H_{oy} & H_{oz} \end{vmatrix} = \begin{pmatrix} 0 \\ -kH_{oz} \\ kH_{oy} \end{pmatrix} = -\omega\epsilon_o \begin{pmatrix} 2.75\,E_{ox} \\ 2.75\,E_{oy} \\ 2.21\,E_{oz} \end{pmatrix}.$$

Thus $E_{ox} = 0$ and the electric field is in the y–z plane. We can break the electric field vector into components with $\vec{E}_o = (E_{oy}\hat{y} + E_{oz}\hat{z})$, solve the wave equation separately for each component, and then add the components to find the total field at any value of x and t. For each component, $\vec{D}$ is just a constant (2.75 ϵ_o or 2.21 ϵ_o) multiplied by $\vec{E}$, so $\vec{\nabla} \cdot \vec{E} = 0$ and we have two equations for the y- and z-component fields $k^2 E_{oy} = \mu_o \epsilon_o$ 2.75 $\omega^2 E_{oy}$ and $k^2 E_{oz} = \mu_o \epsilon_o$ 2.21 $\omega^2 E_{oz}$ with different solutions for k. Combining the component fields we find $\vec{E}(\vec{r}, t) = \left(E_{oy} e^{i\left(\frac{\omega}{c}n_o x\right)}\hat{y} + E_{oz} e^{i\left(\frac{\omega}{c}n_e x\right)}\hat{z} \right) e^{-i\omega t}$ where $n_e = \sqrt{2.21} = 1.49$ is the *extraordinary index of refraction*. The y-polarized and z-polarized components will develop a phase shift as they propagate through the crystal. Propagation perpendicular to the optic axis through a *quarter-wave plate* of selected thickness will create a phase shift of 90° between two equal perpendicular polarization components, so-called *circular polarization*. The two polarizations will also have different angles of refraction if the beam is obliquely incident on an interface, leading to the phenomenon of *birefringence*.

15.3   EM wave propagation in conductors

A conductor, as we have seen, is a material with charges that are free to move and carry current when an electric field is applied, as characterized by the constitutive relation $\vec{J} = \sigma\vec{E}$. In general, the conductivity can be anisotropic and represented by a tensor, but in what follows we will assume for simplicity that the conductivity is isotropic and can be represented by a scalar quantity.

In the presence of conduction currents, the development of the electromagnetic wave equation proceeds from Maxwell's equations as before, except that we cannot neglect the current density $\vec{J}$ in the $\vec{\nabla} \times \vec{H}$ equation. We thus have an equation for the electric field $\vec{E}$

$$\vec{\nabla} \times \left(\vec{\nabla} \times \vec{E} \right) = -\vec{\nabla} \times \frac{\partial \vec{B}}{\partial t} = \frac{\partial}{\partial t} \left(\vec{\nabla} \times \mu \vec{H} \right) = \mu \frac{\partial}{\partial t} \left(\vec{J} + \frac{\partial \vec{D}}{\partial t} \right) = \mu \frac{\partial}{\partial t} \left(\sigma \vec{E} + \varepsilon \frac{\partial \vec{E}}{\partial t} \right).$$

$$(15.17)$$

If the medium is effectively electrically neutral and isotropic, the left side reduces as before: $\vec{\nabla} \times \left(\vec{\nabla} \times \vec{E} \right) = \vec{\nabla}^2 \vec{E}$. Assuming a wave solution of the form $\vec{E}_o e^{i(\vec{k} \cdot \vec{r} - \omega t)}$ (where we allow the possibility that $\vec{k}$ has both real and imaginary parts), the equation for the electric fields leads to the dispersion relation for $\vec{k}$

$$k^2 = \mu \varepsilon \omega^2 + i \mu \sigma \omega. \qquad (15.18)$$

Question 15.2

Show that Eq. (15.17) for $\vec{E}_o e^{i(\vec{k} \cdot \vec{r} - \omega t)}$ implies the given dispersion relation between k and ω.

Thus, for a material with non-zero conductivity the wave vector is complex with an imaginary part, even if ε and μ are real quantities. Materials with a finite conductivity are lossy for electromagnetic wave propagation. At microwave frequencies and below in materials with high conductivity (metals), the wave is essentially damped out within one wavelength in the material.

Example 15.3 Consider 1 GHz microwave radiation incident on an aluminum plate with conductivity $\sigma = 3.5 \times 10^7$ S/m. With $\omega = 2\pi f = 6.28 \times 10^9$ s^{-1}, $\mu \simeq \mu_o = 4\pi \times 10^{-7}$ H/m, and $\varepsilon \sim \epsilon_o = 8.85 \times 10^{-12}$ F/m, the first term in the dispersion equation (Eq. (15.18)) is approximately 10^9 times smaller than the second term and can be neglected. Thus, recalling that $\sqrt{i} = \sqrt{\frac{1}{2}(1 + i)}$, we have $k = \sqrt{i \mu \sigma \omega} = \sqrt{\mu \sigma \omega / 2} + i \sqrt{\mu \sigma \omega / 2}$. The real and imaginary parts of the wave vector are thus equal (and large!) and the wave is damped out in a distance comparable to one wavelength *in the metal*, on the order of microns. The *skin depth*, the distance where the imaginary part of k times the z is 1, is given by $\delta = \sqrt{2 / \mu \sigma \omega}$. A snap-shot of the real electric field of the wave as a function of z at $t = 0$ is shown in Figure 15.3 below.

We can extend the definition of the complex index of refraction to allow for conducting media by using the relation $k = \frac{\omega}{c} \tilde{n}$ yielding

$$\tilde{n} = \sqrt{\mu_r \left(\epsilon_r + \frac{i \sigma}{\omega \epsilon_o} \right)}. \qquad (15.19)$$

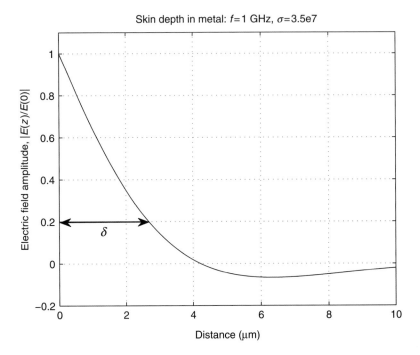

Figure 15.3 The electric field, normalized to the value at the surface, of 1 GHz microwave radiation inside the surface of an aluminum plate at $t = 0$. The skin depth δ is shown.

Question 15.3

Derive Eq. (15.19) from Eq. (15.18) remembering that $c = 1 / \sqrt{\epsilon_0 \mu_0}$.

If we consider our copper screen room with which we began the chapter, and if the screen wire is 1 mm in diameter and copper has conductivity $\sigma = 5.96 \times 10^7$ S/m, we find the skin depth at AM radio frequencies ($f = 500$ kHz) to be $\delta = \sqrt{2/\mu\sigma\omega} =$

$$\sqrt{\frac{2}{(4\pi \times 10^{-7})(5.96 \times 10^7)(2\pi\,500 \times 10^3)}} = 9.2 \times 10^{-5}\text{m}.$$ The propagation through the 1mm thick wire is $e^{-\frac{x}{\delta}} = e^{-1 \times 10^{-3}/9.2 \times 10^{-5}} = 1.9 \times 10^{-5}$; the radio signal is reduced by a factor of 200,000! Note that the wavelength at 500 kHz is about 600 meters, so the small millimeter-scale gaps in the screen are even less resolvable to radio waves than the atomic structure on the order of 0.1 nm is to visible light with a wavelength around 500 nm. FM radio with frequencies 200 times higher would have a skin depth 15 times smaller.

Nuclear Magnetic Resonance (NMR) is being considered as a probe for explosives inside aluminum cans. A soft drink can is made of aluminum with a conductivity $\sigma = 3.5 \times 10^7 \frac{\text{S}}{\text{m}}$ and a thickness around 100 μm. An NMR probe electromagnetic wave at a frequency of 4 kHz has a skin depth in aluminum of around 1 mm, giving an electric field transmission $e^{-\frac{x}{\delta}} = e^{-100 \times 10^{-6}/1 \times 10^{-3}} = 0.90$. A 90% transmission signal is certainly adequate to do measurements inside the can, but we also need to take into account how much of the incident wave is reflected at the interface between the aluminum and the liquid inside or the air outside. To find out the reflection at the interface, we will need to consider the boundary conditions for the electric and magnetic fields at the boundary between two materials in the next sections.

15.4   Boundary conditions at an interface

So far, we have seen how an electromagnetic wave propagates in different media. When an electromagnetic wave is incident on the plane interface between two different materials some of the wave is transmitted into the second material and some of the wave is reflected back into the first medium. The incident, reflected, and transmitted waves at normal incidence are shown in Figure 15.4. For *normal incidence*, as shown, the waves are propagating perpendicular to the interface and the electric and magnetic fields are parallel to the interface and either in (E fields) or perpendicular to (H fields) the plane of the paper. To determine the relative amplitudes and phases of the incident (E_i, H_i), reflected (E_r, H_r), and transmitted (E_t, H_t) fields, we have to apply the *boundary conditions* on the electric and magnetic fields.

   The boundary conditions follow directly from the four Maxwell's equations and the Gauss's and Stokes' theorem integral relations. Considering first the electric field, we consider the integral of Maxwell's "curl E" equation around the path shown by the dotted line in Figure 15.5.

$$\int (\vec{\nabla} \times \vec{E}) \cdot d\vec{S} = \oint \vec{E} \cdot d\vec{\ell} = \int -\frac{\partial \vec{B}}{\partial t} \cdot d\vec{S}. \qquad (15.20)$$

The surface integrals are over the surface of length ℓ and width Δ, while the closed loop line integral is over the perimeter of that surface. Assuming that the width Δ of the area is small and allowed to shrink to zero, the only contribution to the line integral is from

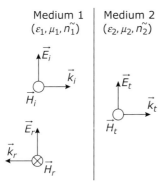

Figure 15.4 Incident (i), relected (r), and transmitted (t) fields and wave vectors for a wave normally incident on an interface between two media.

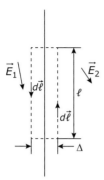

Figure 15.5 Path of integration for Eq. (15.20) along both sides of the interface between the two media.

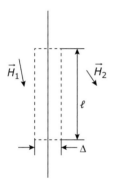

Figure 15.6 Path of integration for the curl H equation on both sides of the interface.

the vertical sides of the path. Dotting electric fields into the length elements parallel to the surface $d\vec{\ell}$ and letting $\Delta \to 0$ we find

$$\vec{E}_1 \cdot \vec{\ell} - \vec{E}_2 \cdot \vec{\ell} = \lim_{\Delta \to 0} \int -\frac{\partial \vec{B}}{\partial t} \cdot d\vec{S} = 0 \qquad (15.21)$$

since the area integral goes to zero with Δ. The electric field dotted into $\vec{\ell}$ just gives the *tangential component* of the fields and the lengths of the sides are equal, giving the boundary condition on the component of the electric fields tangential to the surface

$$E_{1\,tan} = E_{2\,tan}. \qquad (15.22)$$

Similarly, integrating Maxwell's "curl H" equation around a similar path shown in Figure 15.6 gives a similar result

$$\vec{H}_1 \cdot \vec{\ell} - \vec{H}_2 \cdot \vec{\ell} = \lim_{\Delta \to 0} \int \left(\vec{J} - \frac{\partial \vec{D}}{\partial t} \right) \cdot d\vec{S}. \qquad (15.23)$$

The integral over the $\vec{D}$ field on the right side goes to zero as the area goes to zero by the same reasoning as before. If the two materials are insulators, which means that the

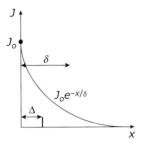

Figure 15.7 The current density associated with an electromagnetic wave decays exponentially into the conductor.

current $\vec{J} = 0$, we have the same result as for the electric field: the tangential component of the magnetic field H_{tan} must be continuous across the interface

$$H_{1tan} = H_{2tan}. \qquad (15.24)$$

The case where one of the materials is a conductor appears at first sight to be different. If there is a current confined to the surface of the conductor, the right-hand side of Equation (15.23) could remain finite because the surface current $\vec{J}_s$ could pass through the contour in Figure 15.5 and create a discontinuity between H_{1tan} and H_{2tan}. However, if we look on a microscopic length scale, we know from our discussion of wave propagation in conductors that the electric field will decay exponentially into the conducting material as shown in Figure 15.7. The current density, given by our usual constitutive relation $\vec{J} = \sigma\vec{E}$, will decay in the same way as shown in Figure 15.7. If we go to the limit where the width of the line integral Δ is much less than the skin depth δ and let $\Delta \to 0$, we see that the current passing through the line contour goes to zero the same as the integral of the $\vec{D}$ field over the surface does and the condition of the continuity of the tangential $\vec{H}$ field is maintained. The surface current is accounted for by the electric field penetration into the skin depth of the conductor and the constitutive relation $\vec{J} = \sigma\vec{E}$ just as we did in finding the wave in a conductor. Hence the condition expressed in Eq. (15.24) holds for all materials.

For completeness we will also consider the boundary conditions on the displacement field $\vec{D}$ and the magnetic induction $\vec{B}$. Integrating the Maxwell's equation $\nabla \cdot \vec{D} = \rho$ over the *Gaussian pillbox* shown in Figure 15.8 and applying Gauss's integral theorem to convert the volume integral into a surface integral, we have

$$\int_V \nabla \cdot \vec{D} \, dV = \oint_S \vec{D} \cdot d\vec{S} = \int_V \rho dV = q_{enclosed}. \qquad (15.25)$$

The integral over the curved surface goes to zero as $\Delta \to 0$ and the integral over the end caps gives us

$$A(D_{1norm} - D_{2norm}) = q_{enclosed} \qquad (15.26)$$

where D_{1norm} and D_{2norm} are the components of the displacement field normal to the interface (i.e., those fields that are parallel to the $d\vec{S}$ surface elements perpendicular to

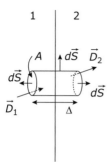

Figure 15.8 Gaussian surface for integrating Maxwell's two divergence equations over both sides of an interface.

the end caps of the cylinder). **If there is no free charge at the interface, $q_{enclosed} = 0$, and the normal component of $\vec{D}$ is continuous across the interface.** The bound charge at the interface due to the difference in the polarizations of the two media is taken into account through the constitutive relation $\vec{D} = \varepsilon\vec{E}$

$$D_{1norm} = \varepsilon_1 E_{1norm} = D_{2norm} = \varepsilon_2 E_{2norm}. \tag{15.27}$$

Thus the normal component of the electric field at an interface without free charge is discontinuous with the discontinuity given by the ratio of the dielectric constants of the two materials

$$E_{1norm} = \frac{\varepsilon_2}{\varepsilon_1} E_{2norm}. \tag{15.28}$$

A similar argument gives us the boundary condition on the magnetic induction

$$B_{1norm} = B_{2norm}, \tag{15.29}$$

or in terms of the magnetic fields and the magnetic permeability of the two media

$$H_{2norm} = \frac{\mu_2}{\mu_1} H_{2norm}. \tag{15.30}$$

15.5   Reflection and transmission at normal incidence

Applying the continuity of the electric field across the interface to an electromagnetic wave normally incident on an interface as illustrated in Figure 15.4, we find

$$E_i + E_r = E_t. \tag{15.31}$$

(At normal incidence the electric fields of the propagating electromagnetic waves are all tangential to the interface.)

Similarly, the magnetic fields of the propagating fields are continuous and, at normal incidence, all parallel to the interface

$$H_i - H_r = H_t. \tag{15.32}$$

The negative sign comes in here because the reflected wave is propagating away from the interface. From the curl E Maxwell's equation and assuming harmonic waves of the form $e^{i(\vec{k}\cdot\vec{r}-\omega t)}$

$$\vec{\nabla}\times\vec{E}=i\vec{k}\times\vec{E}=-\frac{\partial\vec{B}}{\partial t}=i\omega\mu\vec{H}. \tag{15.33}$$

Thus, if all the electric fields are vertical (in the $+\hat{y}$-direction), the reflected wave propagating to the left has $\vec{k}_r||(-\hat{x})$ and $\vec{k}_r\times\vec{E}$ and thus $\vec{H}_r$ must be directed in the $-\hat{z}$-direction (into the page) as shown in Figure 15.4. The incident and transmitted waves have $\vec{k}_i, \vec{k}_t||(+\hat{x})$ and the cross product of the wave vector and the electric field is in the positive $\hat{z}$-direction (out of the page).

Solving Eq. (15.33) for the magnitude of the magnetic field gives

$$|\vec{H}|=\frac{k|\vec{E}|}{\omega\mu}=\frac{\sqrt{\mu\varepsilon\omega^2+i\mu\sigma\omega}|\vec{E}|}{\omega\mu}=\sqrt{\frac{\varepsilon+i\sigma/\omega}{\mu}}|\vec{E}|=\frac{1}{\tilde{Z}}|\vec{E}| \tag{15.34}$$

using Eq. (15.34) and defining the complex characteristic impedance of the medium, including the conductivity σ, $\tilde{Z}=\sqrt{\frac{\mu}{\varepsilon+i\sigma/\omega}}$. Using this relation we can rewrite Eq. (15.32) for the continuity of the magnetic field in terms of the electric field amplitudes

$$\frac{1}{\tilde{Z}_1}(E_i-E_r)=\frac{1}{\tilde{Z}_2}E_t. \tag{15.35}$$

Substituting from the equation of continuity of the electric field $E_i+E_r=E_t$ to eliminate E_t, we find the *field reflection coefficient*

$$r=\frac{E_r}{E_i}=\frac{H_r}{H_i}=\frac{1/\tilde{Z}_1-1/\tilde{Z}_2}{1/\tilde{Z}_1+1/\tilde{Z}_2}=\frac{\tilde{Z}_2-\tilde{Z}_1}{\tilde{Z}_2+\tilde{Z}_1}. \tag{15.36}$$

In the optical spectra region where magnetic effects are negligible so $\mu=\mu_o$, we can express this in terms of the complex indices of refraction $\tilde{n}_1$ and $\tilde{n}_2$

$$r=\frac{\tilde{n}_1-\tilde{n}_2}{\tilde{n}_1+\tilde{n}_2}. \tag{15.37}$$

Similarly, the *field transmission coefficient* is found by eliminating E_r

$$t=\frac{E_t}{E_i}=\frac{H_t}{H_i}=\frac{2\tilde{Z}_2}{\tilde{Z}_2+\tilde{Z}_1} \tag{15.38}$$

or at optical frequencies,

$$t=\frac{2\tilde{n}_1}{\tilde{n}_1+\tilde{n}_2}. \tag{15.39}$$

The *power reflection and transmission coefficients* are found using the time average Poynting vector $\vec{S}=\frac{1}{2}\vec{E}\times\vec{H}^*$. The ratio of the reflected power to the incident power is given by

$$R = \frac{|\vec{S_r}|}{|\vec{S_i}|} = rr^* = \left|\frac{\tilde{Z}_2 - \tilde{Z}_1}{\tilde{Z}_2 + \tilde{Z}_1}\right|^2 \rightarrow \left|\frac{\tilde{n}_1 - \tilde{n}_2}{\tilde{n}_1 + \tilde{n}_2}\right|^2 \tag{15.40}$$

where the last relation is the limit at optical frequencies. Since the incident and reflected waves are in the same medium they have the same velocity and thus the ratio of the reflected and incident field amplitudes can be converted directly into power reflection. This is not true for the transmitted wave, which leaves the interface at a different speed $v = c/n$. For non-normal incidence the angle of propagation also changes on transmission, so the perpendicular cross-section area of the beam is different in the transmitted medium. However, the interface has a vanishing small thickness and can't dissipate any power, so the power arriving at the interface must be the same as the power leaving the interface from conservation of energy. This gives the relationship between the power reflection coefficient and power transmission coefficients $T + R = 1$. The transmitted power ratio is then

$$T = \frac{|\vec{S_t}|}{|\vec{S_i}|} = 1 - R = 1 - \left|\frac{\tilde{Z}_2 - \tilde{Z}_1}{\tilde{Z}_2 + \tilde{Z}_1}\right|^2 = \frac{4Re(\tilde{Z}_1\tilde{Z}_2)}{|\tilde{Z}_2 + \tilde{Z}_1|^2} \rightarrow \frac{4Re(\tilde{n}_1\tilde{n}_2)}{|\tilde{n}_1 + \tilde{n}_2|^2}. \tag{15.41}$$

15.6   Example: reflection and transmission from a thin metal sheet

So, with what we know of reflection and transmission of electromagnetic waves, let us calculate the electric field transmission into a soda can at 4 kHz if we assume the can is 100 μm thick aluminum with conductivity $\sigma = 3.5 \times 10^7$ S/m and the liquid on the inside has a permittivity close to that of water $\epsilon_r \approx 120$ at 4 kHz. Assume that all the materials are non-magnetic with $\mu = \mu_o$ and the background permittivity of aluminum is $\varepsilon_r = \varepsilon/\epsilon_o \simeq 5$.

We will do this example in two ways. First we will apply a straightforward calculation of the multiply-reflected wave as we did in Chapter 7. We will compare this with a calculation where we consider the can as a sheet of current between the air outside and the liquid inside the can. Following the technique of Chapter 7, we add up the multiply-reflected beams with a phase shift and attenuation of $e^{ik_1 d}$ for each pass through the aluminum as in Figure 15.9. This gives

$$t = \frac{E_t}{E_i} = t_{01}e^{ik_1 d}t_{12} + t_{01}e^{ik_1 d}r_{12}e^{ik_1 d}r_{10}e^{ik_1 d}t_{12} + \cdots$$
$$= t_{01}e^{ik_1 d}t_{12}(1 + r_{12}e^{i2k_1 d}r_{10} + [r_{12}e^{i2k_1 d}r_{10}]^2 + \cdots). \tag{15.42}$$

Using the binomial expansion $\frac{1}{1-x} = 1 + x + x^2 + \cdots$ we find

$$t = \frac{t_{01}e^{ik_1 d}t_{12}}{1 - r_{12}e^{i2k_1 d}r_{10}}. \tag{15.43}$$

For the aluminum, we find $k_1 = \frac{\omega}{c}\sqrt{\varepsilon_r + \frac{i\sigma}{\omega\epsilon_o}} \simeq \frac{\omega}{c}\sqrt{5 + \frac{i3.5 \times 10^7}{(2\pi \times 4000)(8.85 \times 10^{-12})}} = 743.4 +$
$743.4i$ and $\tilde{n}_1 = \frac{ck_1}{\omega} = 8.87 \times 10^6 + 8.87 \times 10^6 i$. Using $\tilde{n}_0 = 1$ and $\tilde{n}_2 = \sqrt{120}$

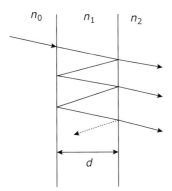

Figure 15.9 Multiply-reflected beams that contribute to the transmission of an EM wave through a thin sheet.

$= 10.95$ and using Equations (15.37) and (15.39) above for t_{01}, t_{12}, r_{12}, and r_{10} and $e^{ik_1 d} = 0.9406 + 0.0560i$, we find

$$t = 1.52 \times 10^{-6} + 2.79 \times 10^{-9} i. \tag{15.44}$$

The impedance mismatch between the free space/liquid and the aluminum reduces the transmission far below the estimate we determined from $e^{-d/\delta}$ before. We note that if we use just the numerator of Eq. (15.43) (ignoring the multiple reflections) we find a significantly different answer: $t_{01} e^{ik_1 d} t_{12} = 2.24 \times 10^{-7} - 1.03 \times 10^{-7} i$.

Now consider the aluminum to be a sheet of current between the air and the liquid (certainly an appropriate approximation at 4 kHz where the free-space wavelength is approximately 75 km). If we integrate Maxwell's "curl H" equation over a rectangle across the interface of length ℓ and vanishing width Δ as before, we find the integral form of Maxwell's "curl H" equation yields

$$\oint_c \vec{H} \cdot d\vec{\ell} = \int_s \left(\vec{J} + \frac{\partial \vec{D}}{\partial t} \right) \cdot d\vec{S} = \lim_{\Delta \to 0} \left(J\ell\Delta + \frac{\partial D}{\partial t} \ell\Delta \right). \tag{15.45}$$

At normal incidence with the fields $H_i - H_r$ in Medium 1 and H_t and E_t in Medium 2 parallel to the interface and using the constitutive relation $\vec{J} = \sigma \vec{E}$ we find

$$(H_i - H_r - H_t)\ell = \lim_{\Delta \to 0} \left(\sigma E_t \ell\Delta + \frac{\partial D}{\partial t} \ell\Delta \right). \tag{15.46}$$

As $\Delta \to 0$ the displacement current flux vanishes but the surface current remains to contribute to a discontinuity in the magnetic fields (but not the electric field)

$$H_i - H_r - H_t = \sigma \Delta E_t. \tag{15.47}$$

The "curl E" Maxwell's equation gives us $H = \frac{kE}{\omega\mu}$ for the incident, reflected, and transmitted fields. Substituting $E_r = E_t - E_i$ from the continuity of the electric field across the interface, we find

$$\frac{k_1}{\omega\mu_1}E_i - \frac{k_1}{\omega\mu_1}(E_t - E_i) - \frac{k_2}{\omega\mu_2}E_t = \sigma\Delta E_t. \tag{15.48}$$

Solving for the transmitted field we find

$$t = \frac{E_t}{E_i} = \frac{2\frac{k_1}{\mu_1}}{\left(\frac{k_1}{\mu_1} + \frac{k_2}{\mu_2} + \omega\sigma\Delta\right)}. \tag{15.49}$$

If the permeabilities $\mu_1 = \mu_2 = \mu_o$ and we express the wave vectors in terms of the complex indices of refraction $k = \frac{\omega}{c}\tilde{n}$, we find

$$t = \frac{2\tilde{n}_1}{(\tilde{n}_1 + \tilde{n}_2 + \mu_o c\sigma\Delta)} \tag{15.50}$$

where the surface conductance $\sigma\Delta$ (in siemens) is normalized by the product $\mu_o c = 376.73$ Ω, known as the *impedance of free space*. Applying this to our problem with $\tilde{n}_1 = 1$, $\tilde{n}_2 = 10.95$, and $\sigma\Delta = (3.5 \times 10^7)(100 \times 10^{-6}) = 3500$ S, we find $t = 1.52 \times 10^{-6}$, which matches our previous calculation to three decimal places if we neglect the imaginary part (which is three orders of magnitude smaller). The previous calculation leading to Eq. (15.44) is the exact solution, but the treatment leading to Eq. (15.50) gives a very close approximation without considering the complications of multiple reflections in the strongly absorbing metal film.

15.7   Reflection and transmission at non-normal incidence on an interface

When an electromagnetic wave is incident at an arbitrary angle on an interface, the boundary conditions developed in Section 15.4 must hold at all points on the interface. Consider a plane wave $\vec{E}_i e^{i(\vec{k}_i \cdot \vec{r} - \omega t)}$ incident on the interface between Medium 1 and Medium 2, as shown in Figure 15.10. Assuming the incident wave vector is in the x–y plane, there will be two polarization cases to consider: (1) the electric field $\vec{E}_i$ is perpendicular to the x–y plane in the z-direction while the magnetic field $\vec{H}_i$ is in the

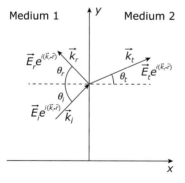

Figure 15.10 Plane wave non-normally incident on a interface between two media showing angle of incidence, angle of reflection, and angle of transmission.

x–y plane and (2) the magnetic field $\vec{H}_i$ is perpendicular to the *x–y* plane in the *z*-direction while the electric field $\vec{E}_i$ is in the *x–y* plane. If we define the *plane of incidence* as the plane which contains the incident, reflected, and transmitted wave vectors $\vec{k}_i$, $\vec{k}_r$, and $\vec{k}_t$ (the *x–y* plane in this case), the geometry in case (1) is referred to as *s-polarization* where the electric field is perpendicular to the plane of incidence and the magnetic field is in the plane of incidence, while case (2) with $\vec{E}_i$ in the plane of incidence and $\vec{H}_i$ perpendicular to the plane of incidence is referred to as *p-polarization*. In both polarizations, the boundary conditions must hold at all times and all points on the interface. The condition that the boundary conditions must be satisfied at all times implies that the frequency of the reflected and transmitted waves must be the same as the frequency ω of the incident wave.

Putting the origin of the coordinate system on the interface as shown, implies that every point $\vec{r}$ on the interface in the plane of the figure can be written as $\vec{r} = y\hat{y}$. The condition that the boundary conditions hold at all points on the interface thus yields $e^{ik_{iy}y} = e^{ik_{ry}y} = e^{ik_{ty}y}$. For this to hold at all values of *y* implies that the *y*-components of the wave vectors of the incident, reflected, and refracted waves are equal

$$k_{iy} = k_{ry} = k_{ty}. \tag{15.51}$$

If we assume that the incident wave has real wave vector $\vec{k}_i$, as will be the case when effective dielectric constant $\tilde{\varepsilon}_1$ and permeability μ_1 are real, the first equation is

$$|k_i| \sin \theta_i = |k_r| \sin \theta_r. \tag{15.52}$$

Since the medium for the incident and reflected wave is the same, $|k_i| = |k_r|$, we have the result that **the angle of reflection θ_r equals the angle of incidence θ_i.**

If Medium 2 also has a real wave vector, then we have Snell's Law for the relationship between the angle of incidence θ_i and the angle of refraction θ_t

$$|k_1| \sin \theta_i = |k_2| \sin \theta_t. \tag{15.53}$$

If $|k_2| < |k_1|$, there are angles θ_i for which there is no physical θ_t ($\sin \theta_t > 1$). At these θ_i there is no transmitted wave and the incident wave is totally reflected.

Note: If Medium 2 has a complex index of refraction $\tilde{n}$, the propagating wave is *inhomogeneous* with the real and imaginary parts of the wave vector not parallel. In this case $\vec{k}_t$ is complex $\vec{k}_t = \vec{k}_t' + i\vec{k}_t''$. The real part $\vec{k}_t'$ (perpendicular to planes of constant phase) must have a *y*-component equal to the *y*-component of the incident wave as given in Eqs. (15.51) and (15.53). Since the incident wave is real, however, the *y*-component of the imaginary part $\vec{k}_t''$ (perpendicular to planes of constant intensity) must be zero, i.e., $\vec{k}_t''$ is entirely in the *x*-direction perpendicular to the interface. Thus the wave in Medium 2 propagates in the direction predicted by Snell's Law, but decays with an intensity $e^{-k_t'' d}$ where *d* is the perpendicular distance from the interface.

The reflection coefficients at non-normal incidence are different in *s-polarization* and *p-polarization*. In **s-polarization**, the electric field is in the $\hat{z}$-direction, perpendicular to the plane of incidence which contains the incident, reflected, and transmitted $\vec{k}$ vectors, and tangent to the interface. Thus we have $E_i + E_r = E_t$ from the condition of the continuity of the tangential component of the electric field. By Maxwell's "curl E" equation, we have $\vec{k} \times \vec{E} = \mu\omega\vec{H}$ implying that $\vec{H}$ is perpendicular to $\vec{E}$ and thus in the plane of incidence.[7] We can find the tangential component of $\vec{H}$ by taking the dot product of $\hat{y}$ with $\vec{k} \times \vec{E}$. Using the vector identity $\vec{A} \cdot (\vec{B} \times \vec{C}) = \vec{B} \cdot (\vec{C} \times \vec{A})$ we find

$$H_{tan} = \frac{1}{\mu\omega}\hat{y} \cdot (\vec{k} \times \vec{E}) = \frac{1}{\mu\omega}\vec{k} \cdot (\vec{E} \times \hat{y}) = \frac{1}{\mu\omega}\vec{k} \cdot (E\hat{z} \times \hat{y}) = -\frac{E}{\mu\omega}\vec{k} \cdot \hat{x} = -\frac{E}{\mu\omega}k_x.$$

$$(15.54)$$

This follows from the condition that in s-polarization $\vec{E}$ is in the $\hat{z}$-direction and $\hat{z} \times \hat{y} = -\hat{x}$. Using the continuity of H_{tan} and the fact that the x-component of the reflected wave vector is opposite to the x-component of the incident wave $k_{rx} = -k_{ix}$ (because the reflected wave is traveling away from the interface, while the incident wave is traveling toward the interface), we find

$$\frac{1}{\mu_1}(E_ik_{ix} + E_rk_{rx}) = \frac{1}{\mu_1}(E_ik_{ix} - E_rk_{ix}) = \frac{1}{\mu_2}E_tk_{tx}. \qquad (15.55)$$

Substituting $E_t = E_i + E_r$ and using Snell's Law, $k_{ty} = k_{iy}$, we find the field reflection coefficient is given by

$$r_s = \frac{E_r}{E_i} = \frac{\dfrac{k_{ix}}{\mu_1} - \dfrac{k_{tx}}{\mu_2}}{\dfrac{k_{ix}}{\mu_1} + \dfrac{k_{tx}}{\mu_2}} = \frac{\dfrac{k_{ix}}{\mu_1} - \dfrac{1}{\mu_2}\sqrt{k_2{}^2 - k_{iy}{}^2}}{\dfrac{k_{ix}}{\mu_1} + \dfrac{1}{\mu_2}\sqrt{k_2{}^2 - k_{iy}{}^2}} \qquad (s\text{-polarization}). \qquad (15.56)$$

If the incident wave is in a medium with real k, $k_{iy} = k_1 \sin \theta_i$ and $k_{ix} = k_1 \cos \theta_i$. The angle θ_i is the angle of propagation of the incident plane wave, defined as the angle between $\vec{k}_i$ and the normal to the interface, as shown in Figure 15.10. In this case, the reflection coefficient takes the form

$$r_s = \frac{E_r}{E_i} = \frac{\dfrac{k_1 \cos \theta_i}{\mu_1} - \dfrac{1}{\mu_2}\sqrt{k_2{}^2 - k_1{}^2 \sin^2 \theta_i}}{\dfrac{k_1 \cos \theta_i}{\mu_1} + \dfrac{1}{\mu_2}\sqrt{k_2{}^2 - k_1{}^2 \sin^2 \theta_i}}. \qquad (15.57)$$

The transmission coefficient follows from the condition on the continuity of the electric field $E_i + E_r = E_t$,

$$t_s = \frac{E_t}{E_i} = 1 + r_s. \qquad (15.58)$$

[7] Unless μ is a tensor with $\vec{H}$ not along a principal axis of the permeability. At infrared and optical frequencies μ is a constant $\mu = \mu_o = 4\pi \times 10^{-7}\text{F/m}$.

The power reflection coefficient follows from the ratio of incident and reflected Poynting vector

$$R_s = \frac{|\vec{S}_r|}{|\vec{S}_i|} = \frac{|\vec{E}_r|^2}{|\vec{E}_i|^2} = |r_s|^2 \tag{15.59}$$

and the power transmission coefficient follows from conservation of energy

$$T_s = \frac{|\vec{S}_t|}{|\vec{S}_i|} = 1 - R_s. \tag{15.60}$$

For **p-polarization**, where the $\vec{H}$ fields are parallel to the interface in the $\hat{z}$-direction and the $\vec{E}$ fields are in the $\hat{x}-\hat{y}$ plane of incidence, the application of the boundary condition on the magnetic field gives us $\vec{H}_i + \vec{H}_r = \vec{H}_t$. Taking the dot product of $\hat{y}$ with the $\vec{E}$ fields in the plane of incidence to find the tangential components of the incident, reflected, and transmitted waves and applying the continuity of the tangential electric field, we find

$$\frac{H_r}{H_i} = \frac{\dfrac{\mu_1}{k_1^2}k_{ix} - \dfrac{\mu_2}{k_2^2}k_{tx}}{\dfrac{\mu_1}{k_1^2}k_{ix} + \dfrac{\mu_2}{k_2^2}k_{tx}}. \tag{15.61}$$

The reflection coefficient r_p for p-polarization is defined differently in different books and papers. Setting $r_p = {}^{H_r}/_{H_i}$ while $r_s = {}^{E_r}/_{E_i}$ has the disadvantage that $r_p = -r_s$ at normal incidence $\theta_i = 90°$ when there is no distinction between s- and p-polarization. **We will use the convention where r_p represents the ratio of the reflected and incident *electric fields*.** This convention with the application of Snell's Law, $k_{ty} = k_{iy}$, gives us

$$r_p = -\frac{H_r}{H_i} = -\frac{-\dfrac{\mu_1}{k_1^2}k_{ix} + \dfrac{\mu_2}{k_2^2}\sqrt{k_2^2 - k_{iy}^2}}{\dfrac{\mu_1}{k_1^2}k_{ix} + \dfrac{\mu_2}{k_2^2}\sqrt{k_2^2 - k_{iy}^2}} \qquad (p\text{-}polarization) \tag{15.62}$$

and

$$t_p = \frac{E_t}{E_i} = 1 - r_p. \tag{15.63}$$

If the incident medium is real with a direction of propagation θ_i, we find

$$r_p = -\frac{H_r}{H_i} = \frac{-\dfrac{\mu_1}{k_1^2}k_{ix} + \dfrac{\mu_2}{k_2^2}k_{tx}}{\dfrac{\mu_1}{k_1^2}k_{ix} + \dfrac{\mu_2}{k_2^2}k_{tx}} = \frac{-\dfrac{\mu_1}{k_1}\cos\theta_i + \dfrac{\mu_2}{k_2^2}\sqrt{k_2^2 - k_1^2\sin^2\theta_i}}{\dfrac{\mu_1}{k_1}\cos\theta_i + \dfrac{\mu_2}{k_2^2}\sqrt{k_2^2 - k_1^2\sin^2\theta_i}}. \tag{15.64}$$

The power reflection and transmission coefficients follow from the same logic as s-polarization

$$R_p = \frac{|\vec{S}_r|}{|\vec{S}_i|} = \frac{|\vec{E}_r|^2}{|\vec{E}_i|^2} = |r_p|^2 \qquad (15.65)$$

and

$$T_p = \frac{|\vec{S}_t|}{|\vec{S}_i|} = 1 - R_p. \qquad (15.66)$$

In Figure 15.11(a), we have plotted the power reflection coefficients R_s and R_p for reflection at an interface when the incident medium 1 has a refractive index $\tilde{n}_1 = 1$ ($\Rightarrow k_1 = \frac{\omega}{c}$) and the transmitted medium 2 has a refractive index $\tilde{n}_2 = 1.5$ ($\Rightarrow k_2 = 1.5 \frac{\omega}{c}$). When the propagation is from the material with the greater index into the material with the smaller index (in Figure 15.11(b)) we see the total internal reflection at the critical angle predicted by Snell's Law

$$\theta_c = \sin^{-1}(\tilde{n}_2/\tilde{n}_1). \qquad (15.67)$$

In Figure 15.11(a), we note that in s-polarization the reflection increases uniformly from $R_s = 0.04$ to $R_s = 1$ as the angle of incidence increases from $0°$ (normal incidence) to $90°$ (grazing incidence). In p-polarization, however, we note that the reflection decreases from $R_p = 0.04$ at $0°$ to zero at $\theta_i = 56.3°$ and then increases to $R_p = 1$ at $\theta_i = 90°$. The angle at which the polarization reflection goes to zero is the **Brewster angle** and can be shown to be given by

$$\theta_B = \tan^{-1}(\tilde{n}_2/\tilde{n}_1). \qquad (15.68)$$

At the Brewster angle, the normal component of the p-polarized optical electrical field is exactly canceled out by the polarization at the interface between the two media and the reflection coefficient becomes zero. This does not occur in s-polarization where the electric field has no component normal to the interface. One implication of this is that for p-polarization there is 100% transmission for electromagnetic radiation incident on an interface at the Brewster angle. In fact, it is easy to show that if light is incident on a plane-parallel slab of transparent material in air at the "air-to-glass" Brewster angle, the transmitted light inside the material will be refracted so it will be incident on the interface at the other side of the slab at the correct Brewster angle for 100% "glass-to-air" transmission. Thus the window will be 100% transmitting for p-polarized light. Such *Brewster windows* are commonly used in gas lasers with feedback mirrors that need to be outside of the discharge tube where the stimulated emission occurs, but the gain is not sufficient to make up for any reflection losses out of the beam. For the p-polarized light, the windows are perfectly transmitting with no reflection losses. Thus the output light of the laser is nearly all p-polarized.

(a)

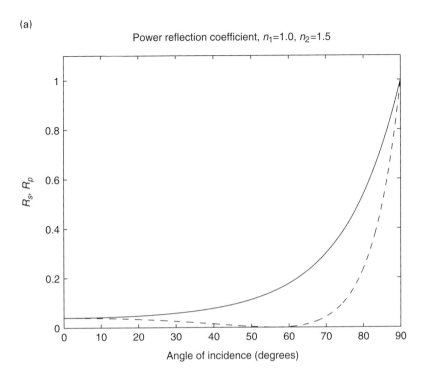

(b)

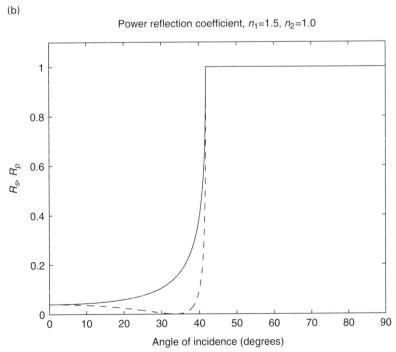

Figure 15.11 Reflection of s-polarized (solid line) and p-polarized (dashed line) optical radiation from an interface as a function of angle of incidence illustrating the Brewster angle of zero reflectivity in p-polarization, shown both for the case (a) when $n_2 > n_1$ and (b) when $n_2 < n_1$ where the critical angle marking the beginning of total internal reflection is evident.

Chapter 15 problems

1. As an EM wave propagates in a medium, discuss what happens to its characteristics (amplitude, frequency, wavelength, and phase velocity). Consider different type of media (isotropic, non-isotropic, lossy, and non-lossy).

2. An EM wave traveling in air (vacuum) is incident on a material. Compare the characteristics of the transmitted and reflected waves to those of the incident wave. Consider cases for different types of material as in the previous problem.

3. Maxwell's equations in a conducting solid yield a dispersion relation $k^2 = \mu\varepsilon\omega^2 + i\mu\sigma\omega$.

    Show that the fields for a 1 GHz plane wave propagating in the z-direction decay into aluminum as $E(z) = E_o e^{-z/\delta}$ where δ is the *skin depth*. Find a value for the skin depth of aluminum assuming that ε and μ are close to their free-space values.

4. For an EM wave in vacuum,
    a. show that $E = c\,B$

    b. compare the energy associated with the electric field E with the energy associated with the magnetic field B, if the energy density within an electric field is $u_E = \frac{1}{2}\epsilon_o E^2$ and within a magnetic field $u_B = \frac{B^2}{2\mu_o}$.

    c. From part (a), can we conclude that the electric field component of an EM wave is stronger than the magnetic field one? Justify your answer.

5. In Example 15.1, what will happen to the wavelength, propagation velocity, and $1/e$ damping distance of a wave with a free-space wavelength of 1100 nm (twice that of the example)? Assume that the permeability μ and permittivity ε remain the same as the example. Predict whether the above quantities will increase or decrease or stay the same using reasoning and then perform the calculations.

6. What thickness of calcite crystal (see Example 15.2) would convert an incident linearly polarized laser beam with $\lambda = 633$ nm propagating perpendicular to the optic axis, with its electric field polarized at $45°$ to the optic axis, into a circularly polarized beam with $\vec{E}\,(\vec{r},t) = E_o(e^{i\pi/2}\hat{y} + \hat{z})e^{i(kx-\omega t)}$?

7. The concept of the polarization on reflection from an interface is also part of the reason that polarized sunglasses allow you to see through glare. Why would glass that only transmits one polarization of light be more helpful in sunglasses than just a neutral filter that attenuates both polarizations equally? What orientation should the plane of polarization be in sunglasses? How could you check with your sunglasses to see what the effect of the polarization is?

8. Calculate the intensity of an EM wave propagating in vacuum in terms of the maximum value of an oscillating electric field. Also express it as a function of the maximum magnetic field.

9. If the maximum electric field at 20 m from a point source in an isotropic medium is 20 V/m, calculate:

    a. the maximum magnetic field at the point

    b. the average intensity of the light there and

    c. the power of the source.

10. Consider an electromagnetic wave traveling from a dielectric medium to a perfect electric conductor ($\sigma = \infty$). Examine the boundary conditions at the interface between the two media.

11. In the discussion of *Brewster windows* in the last section, it is asserted that "it is easy to show that if light is incident on a plane-parallel slab of transparent material in air at the 'air-to-glass' Brewster angle, the transmitted light inside the material will be refracted so it will be incident on the interface at the other side of the slab at the correct Brewster angle for 100% 'glass-to-air' transmission. Thus the window will be 100% transmitting for p-polarized light." In the diagram below the slab of transparent glass with an index of refraction $n_1 = 2.5$ is in air with an index of $n_0 = n_2 = 1.0$. What is the Brewster angle for "air-to-glass" transmission? What is the Brewster angle for "glass-to-air" transmission? Show that the angle β is the same as the angle α, and that if α is the Brewster angle for transmission from the air to the glass, then γ is the Brewster angle for transmission from the glass to air.

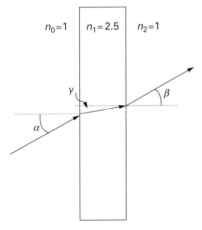

16 Electromagnetic materials

Late at night AM radio signals broadcasting in the band from 531–1710 kHz (1 kHz = 1000 cycles/s) can be picked up over long distances from distant cities. AM radio stations from Chicago can sometimes be received at night in cities as far away as New York or Washington DC. Because of this long-distance interference, many AM radio stations have a license to broadcast only during daylight hours. Similarly, "short-wave" radio hobbyists, broadcasting in the 1.6–30 MHz range (1 MHz = 10^6 cycles/s), can communicate with other operators even on different continents. On the other hand, FM radio stations broadcasting in the 87.5 to 108.0 MHz band, in common with television stations, can only be received in a single metropolitan market – essentially reaching only those receivers that have direct line of sight from a tall transmission tower. These differences in transmission properties are due to the frequency dependence of the electromagnetic properties of the upper atmosphere. Similarly, the different optical transmission and reflection characteristics that we perceive as different colors of objects are the result of frequency-dependent electromagnetic properties, in this case at optical frequencies from about 400–750 THz (1 THz = 10^{12} cycles/s). What causes these kinds of frequency-dependent effects on the propagation of electromagnetic waves and how we can model them are the topic of this chapter.

We note that Maxwell's equations say nothing explicitly about the effect of materials on electromagnetic wave propagation. All of the electromagnetic properties are contained in the constitutive relations that relate (i) the displacement field $\vec{D}$ to the electric field $\vec{E}$ through the *permittivity* ε: $\vec{D} = \varepsilon \vec{E}$, (ii) the magnetic flux density $\vec{B}$ to the magnetic field $\vec{H}$ through the *permeability* μ: $\vec{B} = \mu \vec{H}$, and (iii) the current density $\vec{J}$ to the electric field $\vec{E}$ through the *conductivity* σ: $\vec{J} = \sigma \vec{E}$. *All of the differences in electromagnetic properties of materials are contained in the details of ε, μ, and σ.*

We have already noted that in general ε, μ, and σ are *tensors* with different values along different directions in materials that are *anisotropic*. At the very least, electromagnetic waves in anisotropic materials have different velocities and damping for different polarizations when propagating along high-symmetry directions. For the most general propagation directions and polarizations in anisotropic materials, we would have to revisit our derivation of the wave equation since $\vec{\nabla} \cdot \vec{D} = 0$ does not imply that $\vec{\nabla} \cdot \vec{E} = 0$, and the dispersion relation that connects the wave number k and frequency ω becomes more complicated.

In addition, we have implicitly assumed that our materials have a *linear* electromagnetic response, i.e., the values of ε, μ, and σ do not depend on the strength of the fields.

This is a convenient assumption because it makes Maxwell's equations and the dispersion equations that result from them a linear system of equations with the fundamental property that for a single frequency excitation of the form $e^{-i\omega t}$ the response is also of the form $e^{-i\omega t}$ with the same frequency as the excitation. This fundamental property is what unleashes the power of Fourier decomposition that allows us to predict the response of a material to any arbitrary excitation by combining the responses at all of the component frequencies. In actual fact, non-linear effects occur in all materials at high enough excitation fields, and non-linear materials and devices are commonly used in technologies for frequency mixing in electromagnetic communications and frequency doubling in optical instrumentation. Non-linear materials and their important and useful effects are beyond the scope of phenomena that we can explore in detail.

Within the subset of linear and isotropic materials, the electromagnetic properties are always frequency dependent: no medium, except free space, has the same properties from radio frequencies to optical, X-ray, and gamma-ray frequencies. In addition, in some cases, the material response to an electromagnetic wave $e^{i(\vec{k}\cdot\vec{r}-\omega t)}$ is dependent not only on the frequency ω but also separately on the wave vector $\vec{k}$ in a way that cannot be simply deduced from the frequency response and the relationship between ω and $\vec{k}$ given by the dispersion relation. *Non-local* electrodynamics result from electromagnetic effects at one position in a material affecting the electromagnetic response at another position. Since the electromagnetic wavelength (~0.5 μm for optical frequencies, for example) is usually much longer than the length scale of material phenomena (typically of the order of the lattice constant or the scattering length), most electromagnetic phenomena are well-approximated by the $k = 0$ local limit. For frequencies well above the optical spectral range, however, as we have already seen in *Compton scattering* of gamma rays from electrons, energy $\hbar\omega$ and momentum $\hbar\vec{k}$ must be independently considered and conserved. In what follows we will ignore any non-local effects and consider models for the frequency dependence of electromagnetic effects (*dispersion*) in materials where the fields at any given point in space are related only to the material properties at that point.

16.1 The Drude frequency-dependent conductivity

The electromagnetic properties of materials are specified by the response functions in the constitutive relations, the permittivity ε, the permeability μ, and the conductivity σ. Probably the easiest electromagnetic response to model is the conductivity of a *plasma* – a material with free charges that can respond to the electromagnetic field. Plasmas can be found in ionized gases, where both the positive ions and the negative electrons are free to move, and inside metals and semiconductors where only the electrons are free to move while the positive ions are fixed with the atomic lattice.

The current density $\vec{J} = \sigma\vec{E}$ is the current per unit area crossing a cross-section plane in the sample (see Figure 16.1). If the average velocity of the current-carrying charges is $\vec{v}$, the total charge crossing a plane in a sample in a time interval dt is $Q = qnAvdt$,

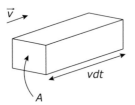

Figure 16.1 Volume element for the calculation of the current density $\vec{J}$.

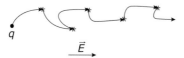

Figure 16.2 Motion of a positive charged particle subject to an electric field $\vec{E}$ and random scattering events. The model also applies to electrons in a metal with the motion opposite to (instead of parallel to) $\vec{E}$.

where n is the volume density of charge carriers and q is the unit charge of the charge carriers (usually the charge of an electron). Since $I = dQ/dt$ is the current we have

$$\vec{J} = \vec{I}/A = qn\vec{v}. \tag{16.1}$$

Finding the expression for the conductivity σ, thus, is a matter of finding how the average velocity $\vec{v}$ depends on the electric field $\vec{E}$.

The simplest and most robust model for a plasma is the *Drude model* of a gas of charged particles of mass m and charge q that respond to the electric field with acceleration given by $\vec{a} = \frac{\vec{F}}{m} = q\vec{E}/m$ between discrete point scattering events that (1) occur with an average frequency $1/\tau$ and that (2) completely randomize the particle momentum. The parameter τ is known as the *scattering time* which, along with the particle density n, charge, and mass completely characterizes the response of the plasma in this approximation. Figure 16.2 shows the trajectory of a charged positive particle under the influence of a constant electric field and random scattering events. The equation of motion for such a particle is given by

$$m\vec{a} = q\vec{E} - \vec{F}_d = q\vec{E} - \frac{m\vec{v}}{\tau}, \tag{16.2}$$

where the effective damping force due to the random collisions $\vec{F}_d = d\vec{p}/dt$ is given by the change in momentum divided by the time between collisions. We assume that, since the collisions randomize the momentum, the average momentum immediately after a collision is 0 while the momentum before the collision is mv where v is the average velocity of the charges, so $d\vec{p} = m\vec{v}$ while $dt \approx \tau$, leading to the form for the damping force in the last term in Eq. (16.2).

Assuming that the electromagnetic field $\vec{E} = \vec{E}_o\, e^{-i\omega t}$ is in steady state, the acceleration and velocity of the charges should have the same $e^{-i\omega t}$ time dependence, with complex amplitudes that express the amplitude and relative phase of the oscillatory motion $\vec{v}(t) = \vec{v}_o e^{-i\omega t}$ and $\vec{a}(t) = a_o e^{-i\omega t} = \frac{d}{dt}\vec{v}(t) = (-i\omega)\vec{v}_o e^{-i\omega t}$. Putting all this together in the equation of motion Eq. (16.2), we have

$$-i\omega m \vec{v}_o e^{-i\omega t} = q\vec{E}_o e^{-i\omega t} - \frac{m}{\tau}\vec{v}_o e^{-i\omega t}. \tag{16.3}$$

Canceling out the $e^{-i\omega t}$ factors and solving for $\vec{v}_o$ we have,

$$\vec{v}_o = \frac{q\tau\vec{E}_o}{m}\left(\frac{1}{1 - i\omega\tau}\right). \tag{16.4}$$

Substituting this into the expression for current density $\vec{J}$ in terms of carrier velocity, we find

$$\vec{J} = nq\vec{v} = \frac{nq^2\tau\vec{E}_o}{m}\left(\frac{1}{1 - i\omega\tau}\right)e^{-i\omega t}, \tag{16.5}$$

leading to an expression for the Drude frequency-dependent conductivity

$$\sigma(\omega) = \frac{nq^2\tau}{m(1 - i\omega\tau)}. \tag{16.6}$$

We note that at DC or low frequencies where $\omega\tau \ll 1$ we have $\sigma_{DC} = \frac{nq^2\tau}{m}$, while in the opposite high-frequency limit ($\omega\tau \gg 1$), we find $\sigma \to i\frac{n q^2}{m \omega}$. The factor of i in this equation implies that the current density at high frequencies is $90°$ out-of-phase with the driving $\vec{E}$ field.

Question 16.1

Using Euler's relation $e^{i\delta} = \cos\delta + i\sin\delta$, can you show that a factor of i corresponds to a phase shift of $\delta = 90°$?

Combining the Drude conductivity with the dispersion relation for electromagnetic wave propagation in a conducting medium (see Section 15.3), we find for a Drude plasma the wave vector k is related to the frequency by

$$k = \sqrt{\mu\varepsilon\,\omega^2 + i\mu\sigma\,\omega} = \frac{\omega}{c}\sqrt{\mu_r\varepsilon_r + \frac{i\mu_r}{m\epsilon_o\omega}\frac{n\,q^2\tau}{(1 - i\omega\tau)}}, \tag{16.7}$$

leading to an effective index of refraction

$$\tilde{n} = \frac{ck}{\omega} = \sqrt{\mu_r\varepsilon_r + \frac{i\mu_r}{m\epsilon_o\omega}\frac{n\,q^2\tau}{(1 - i\omega\tau)}}, \tag{16.8}$$

where $\mu_r = \frac{\mu}{\mu_0}$ and $\varepsilon_r = \frac{\varepsilon}{\epsilon_o}$ are the relative magnetic permeability and electric permittivity, respectively. In the high-frequency (optical) limit where $\omega\tau \gg 1$ and the relative permeability $\mu_r \cong 1$, we find the complex index of refraction $\tilde{n}$

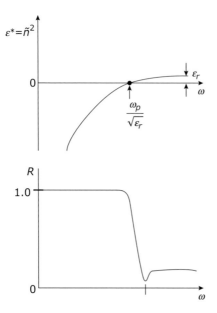

Figure 16.3 The real part of the effective permittivity for a material with mobile carriers and the corresponding reflectivity.

$$\tilde{n} = \sqrt{\varepsilon_r - \frac{n\,q^2}{m\epsilon_o\omega^2}} = \sqrt{\varepsilon_r - \frac{\omega_p^{\,2}}{\omega^2}} \tag{16.9}$$

where we have introduced the *plasma frequency* $\omega_p = \sqrt{\frac{nq^2}{m\epsilon_o}}$.

In Figure 16.3, we plot the effective dielectric permittivity $\varepsilon^* = \frac{(\varepsilon + i\sigma/\omega)}{\epsilon_o} = \tilde{n}^2$ against frequency for a plasma in the high-frequency limit discussed here. For frequencies below $\omega = \omega_p/\sqrt{\varepsilon_r}$, the effective permittivity is negative, which makes the index of refraction imaginary. This makes the normal incidence reflection $R = |\frac{\tilde{n}-1}{\tilde{n}+1}|^2 \to 1$ and the plasma has a very high reflectivity, which drops dramatically when $\omega > \omega_p/\sqrt{\varepsilon_r}$ and $\tilde{n}$ passes through 0 and then gradually approaches $\sqrt{\varepsilon_r}$ as $\omega \to \infty$, as shown in the bottom part of Figure 16.3. In gas plasmas, a similar model can be used to find the contribution to the conductivity from the ion cores in addition to the electron conductivity. Since the mass of the ion cores is orders of magnitude higher than the electron mass, the plasma frequency associated with the ions is much lower than the electron plasma frequency.

Question 16.2

If $\varepsilon_r = \varepsilon/\epsilon_o = -9$ for a plasma at a frequency $< \omega_p/\sqrt{\varepsilon}$, find $\tilde{n} = n + i\kappa$. What is n? What is κ? What is the power reflection between vacuum ($\tilde{n} = 1$) and the plasma $R = |\frac{\tilde{n}-1}{\tilde{n}+1}|^2$? (Hint: $|z|^2 = z \cdot \text{conj}(z)$.)

Example 16.1 What is the plasma frequency associated with the ions and the electrons in a hydrogen gas plasma with the electron and ion density given by $n_e = n_{ion} = 1.0 \times 10^{18}$ m^{-3}?

Solution: Using the free-electron mass $m_e = 9.11 \times 10^{-31}$ kg and an ion mass equal to the proton mass $m_p = 1.67 \times 10^{-27}$ kg (assuming that the H$_2$ molecules are completely dissociated) the electron plasma frequency is given by

$$\omega_{pe} = \sqrt{\frac{(1.0 \times 10^{18})(1.6 \times 10^{-19})^2}{(9.11 \times 10^{-31})(8.85 \times 10^{-12})}} = 5.63 \times 10^{10} \tfrac{\text{rad}}{\text{s}} = 8.97 \text{ GHz} \quad \text{in the microwave-}$$

frequency range. The ion plasma frequency is smaller by a factor of $\sqrt{m_p/m_e} = 42.5$ giving a plasma frequency $\omega_{pi} = 1.32 \times 10^9 \tfrac{\text{rad}}{\text{s}} = 211$ MHz in the radio-frequency range.

The *plasma reflectivity edge* is observed not only in gaseous plasmas, but also in semiconductors, where the free electrons can be modeled by a Drude theory as above. In metals, the plasma frequency is at such high frequency that $\hbar\omega_p$ is at energies where the pure conductivity response is obscured by *interband transitions*. Nevertheless, some metals show indications of *ultraviolet transparency* at frequencies above the plasma frequency.

Example 16.2 (a) Find the plasma frequency for silicon with $n = 10^{24}$ electrons/m^3, a conductivity effective mass $m^* = 0.26m_o$, and a low-frequency dielectric permittivity $\varepsilon_s = 11.8\,\epsilon_o$. Compare the photon energy at this frequency with the band gap of silicon $E_g = 1.1$ eV. (b) What is the plasma frequency in copper with $n = 8.5 \times 10^{28}$ m^{-3}, effective mass $m^* \simeq m_o$, and background permittivity $\varepsilon \approx \epsilon_o$.

Solution: (a) For silicon with n-type doping of 10^{24} electrons/m^3 we find

$$\omega_p = \sqrt{\frac{ne^2}{m\epsilon_o}} = \sqrt{\frac{10^{24}(1.6 \times 10^{-19})^2}{0.26(9.11 \times 10^{-31})11.8(8.85 \times 10^{-12})}} \simeq 3.2 \times 10^{13} \text{ s}^{-1}.$$

This frequency corresponds to a wavelength $\lambda_p = \tfrac{2\pi c}{\omega_p} = 58$ μm. The photon energy at this energy is $E_p = \hbar\omega_p = (1.054 \times 10^{-34})(3.2 \times 10^{13}) = 3.4 \times 10^{-21}$ J $= 0.02$ eV, much less than the band-gap energy. The decrease in infrared reflectivity above the "plasma edge" at $\omega = \omega_p/\sqrt{\varepsilon}$ is quite dramatic in silicon and can be used to estimate the doping level of the material. (b) For copper, the same calculation gives the plasma frequency $\omega_p = 1.64 \times 10^{16}$ s^{-1} corresponding to a photon wavelength of around 115 nm in the ultraviolet. This result is typical for metals and explains the high reflectivity of metals for all EM waves with wavelengths in the visible range and above. The equivalent photon energy is $\hbar\omega_p = 10.8$ eV, which is greater than band-to-band energy separations in most materials.

How does this concept of the frequency-dependent conductivity of a plasma explain the different propagation characteristics of AM and FM radio that were noted in the introduction to this chapter? The ionosphere is the region of the atmosphere between 85 and 600 km from the surface of the Earth where the ultraviolet solar radiation ionizes the atmospheric gases and creates a layer of free electrons with a density of $10^{11}–10^{12}$ m^{-3}. The electron plasma frequency is thus in the range

$$\omega_p = \sqrt{\frac{ne^2}{m\epsilon_o}} = \sqrt{\frac{10^{12}\,(1.6 \times 10^{-19})^2}{(9.11 \times 10^{-31})(8.85 \times 10^{-12})}} = 5.6 \times 10^7\,\text{s}^{-1} \approx 9\,\text{MHz}. \quad (16.10)$$

This is well below the frequency range of FM radio band (87.5–108 MHz), so FM radio (and television) signals pass through the ionosphere without reflection. On the other hand, the AM radio band (535–1700 kHz) and the low-frequency end of the short-wave radio band (1.6–30 MHz) will reflect off the ionosphere plasma and can be received at sites on the Earth's surface beyond the horizon. During the day, the sunlight creates an ionization layer at low altitudes (under 80 km above the surface) which attenuates the signals at relatively short ranges after one or two "bounces" off the ionosphere and the Earth's surface. At night, when the Sun is not directly shining, the lower layers of the ionosphere greatly reduce their ionization level, leaving only the highly ionized layer about 200 km above the surface. Bounces off this high-altitude, high-reflectivity plasma layer can be picked up thousands of miles from the transmitting antenna.

16.2   Optical phonons: infrared optical properties of ionic crystals

Ionic crystals like NaCl, CdTe, PbS, etc. have a sublattice of positive ions – metallic atoms that have weakly bound outer electrons – and a sublattice of negative ions – atoms from column VI or VII on the Periodic Table with nearly filled shells that are strongly electro-negative. Each atom is balanced by surrounding atoms of opposite charge, but under the effect of an applied electric field $\vec{E}$ the sublattices will move apart, the positive sublattice moving slightly in the direction of the electric field and the negative sublattice opposite to the direction of the field as shown in Figure 16.4. The ions also experience an elastic restoring force $\vec{F}_{el} = -k\,\Delta\vec{d}$ with the equilibrium displacement representing a balance between the electric force and the elastic restoring force.

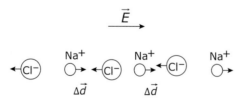

Figure 16.4 Charge separation in a polarizable medium (a NaCl crystal) caused by an applied electric field $\vec{E}$.

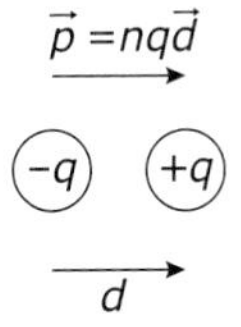

Figure 16.5 Dipole moment $\vec{p}$ created by a charge separation $\vec{d}$ of charges of $+q$ and $-q$.

The separation of charge causes a *polarization* of the medium. The polarization of an *electric dipole*, defined as a positive charge of $+q$ separated by a distance of d from a negative charge of $-q$ as shown in Figure 16.5, is $\vec{p} = q\,\vec{d}$. In the case of the ionic crystal where each charge is balanced by opposite charges on either side, we can consider the dipole moment to be $\vec{p} = q\,\Delta\vec{d}$ where $\Delta\vec{d}$ is the displacement from the atomic positions at $\vec{E} = 0$. If there are n dipoles per unit volume, the medium has a polarization field $\vec{P} = nq\Delta\vec{d}$, which depends on the electric field through the *electric susceptibility* of the medium χ_e: $\vec{P} = \epsilon_o \chi_e \vec{E}$, where $\epsilon_o = 8.85 \times 10^{-12}$ F/m (farad per meter) is the *permittivity of free space*. The polarization field $\vec{P}$, in units of C/m^2, is the volume density of the aligned electric dipoles. The effect of the polarization on the electric field in the material is given by the constitutive relation

$$\vec{D} = \epsilon_o \vec{E} \; + \; \vec{P} = \epsilon_o \left(1 + \chi_e\right) \vec{E} = \varepsilon \vec{E} \qquad (16.11)$$

where ε is the *permittivity* of the medium, which describes the effect of the lattice polarization on the electric field $\vec{E}$ and the displacement field $\vec{D}$. *In general, electric susceptibility measures how easily a dielectric will polarize in response to an electric field, while permittivity measures how easily an electric field will form in a medium.*

If we apply the equation of motion method to an ion under the influence of a sinusoidal oscillating electric field at frequency ω, $\vec{E} = \vec{E}_o\, e^{-i\omega t}$, and an elastic restoring force proportional to the displacement Δd with a linear damping parameter Γ, we find a permittivity given by a constant term plus a Lorentzian oscillator

$$\varepsilon(\omega) = \varepsilon_\infty + \frac{\omega_T^2\left(\varepsilon_s - \varepsilon_\infty\right)}{\omega_T^2 - \omega^2 - i\Gamma\omega}. \qquad (16.12)$$

Question 16.3

What is the oscillator strength of the Lorentzian oscillator?

Here ω_T is the *transverse optical phonon frequency*. A *phonon* is a lattice vibration, and the transverse optical phonon frequency is the resonant frequency at which the positively charged sublattice and the negatively charged sublattice will vibrate against each other if one lattice is displaced with respect to the other. The constants ε_s and ε_∞ are the static (DC) and high-frequency dielectric permittivities. The damping parameter Γ

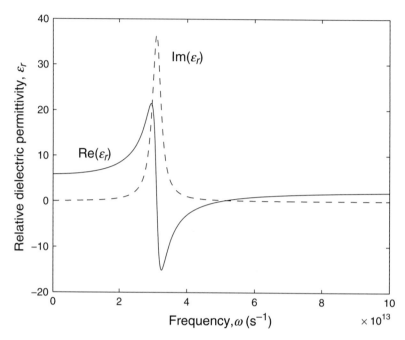

Figure 16.6 Real (solid line) and imaginary (dashed line) permittivities of NaCl with an intrinsic transverse mode frequency $\omega_T = 3.1 \times 10^{13} \text{ s}^{-1}$.

depends on the temperature and how perfect the crystal is, but at room temperature it is typically about ten times less than ω_T: $\Gamma \simeq 0.1 \, \omega_T$.

For NaCl in the infrared spectral region near $\lambda = 50 \, \mu\text{m}$, for example, $\varepsilon_s = 5.9 \, \epsilon_o$, $\varepsilon_\infty = 2.25 \, \epsilon_o$, $\omega_T = 3.1 \times 10^{13}$ radians/s, and $\Gamma \simeq 3.1 \times 10^{12}$ radians/s. If we plot the *relative dielectric permittivity* $\varepsilon_r = \varepsilon/\epsilon_o$ vs. ω we get the plot shown in Figure 16.6.

The *index of refraction* is the square root of the relative dielectric permittivity

$$\tilde{n} = \sqrt{\varepsilon_r} = \sqrt{\varepsilon/\epsilon_o} \tag{16.13}$$

and the normal incidence reflectivity of a material is given by

$$R = \left| \frac{\tilde{n} - 1}{\tilde{n} + 1} \right|^2. \tag{16.14}$$

If we plot the reflectivity of NaCl, which results from the permittivity above, we find the plot shown in Figure 16.7.

The high reflectivity of ionic crystals in a band just above the transverse optical phonon frequency is known as the *restrahlen* (German for "residual ray"), so-named because early infrared spectroscopists observed that only a narrow band of infrared wavelengths (only those wavelengths corresponding to frequencies where the reflectivity is high) would transmit between two ionic crystals. The reflectivity peak occurs in the region where the real part of the dielectric permittivity goes negative and the refractive index becomes almost completely imaginary, so the infrared light doesn't transmit in the crystal.

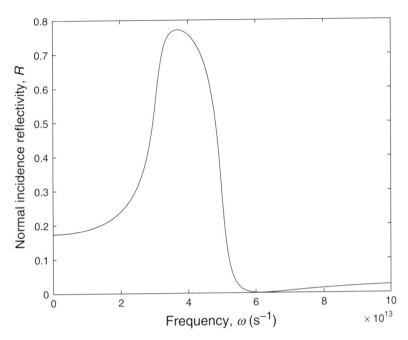

Figure 16.7 Infrared reflectivity of NaCl calculated from the $\varepsilon(\omega)$ in Figure 16.6 showing the large increase in reflectivity in the *restrahlen band* where the dielectric permittivity is negative.

16.3   Optical absorption of an atomic resonance in a gas

Another phenomenon described by a Lorentzian oscillator is visible light absorption by an electronic transition in a dilute gas. In this case, light can be absorbed by exciting an electron from a lower energy state to a higher energy state (see Figure 16.8). According to quantum theory, the frequency of the absorbed light will be

$$\omega_o = \Delta E/\hbar = (E_2 - E_1)/\hbar. \tag{16.15}$$

In contrast to the optical phonon where every lattice site was part of a dipole that interacted with the electric field, in a dilute gas the density of absorbing centers is much less and the oscillator strength S is consequently much smaller. We will model the effect on the dielectric permittivity due to the absorption of the photon and the resulting electronic transition in the gas. The frequency corresponding to the transition energy $\omega_o = \Delta E/\hbar$ is equivalent to the resonant frequency of the transition, and the damping γ is related to the lifetime τ of the electron in the upper state (how long the electron remains in the excited state E_2 before relaxing back to *ground state* E_1) by $\gamma = 1/\tau$.

In terms of these parameters, the permittivity of the dilute gas with the resonant absorption is described by a Lorentzian oscillator with an oscillator strength S, a resonant frequency ω_o, and a damping parameter γ:

$$\varepsilon(\omega) = \epsilon_o + \frac{S\,\omega_o^2}{\omega_o^2 - \omega^2 - i\gamma\omega}. \tag{16.16}$$

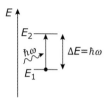

Figure 16.8 Absorption of a photon with a frequency $\hbar\omega_o$ in a material with two quantum energy levels separated by $\Delta E = E_2 - E_1$.

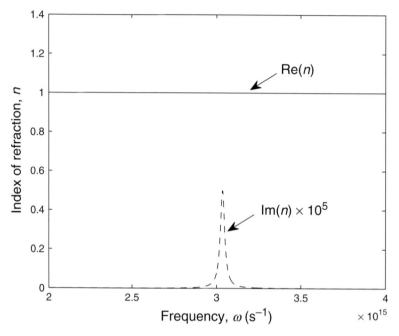

Figure 16.9 Real and imaginary parts of the index of refraction vs. frequency for a gas with a weak quantum transition with $\Delta E = 2.0$ eV.

Again, the refractive index is given by the square root of $\varepsilon_r = \varepsilon(\omega)/\varepsilon_o$:

$$\tilde{n} = \sqrt{1 + \frac{S\omega_o^2/\epsilon_o}{\omega_o^2 - \omega^2 - i\gamma\omega}}. \tag{16.17}$$

The optical wave propagating a distance x through the gas will have a relative transmission $T = |e^{i\frac{\omega}{c}\tilde{n}x}|^2$. We will model the transmission with $S/\epsilon_o = 1 \times 10^{-7}$, and $\Delta E = 2$ eV $= 3.20 \times 10^{-19}$ J, which gives $\omega_o = 3.04 \times 10^{15}$ s^{-1}. We will select $\gamma = 0.01\omega_o = 3.04 \times 10^{13}$ s^{-1}. Because the oscillator strength is so small, as shown in Figure 16.9, the real part of $\tilde{n}$ has no effect on the transmission. The imaginary part of $\tilde{n}$, however, gives rise to an exponential decrease in the transmission centered at ω_o. The plot is for propagation through 0.01 m of the gas.

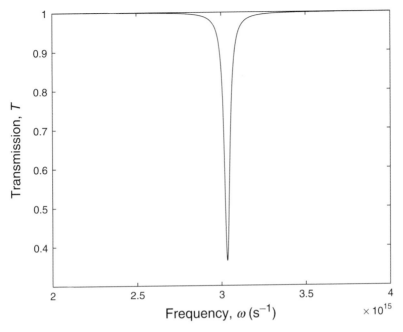

Figure 16.10 Transmission of light vs. frequency through 1 cm of gas with a complex index of refraction as illustrated in Figure 16.9.

Question 16.4

What changes would you expect in the transmission spectrum if the energy of the transition ΔE increases from 2.0 eV to 2.2 eV? How do you think the shape of the absorption line in the transmission of the gas will change if the Lorentzian damping parameter γ increases (the lifetime of the state τ decreases)? How about if the oscillator strength S increases?

The precise measurement of electronic transition energies in gases and liquids by the absorption or emission spectra makes spectroscopy a powerful tool for identification of chemical species. The frequency of maximum absorption in a dilute gas or liquid solution of a material is usually a measure of the energy of the electronic transition that is being optically excited, but the strong colors in empirically discovered natural and synthetic dyes are often the result of complex electronic processes that can't necessarily be identified with a single electron transition.

In solids, the optical properties in the visible are usually dominated by broader spectral features due to continuous electronic band-to-band transitions. Transparent material like silicate glasses (SiO_2), alumina (Al_2O_3), and diamond have band-gap energies that are above the range of visible light ($E_g \gtrsim 3.1$ eV), but impurity states in the gap can give the materials a strong color. Ruby and sapphire, for example, are both mostly aluminum oxide Al_2O_3. In the case of ruby, a chromium impurity Cr^{3+} absorbs in the yellow-green portion of the spectra leaving the transmitted light in the red portion of the spectra. In addition, the yellow-green light that is absorbed is re-emitted after a lattice relaxation in the red portion of the spectra. The 694 nm emission of the excited Cr^{3+} ions in ruby was used by Theodore Maiman in 1960 to make the first solid state laser. In sapphire, electronic charge transfer between titanium and iron impurities Fe^{2+} and Ti^{4+} in the Al_2O_3 lattice absorbs yellow light, resulting in a deep blue or purple color. Electronic transitions in the bands of solids or in the electronic states of localized impurities create the colors of materials that we see all around us. For example, chlorophyll which is a key compound for photosynthesis in plants absorbs light in the red and blue regions of the spectra; the residual reflected and transmitted green light gives us the color of leaves and plants.

Note that strong electronic absorption at a particular wavelength does not necessarily mean that no light is reflected at that wavelength. Electronic absorption results in a peak in the imaginary parts of the permittivity and index of refraction. If the imaginary part of the complex index of refraction κ is very large, it can create a refractive index mismatch that causes a high reflectivity. An example of this is metals that for frequencies below the plasma frequency have a very large κ but are highly reflective because $R = \frac{(n-1)^2+\kappa^2}{(n+1)^2+\kappa^2} \simeq 1$. Essentially no light gets into the metal to be absorbed. In gold and copper at short wavelengths (high energies) interband electronic transitions come into play that reduce n and κ and allow blue light to propagate into the metal where it is quickly absorbed. The reflected light is thus richer in the red and yellow hues that give the gold and copper colors.

In addition, one must consider the effect of the real part of the permittivity function. For example, the absorption peak of a strong Lorentzian oscillator occurs at the frequency where the permittivity becomes positive again after the highly reflective region at frequencies below the resonant frequency of the system. Thus the absorption peak in $Im\{\varepsilon(\omega)\}$ marks the transition from a region of high reflectivity to a region of transmission, as we saw in the restrahlen band in polar crystals.

16.4   Microwave magnetic resonance

Electromagnetic effects due to magnetic fields and dynamics can be found in the conductivity, in the permittivity, or in the permeability. The frequency- and magnetic-field-dependent effects on the electron conductivity $\sigma(\omega)$ in metals or semiconductors are the basis of a large range of magneto-optical effects at microwave and infrared frequencies. At optical frequencies, magneto-optical effects result from the magnetic-field-induced shift of electronic energy levels – and thus electronic absorption resonance frequencies – with a resultant change in the permittivity $\varepsilon(\omega)$. The parameters of interest for magneto-conductivity and magneto-optical phenomena include the electron *cyclotron resonance frequency* $\omega_{ce} = qB/m$ and the spin-up/spin-down splitting $\Delta E = \mu_B H$ of electronic quantum levels.

Electromagnetic effects in the magnetic permeability μ include a rich range of phenomena, mostly at microwave frequencies and lower. The permeability is related to the magnetization $\vec{M}$ of the medium by

$$\vec{B} = \mu_o \left(\vec{H} + \vec{M} \right) = \mu_o \vec{H} + \mu_o \chi_m \vec{H} = \mu \vec{H} \tag{16.18}$$

where we have introduced the dimensionless *magnetic susceptibility* χ_m, which meas-ures the magnetization $\vec{M}$ of a medium under the application of a magnetic field $\vec{H}$. The magnetization $\vec{M} = \mu_o \chi_m \vec{H}$ in units of A/m is the volume density of the aligned magnetic dipoles caused by net electron orbital motion and unpaired electron spins on the constituent molecules of the material.

A molecule with an unpaired proton or electron has a local magnetic moment $\vec{m}$, which can be modeled as a circulating loop of current. In a magnetic field, the current loop will have a balanced pair of forces due to the Lorentz forces $d\vec{F} = I\,d\vec{\ell} \times \vec{B}$ on the two sides of the current loop. The magnetic moment is related to the effective circulating current I and differential area $d\vec{S}$ encircled by it by $\vec{m} = I\,d\vec{S}$. The torque created by the Lorentz force on the magnetic moment is given by

$$\vec{\tau} = \vec{m} \times \vec{B} . \tag{16.19}$$

From quantum mechanics, the magnetic moment of an electron, proton, or molecule is parallel to the angular momentum $\hbar\vec{L}$ with $\vec{m} = \gamma\hbar\vec{L}$. For electrons the constant of proportionality $\gamma = g\frac{e}{2m}$ where e and m are the electronic charge and mass and g is the *Landé spectroscopic splitting factor*. For most molecules where the magnetic moment is due to unpaired electrons $g \simeq 2$ and $\gamma \simeq 1.76 \times 10^{11}$ rad/s $\cdot$ T. Since the torque gives the time rate of change of the angular momentum, we have the fundamental equation of magnetic dynamics for atomic magnetizations

$$\frac{d(\hbar\vec{L})}{dt} = \frac{1}{\gamma}\frac{d\vec{m}}{dt} = \vec{m} \times \vec{B} . \tag{16.20}$$

If we define the material magnetization $\vec{M}$ as the volume average of the molecular magnetic moments, we can write an equation of motion for the magnetization

$$\frac{d\vec{M}}{dt} = \gamma\,\vec{M} \times \vec{B} \tag{16.21}$$

where $\vec{B}$ is the effective field the moment senses – a combination of the applied field, the *demagnetizing field* caused by the surface poles on the sample, and the local field caused by combined fields of all the other moments around the molecular site.

If the material is in a static magnetic induction field $\vec{B} = B_o\,\hat{z}$ and we assume the magnetization has an $e^{-i\omega t}$ time dependence, the magnetization equation of motion gives

$$\begin{pmatrix} -i\omega M_x \\ -i\omega M_y \\ -i\omega M_z \end{pmatrix} = \gamma \begin{vmatrix} \hat{x} & \hat{y} & \hat{z} \\ M_x & M_y & M_z \\ 0 & 0 & B_o \end{vmatrix} = \begin{pmatrix} \gamma M_y\,B_o \\ -\gamma M_x\,B_o \\ 0 \end{pmatrix}. \tag{16.22}$$

The z-component equation gives us that the time-varying part of $M_z = 0$ – the z-component of the magnetization is constant – while combining the x- and y-component equations we find

$$-i\omega M_x = \gamma B_o \left(\frac{\gamma M_x\,B_o}{i\omega} \right) \tag{16.23}$$

which can only be satisfied if $\left(\frac{\gamma B_o}{\omega}\right)^2 = 1$, which corresponds to a solution where the magnetization is rotating (*precessing*) in the x–y plane with the x-component 90° ($= e^{i\pi/2} = i$) out-of-phase with the y-component at a frequency $\omega_o = \gamma B_o$. This is the fundamental electron magnetic resonance frequency and for internal fields on the order of 0.1 T the precessional frequency is in the microwave frequency range

$$f_o = \omega_o/(2\pi) = \frac{(1.76\times10^{11})(0.1)}{(2\pi)} = 2.8\ \text{GHz}.$$

To model the microwave permeability of a material with local magnetic moments, we would apply Eq. (16.21) above with a total applied field including a DC component $\vec{B} = B_o\hat{z}$, which creates a static magnetization $M_o\hat{z}$, and a much smaller AC component $\vec{B}(t) = \mu_o\,\vec{H}e^{-i\omega t}$, which creates a time-dependent magnetization $\vec{M}e^{-i\omega t}$. Substituting $B_o\hat{z} + \mu_o\,\vec{H}e^{-i\omega t}$ and $M_o\hat{z} + \vec{M}e^{-i\omega t}$ into Eq. (16.21) and neglecting the cross products of the (small) AC field and the (small) time-dependent magnetization, we find

$$\begin{pmatrix} -i\omega M_x \\ -i\omega M_y \\ -i\omega M_z \end{pmatrix} = \gamma \begin{pmatrix} M_y B_o - \mu_o H_y M_o \\ \mu_o H_x M_o - M_x B_o \\ 0 \end{pmatrix}. \tag{16.24}$$

Solving the first two simultaneous equations for M_x and M_y in terms of H_x and H_y we find

$$\begin{pmatrix} M_x \\ M_y \\ M_z \end{pmatrix} = \begin{pmatrix} \chi_{xx} & \chi_{xy} & 0 \\ \chi_{yx} & \chi_{yy} & 0 \\ 0 & 0 & 0 \end{pmatrix} \begin{pmatrix} H_x \\ H_y \\ H_z \end{pmatrix} \tag{16.25}$$

where

$$\chi_{xx} = \chi_{yy} = \frac{\gamma^2 B_o \mu_o M_o}{\gamma^2 B_o^2 - \omega^2} \tag{16.26}$$

and

$$\chi_{xy} = -\chi_{yx} = \frac{i\omega\gamma\mu_o M_o}{\gamma^2 B_o^2 - \omega^2}. \tag{16.27}$$

We recognize this as a Lorentzian oscillator with a resonant frequency B_o and zero damping. Identifying $\omega_o = \gamma B_o$ and $\omega_M = \gamma\mu_o M_o$ and adding a phenomenological damping term $i\alpha\omega$ to represent the relaxation of the magnetic moment through loss processes, we find the (dimensionless) *Polder susceptibility tensor*

$$\begin{pmatrix} M_x \\ M_y \\ M_z \end{pmatrix} = \begin{pmatrix} \dfrac{\omega_o\omega_M}{\omega_o^2 - \omega^2 - i\alpha\omega} & \dfrac{i\omega\omega_M}{\omega_o^2 - \omega^2 - i\alpha\omega} & 0 \\[2ex] \dfrac{-i\omega\omega_M}{\omega_o^2 - \omega^2 - i\alpha\omega} & \dfrac{\omega_o\omega_M}{\omega_o^2 - \omega^2 - i\alpha\omega} & 0 \\[2ex] 0 & 0 & 0 \end{pmatrix} \begin{pmatrix} H_x \\ H_y \\ H_z \end{pmatrix} = \chi_m \begin{pmatrix} H_x \\ H_y \\ H_z \end{pmatrix}.$$

$$\tag{16.28}$$

Substituting $\mu = \mu_o + \mu_o \chi_m$ we find the *Polder magnetic permeability*

$$\mu = \mu_o \begin{pmatrix} 1 + \dfrac{\omega_o \omega_M}{\omega_o^2 - \omega^2 - i a \omega} & \dfrac{i \omega \omega_M}{\omega_o^2 - \omega^2 - i a \omega} & 0 \\ \dfrac{-i \omega \omega_M}{\omega_o^2 - \omega^2 - i a \omega} & 1 + \dfrac{\omega_o \omega_M}{\omega_o^2 - \omega^2 - i a \omega} & 0 \\ 0 & 0 & 1 \end{pmatrix}. \tag{16.29}$$

We can use this permeability in Maxwell's equations for an electromagnetic wave propagating in the $\hat{z}$-direction with a space and time dependence of the form $e^{i(k_z z - \omega t)}$. In this case, we will take the curl of the Maxwell "curl H" equation and eliminate the $\vec{E}$ fields by substituting $\vec{\nabla} \times \vec{E} = -\frac{\partial \vec{B}}{\partial t} = i \omega \mu \vec{H}$

$$\vec{\nabla} \times (\vec{\nabla} \times \vec{H}) = \vec{\nabla} \times \frac{\partial \vec{D}}{\partial t} = (-i\omega)\varepsilon \vec{\nabla} \times \vec{E} = -(i\omega)^2 \varepsilon \mu \vec{H}. \tag{16.30}$$

Using our vector identity $\vec{\nabla} \times (\vec{\nabla} \times \vec{H}) = \vec{\nabla}(\vec{\nabla} \cdot \vec{H}) - \vec{\nabla}^2 \vec{H}$ and recognizing that, since the fields only depend on z and the permeability tensor gives $B_z = \mu_o H_z$ and thus $\vec{\nabla} \cdot \vec{B} = \frac{\partial B_z}{\partial z} = \mu_o \frac{\partial H_z}{\partial z} = 0$, we have

$$-\vec{\nabla}^2 \vec{H} = -(i k_z)^2 \vec{H} = k_z^2 \vec{H} = \omega^2 \varepsilon \mu \vec{H}. \tag{16.31}$$

Writing the permeability tensor as $\mu = \mu_o \begin{pmatrix} v & i\kappa & 0 \\ -i\kappa & v & 0 \\ 0 & 0 & 1 \end{pmatrix}$, where v and κ are the corresponding matrix elements in Eq. (16.29), and expanding the vector equation into component equations, we find

$$k_z^2 H_x = \omega^2 \varepsilon \mu_o (v H_x + i\kappa H_y)$$
$$k_z^2 H_y = \omega^2 \varepsilon \mu_o (-i\kappa H_x + v H_y). \tag{16.32}$$

Eliminating H_y from the two equations and equating the coefficients of H_x, we get the condition on k_z

$$(k_z^2 - \omega^2 \varepsilon \mu_o v)^2 = \omega^4 \varepsilon^2 \mu_o^2 \kappa. \tag{16.33}$$

Taking the positive and negative square roots of this equation we find

$$k_z^2 = \omega^2 \varepsilon \mu_o (v \pm \kappa). \tag{16.34}$$

The two solutions for k_z correspond to left- and right-circularly polarized modes with $H_x = \pm i H_y$, and in terms of the magnetic parameters, we have

$$k_{z\pm} = \frac{\omega}{c} \sqrt{\varepsilon_r} \sqrt{1 + \frac{\omega_M (\omega_o \pm \omega)}{\omega_o^2 - \omega^2 - i a \omega}}. \tag{16.35}$$

Only the k_+ mode – where the $\vec{H}$ vector is circularly polarized in the same sense as the magnetization M_o rotates at resonance – has a pole at $\omega_o = \gamma B_o$, so the two modes have very different propagation characteristics near the resonance frequency ω_o.

In addition, the k_+ mode propagating along the magnetic field corresponds to the k_- mode propagating opposite to the magnetic field: the propagation is *non-reciprocal* and depends on the direction of propagation with respect to the magnetic field. This property of magnetic materials is very useful in microwave devices. For example, *isolators* will propagate a wave in one direction but attenuate the reflected waves that are traveling in the other direction, or *circulators* that are used in radar systems to direct the signal from the high-power pulse generator circuit to the antenna and direct the returning echo signal arriving at the same antenna to the sensitive detector circuit simultaneously without allowing the transmitted signal to go directly to the detector circuit. The selection of appropriate magnetic materials – usually magnetic insulating materials called *ferrites* – for operation in important radar and communication bands remains an area of technological interest.

Chapter 16 problems

1. The constitutive relation $J = \sigma E$ is the microscopic form of Ohm's Law $I = V/R$. The current density $J = I/A$, the electric field is $E = V/\ell$, and the resistance is $R = \left(\frac{1}{\sigma}\right)\frac{\ell}{A}$. From a microscopic model presented in Chapter 16, show that $\vec{J} = \frac{I}{A} = qn\vec{v}$ where $\vec{v}$ is the electron *drift velocity* – the average velocity of the electrons as they are accelerated and scattered.

2. When the Apollo space craft were returning into the atmosphere they would become surrounded by a plasma of ionized molecules, which would cut off the radio transmission from the capsule to the ground station for several minutes until the capsule slowed down enough that the plasma sheath disappeared and the nervous mission control would again hear from the astronauts that they were through re-entry and descending by parachute toward their ocean landing. Why did the plasma sheath cut off the radio transmission and was there a different frequency that could have been used to transmit and receive under re-entry conditions?

3. Apply the equation of motion method to an ion under the influence of a sinusoidal oscillating electric field at frequency ω, $\vec{E} = \vec{E}_o e^{-i\omega t}$, and an elastic restoring force proportional to the displacement Δd with a linear damping parameter Γ, to show that the permittivity is given by a constant term plus a Lorentzian oscillator:

$$\varepsilon(\omega) = \varepsilon_\infty + \frac{\omega_T^2(\varepsilon_s - \varepsilon_\infty)}{\omega_T^2 - \omega^2 - i\Gamma\omega}.$$

4. Aluminum has a density of 2.699 g/cm^3, a molar mass of 26.98 g/mole, and a valence of three conduction electrons per atom.

    a. Estimate the density of conduction electrons (m^{-3}).

    b. If the DC resistivity of aluminum is $\rho = 1/\sigma = 2.75 \times 10^{-8}$ Ω m, estimate the electron scattering time τ.

    c. Using a Drude model for the frequency-dependent conductivity $\sigma(\omega)$, estimate the conductivity of aluminum at a frequency $f = 100$ GHz.

5. Silicon has a low-frequency dielectric constant $\varepsilon_s = 11.8\epsilon_o$ and an electron conductivity effective mass $m^* = 0.2m_o$ where m_o is the free-electron mass. Using a Drude model for the effective relative dielectric permittivity function:

$$\varepsilon^*(\omega) = \frac{\varepsilon_s}{\epsilon_o} - \frac{\omega_p^2}{\omega(\omega + \frac{i}{\tau})}$$

with an electron scattering time $\tau = 1 \times 10^{-13}$ s and an electron density $n = 1 \times 10^{24}$ m^{-3}, use MATLAB to plot the real and imaginary parts of the effective relative dielectric permittivity $\varepsilon_r = \varepsilon/\epsilon_o$ from $\omega = 0.5 \times 10^{13}$ to 5.0×10^{13} s. The *plasma frequency* ω_p is found from $\omega_p^2 = \frac{n q^2}{\epsilon_o m}$. Plot the normal incidence power reflectivity across the same frequency range. At what wavelength does the minimum reflectivity occur?

6. Lead telluride (PbTe) has a transverse optical phonon frequency $\omega_{TO} = 32$ cm^{-1}, a longitudinal optical phonon frequency $\omega_{LO} = 114$ cm^{-1}, a phonon damping parameter $\Gamma = 10$ cm^{-1} and a high-frequency dielectric constant $\varepsilon_\infty/\epsilon_o = 32$. Using a Lorentz model for the permittivity

$$\varepsilon^*(\omega) = \frac{\varepsilon_\infty}{\epsilon_o}\left(1 + \frac{(\omega_{LO}^2 - \omega_{TO}^2)}{(\omega_{TO}^2 - \omega^2 - i\Gamma\omega)}\right),$$

plot the relative dielectric permittivity and the normal incidence power reflectivity from 0 to 150 cm^{-1}. What wavelength corresponds to the minimum in reflectivity? Note: "wave number" (cm^{-1}) is a unit of frequency used in infrared spectroscopy – it is equal to the reciprocal of the wavelength in centimeters. You can convert to ω in s^{-1} by multiplying by $2\pi \times 2.9979 \times 10^{10}$, or you can leave all units in wave numbers and the units will cancel out!

7. Light propagates through a dilute gas of atoms with an electronic transition $\Delta E = 2.5$ eV. The absorption line has a dimensionless oscillator strength $S = 1 \times 10^{-4}$ and the state has a lifetime $T = 10$ ns. Using a Lorentzian approximation for the dielectric permittivity,

$$\varepsilon^*(\omega) = 1 + \frac{S\omega_o^2}{\omega_o^2 - \omega^2 - \frac{i\omega}{T}}$$

plot the real and imaginary parts of the permittivity and calculate the transmission of light through 1 cm of the gas as a function of frequency from 3.5×10^{15} to 4.0×10^{15} s^{-1}. Compare the plots in the case where the lifetime is 1 ns.

17 Fluids

Water cannot be pumped out of a well deeper than about 34 feet with a hand or mechanical pump located on the surface. Why is this? And if this is true how does water get to the top floors of buildings that are over 40 feet high? And how does the size of the pipe affect the water flow? Do these questions have anything to do with how smoke goes up a chimney or how an airplane flies? These are some of the questions that we will address in this chapter where we examine the physics of fluids and the engineering practices of hydrostatics, hydrodynamics, and aeronautics that derive from these principles.

To begin with, what is a fluid and how are the mechanical properties of fluids different from those of the solids that we studied in Chapter 5? Fluids are materials in which the molecules are free to readjust themselves to flow and conform to the boundaries of any container that they are in. A fluid is a material which cannot support shear stresses and thus moves freely in transverse directions to find its state of lowest energy. Highly *viscous* fluids respond more slowly to shear stresses, but all fluids in equilibrium will have zero shear stress. Fluids include *liquids*, which have a free surface between the liquid and gas phases, and *gases* which have no free surface. *Plasmas* are fluids which include free charges and therefore respond to electromagnetic as well as mechanical forces. Plasmas are often described as a "fourth state of matter" (after solids, liquids, and gases) and play critical roles in systems from fluorescent lights to the interiors of stars.

If we put aside the case of plasmas, we can consider fluids as materials that are described by the scalar fields of mass density $\rho_m(x, y, z)$, temperature $T(x, y, z)$, and pressure $P(x, y, z)$ and a vector field of mass velocity $\vec{v}(x, y, z)$. Each of these fields defines the instantaneous value of the parameter for an infinitesimal mass δm of the fluid at the point (x, y, z). The density field is related to the volume δV occupied by a mass element

$$\rho_m(x, y, z) = \lim_{\delta V \to 0} \frac{\delta m}{\delta V}. \tag{17.1}$$

The infinitesimal mass is small enough to be considered a point, but large compared to the actual molecular structure of the fluid. In cases where there is a distribution of values, we consider the value of the field to be the average value. For example, if V' is a volume centered at the point (x, y, z), we can defined $\vec{V}(x, y, z)$ by

$$\vec{v}(x, y, z) = \lim_{V' \to 0} \frac{1}{V'} \int_{V'} \vec{v}(x', y', z') \, dx' \, dy' \, dz'. \tag{17.2}$$

To put some numbers on these parameters, consider air and water as two common fluids. The density of water is very close to 1000 kg/m³ over a wide temperature and pressure range. The density of air depends on the temperature and pressure, but at room temperature and atmospheric pressure at sea level ($T \cong 300\text{K}, P = 1$ atmosphere $\cong 10^5 \text{N}/_{\text{m}^2} = 10^5$ pascals) the density of air is 1.2 kg/m³, about 1/1000 of the density of water. The fluid velocity of air in hurricanes or tornados can be on the order of 100 miles/hour $= 44.7$ m/s. The velocity of water emitted from a fire hose that can reach the fourth story of a building is about 18 m/s. These numbers are about 1/10 of the molecular velocities that can be found from the equipartition theorem. So the air molecules are having multiple collisions and changes of direction inside the fluid volume as it moves.

Question 17.1

Use equipartition to estimate the velocity of air molecules at 300 K. How would you estimate the random thermal velocity of water molecules at 300 K?

The relations between these fields are given by Newton's Laws of Mechanics, just as the relation between the electric and magnetic vector fields are given by Maxwell's Laws of Electromagnetism. Much of the mathematical treatment of fluids will be reminiscent of our analysis of electromagnetic phenomena.

17.1 The continuity equation

The first fundamental principle of mechanics is the conservation of mass – matter can neither be created nor destroyed if nuclear reactions are not on the table. If we consider a non-deforming test volume fixed in space V_o enclosed by a surface S, we can calculate the rate at which matter leaves the volume by (i) finding the density of fluid at an infinitesimal area of the enclosing surface $d\vec{S}$, (ii) multiplying by the (average) velocity normal to the surface at that point $\vec{v} \cdot d\vec{S}$, and (iii) integrating that product over the entire surface. This integral must equal the rate of decrease of mass within the test volume:

$$\frac{\partial}{\partial t} \int_V \rho_m dV = \int_V \frac{\partial \rho_m}{\partial t} dV = -\int_S \rho_m \, \vec{v} \cdot d\vec{S}. \qquad (17.3)$$

Applying Gauss's theorem to convert the surface integral of the dot product to the volume integral of the divergence of the vector, letting the integral shrink to an infinitesimal volume, and equating the integrands, we find the differential form of the continuity equation

$$\frac{\partial \rho_m}{\partial t} = -\vec{\nabla} \cdot (\rho_m \vec{v}) \qquad (17.4)$$

where the negative sign comes from the fact that if mass is flowing out across the surface, the net mass in the volume must decrease.

If we define a fluid current density at any point by $\vec{J}_m = \rho_m \vec{v}$ (the mass flux), we find the mass conservation criteria gives us a continuity equation of exactly the same form as the charge continuity equation we derived in Chapter 14

$$\frac{\partial \rho_m}{\partial t} + \vec{\nabla} \cdot \vec{J}_m = 0. \tag{17.5}$$

We note that the density of the fluid test mass δm can change due to the effect of pressure or temperature on the volume of the fluid element as given by the *equation of state* of the fluid. In general, the volume is a function of temperature and pressure $V(T, P)$ and

$$dV = -\frac{V}{\rho_m} d\rho_m = \left(\frac{\partial V}{\partial T}\right)_P dT + \left(\frac{\partial V}{\partial P}\right)_T dP = \beta V\, dT - \kappa_T V dP \tag{17.6}$$

where we have introduced the parameters *isobaric thermal expansion coefficient* $\beta = \frac{1}{V}\left(\frac{\partial V}{\partial T}\right)_P = -\frac{1}{\rho_m}\left(\frac{\partial \rho_m}{\partial T}\right)_P$ and the *isothermal compressibility* $\kappa_T = -\frac{1}{V}\left(\frac{\partial V}{\partial P}\right)_T = \frac{1}{\rho_m}\left(\frac{\partial \rho_m}{\partial P}\right)_T$. For an ideal gas where $PV = NRT$ (a reasonable approximation for air) $\beta = \frac{1}{T} \sim 0.003$ K^{-1} and $\kappa_T = \frac{1}{P} \sim 10^{-5}$ m^2/$_N$. For water, κ_T is the reciprocal of the bulk modulus B, $\kappa_T = 1/B \sim 4.6 \times 10^{-10}$ m^2/$_N$, while $\beta \sim 2 \times 10^{-4}$ K^{-1}. This implies that a 1% change in volume or density $\left(-\frac{\Delta V}{V} = \frac{\Delta \rho_m}{\rho_m} = 0.01\right)$ would require a pressure increase of 1000 Pa or 0.01 atmosphere in air, but a pressure increase of 2.2×10^7 Pa or about 220 atmospheres in water. The difference in thermal expansion is not as dramatic but water requires 15 times as much temperature change as air to cause the same percentage change in density. Thus, to first approximation we can often assume that water and other liquids are incompressible with $\frac{\partial \rho_m}{\partial t} = 0$ and, if there are no sources or sinks, the divergence of the mass flow is zero, $\vec{\nabla} \cdot (\rho_m \vec{v}) = 0$.

Example 17.1 A stream of water flowing smoothly from a faucet narrows as it falls as show in the figure below. Using the continuity equation, find the area A_1 of the water stream in terms of the area A_o of the same stream a distance h above A_1.

Solution: Since water is an incompressible fluid, ρ_m is constant, and we must have $\vec{\nabla} \cdot \vec{J}_m = \rho_m \vec{\nabla} \cdot \vec{v} = 0$, which from Gauss's Law implies that $\oint_S \vec{v} \cdot d\vec{S} = 0$ since the density of the stream doesn't change in continuous flow. According to our fundamental kinematic equations, the velocity of a volume of water will increase as it falls with the (constant) acceleration of gravity by $v_1^2 = v_o^2 + 2gh$. If we consider a Gaussian surface S as a truncated cone with horizontal surfaces with a top area A_o and bottom area A_1 the fluid flux over the side is zero since there is no flow to the side. The surface integral over the top and bottom surfaces gives us

Continued

Example 17.1 (*cont.*)

$$-v_o A_o + v_1 A_1 = 0 \Rightarrow v_o^2 A_o^2 = v_1^2 A_1^2 = (v_o^2 + 2gh)A_1^2.$$

Solving for A_1 we find

$$A_1 = A_o \sqrt{\frac{v_o^2}{(v_o^2 + 2gh)}}.$$

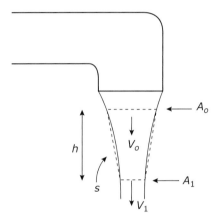

17.2   The Euler equation

The fundamental equations of fluids result from applying Newton's Laws to regions of the fluid described by our scalar pressure, temperature, and density fields and the vector velocity field. Consider an element of an invicid fluid (a fluid without appreciable viscosity) in one-dimensional motion with area A and width Δx shown in Figure 17.1. The total force on the element is the pressure field at x times the area minus the pressure field at $x + \Delta x$ times the area

$$F = PA - (P + \Delta P)A = -\Delta PA. \tag{17.7}$$

We have implicitly used the concept here that pressure is a *scalar*; the pressure at any point in a fluid is exerted in all directions with equal magnitude. This follows from Newton's Third Law.

Using Newton's Second Law, defining the mass of the element $m = \rho_m V$, we find

$$\frac{dp_x}{dt} = \rho_m V \frac{dv_x}{dt} = -\Delta PA, \tag{17.8}$$

where we have implicitly assumed that the fluid is incompressible, so ρ_m is constant. Since the volume $V = A\Delta x$, we have $\rho_m \frac{dv_x}{dt} = -\frac{\partial P}{\partial x}$. Now we need to recognize that the fluid element here is moving with the fluid as described by the vector velocity field,

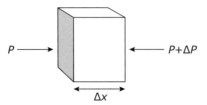

Figure 17.1 Forces on a fluid element due to the pressure change in the fluid.

which changes with position x, and the variable x is also a function of time $x = x_o + v\,dt$. The element velocity changes as the element moves through the fluid as given by the velocity field and, in addition, the pressure gradient changes its acceleration. Using the chain rule for differentiation of a function of x and t we have

$$\frac{dv_x}{dt} = \frac{\partial v_x}{\partial t} + \frac{\partial v_x}{\partial x}\frac{\partial x}{\partial t} = \frac{\partial v_x}{\partial t} + v_x\frac{\partial v_x}{\partial x} \equiv \frac{Dv_x}{Dt} \tag{17.9}$$

where we have introduced the notation $\frac{D}{Dt} = \frac{\partial}{\partial t} + \vec{v}\cdot\vec{\nabla}$ for the *convective derivative*, an operation which is used extensively in fluid mechanics. In three dimensions, we have

$$\frac{d\vec{v}}{dt} = \frac{D\vec{v}}{Dt} = \frac{\partial\vec{v}}{\partial t} + (\vec{v}\cdot\vec{\nabla})\,\vec{v} = -\frac{1}{\rho_m}\vec{\nabla}P. \tag{17.10}$$

If there are other external forces, such as gravity, operating on the volume element we can add the external forces to the right side

$$\frac{\partial\vec{v}}{\partial t} + (\vec{v}\cdot\vec{\nabla})\vec{v} = -\frac{1}{\rho_m}\left(\vec{\nabla}P - \frac{\vec{F}_{ex}}{V}\right). \tag{17.11}$$

This is the Euler equation, which represents the application of Newton's Second Law to a volume element of an invicid (incompressible) fluid.

If we add the viscosity of the fluid to the Euler equation as another force, we will get the Navier–Stokes equation, which is the starting point for most fluid dynamics calculations. (The concept of viscosity will be discussed in some detail later in this chapter. At this point, one can think of viscosity as the quantity that describes the resistance of the fluid to flow.) We note that the $(\vec{v}\cdot\vec{\nabla})\vec{v}$ term on the left-hand side is quadratic in velocity and thus the fluid dynamics equations are inherently non-linear. This represents an increase in complexity over anything we discussed in electromagnetics, where Maxwell's equations and the simplified constitutive relations we considered were all linear equations. But first we will consider the case where the fluid is at rest and the velocity field is zero everywhere, $\vec{v}(x, y, z) = 0$. This is the case of fluid statics or hydrostatics.

17.3   Fluids at rest

The pressure in a fluid increases with depth, a phenomenon well-known to scuba divers and to mountain climbers. Consider the pressure at a depth y under the surface of an

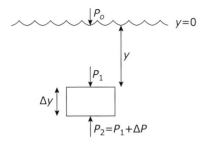

Figure 17.2 Pressures on a rectangular volume a distance y under the surface of a fluid.

incompressible fluid with density ρ_m as shown in Figure 17.2. Euler's equation above for a static fluid with $\vec{v}(x, y, z) = 0$ gives us

$$\vec{\nabla}P - \frac{\vec{F}_{ex}}{V} = 0. \qquad (17.12)$$

The external gravitational force on a test volume V is $F_{ex} = -mg = -\rho_m Vg$, where the acceleration due to gravity is $g \cong 9.8$ m/s^2, so $\frac{\vec{F}_{ex}}{V} = -\rho_m g\,\hat{y}$. Expressing the gradient of pressure as $\frac{dP}{dy}$, we find

$$\frac{dP}{dy} = -\rho_m g. \qquad (17.13)$$

Question 17.2

Show that a force balance equation on the test volume gives us the same equation.

The pressure at any depth under the surface of water can be found by integrating Eq. (17.13)

$$P(y) = \int_{-y}^{0} -\rho_{H_2O}\, g \, dy = P_o + \rho_{H_2O} g y \qquad (17.14)$$

where P_o is the pressure at the surface $P(y = 0)$. The second term in Eq. (17.14) is called the *gage pressure*, because it is the pressure that would be read on a pressure gage which is calibrated to read zero at the surface. The gage pressure is equal to the weight of water supported by a unit area at depth y: $P = W/A = mg/A = \rho_m Vg/A = \rho_m yg$. The density of water $\rho_{H_2O} \cong 1000$ kg/m^3, which implies that the gage pressure increases approximately one atmosphere for every 10 m (~33 feet) below the surface that a diver goes. A diver at 100 feet below the surface is operating at approximately three atmospheres of gage pressure.

The atmospheric pressure variation at any elevation above the Earth's surface can be found from the same analysis except that air is not an incompressible fluid and the density depends on the pressure and temperature. If we neglect the variation of the acceleration of gravity g for the distances within the atmosphere, we find

$$dP = -\rho_{Air} \, g \, dy = -(AW)\frac{N}{V} g \, dy \tag{17.15}$$

where we have set the density of air equal to the number of moles (N) per volume (V) times the atomic weight of air $AW \cong 28 \times 10^{-3}$ kg/mole. Assuming that air can be modeled by an ideal gas law $PV = NRT$ and assuming that T is nearly constant, we find

$$dP = -(AW)\frac{P}{RT} g dy. \tag{17.16}$$

Dividing by P and integrating both sides

$$\ln P = -\frac{AW}{RT} gy \Rightarrow P(y) = P_o \, e^{-\frac{AW}{RT}gy} \tag{17.17}$$

where P_o is the pressure at $y = 0$ ($P_o = 1$ atmosphere $\cong 10^5 \frac{N}{m^2}$) and the pressure goes to zero as $y \to \infty$. By using force balance we can see that this pressure must be equal to the weight of a column of air with cross-section area 1 m^2 extending from the altitude y to the top of the atmosphere.

We note that the pressure at any position y does not depend on the transverse position, only on the value of y representing the vertical distance of the position in the fluid. This implies that the pressure does not depend on the shape of the container. So the pressure at the dashed line is the same in the containers shown in Figure 17.3. At first sight, it seems as if the fluid at Point A at the edge of the container in the center is not supporting the same height of fluid as the fluid at Point B in the center of the container. However, we have to remember that pressure in a fluid is isotropic, equal in all directions. By Newton's Third Law, the upward pressure on the container wall directly above Point A is the same as the downward pressure from the container wall on the fluid (and this is also equal to the pressure at the height of the container lip directly above Point B). The pressure at Point A is equal to the container pressure plus the weight (per unit area) of the fluid directly above Point A. It is apparent that this is equal to the total weight (per unit area) of the fluid above Point B.

17.3.1   Buoyancy

We are all familiar with the concept that objects are "lighter" when they are immersed in a fluid. That is, of course, how objects that are lighter than an equal volume of water float and how we are able to stay on the surface of the water when we swim. Clearly, there is a force which opposes gravity when an object is immersed in a fluid. The name of this force is the *buoyant force* and is due to the increase in pressure with depth in a fluid. As illustrated in Figure 17.4, the forces acting on a rectangular solid submerged in a fluid include the weight of the object $W = mg$, which is the force with which the Earth

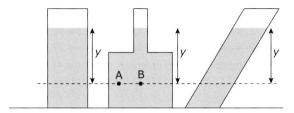

Figure 17.3 The pressure at the dashed line is the same for each container filled with the same liquid at the same ambient pressure. The pressure depends only on the height of the fluid y above the line.

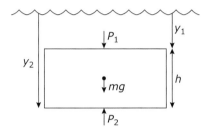

Figure 17.4 Pressures and forces on a rectangular solid of mass m under the surface of a liquid.

is pulling down on the object, the force due to the pressure at the bottom $F_2 = P_2A$ (where A is the area of the bottom rectangule) pushing up, and the force due to the pressure on the top $F_1 = P_1A$ pushing down. Adding these forces gives us the net force on the object

$$F_{net} = P_2A - P_1A - mg \tag{17.18}$$

where we have defined the positive forces pushing up and the negative forces pulling down. (There are also, of course, forces due to the pressure on the sides of the solid, but the forces pushing the object to the left are canceled out by those pushing the object to the right, and similarly for the forces pushing in and out.) As we have just learned, $P_2 = P_1 + \rho_m g h$ where h is the height of the solid. Substituting in the equation above, we find

$$F_{net} = \rho_m g h A - mg \tag{17.19}$$

from which we recognize the first term as $g\rho_m V = gm_{liq}$, which is the weight of the volume of the liquid which is displaced by the solid immersed in it. So, the net force on the object is the weight of the object pushing down minus a force equal to the weight of the displaced liquid pushing up. If the object is denser than the liquid, its weight will be greater than the weight of the displaced water and the object will sink, albeit with less acceleration than g. If the object is less dense than the liquid, the object will rise in the liquid because the buoyant force will be greater than the object's weight when it is completely immersed in it. When an object is floating on the surface the buoyant force, equal to the weight of a volume of liquid equal to the volume of the submerged portion of the object, is exactly equal to the weight of the object. We have shown this for a rectangular solid, but in fact it can be demonstrated that it is true regardless of the shape of the object, a result know as Archimedes' Principle: *When an object is in a fluid, it is acted on by a buoyant force, which opposes the force of gravity on the object and is equal to the weight of the displaced fluid.*[8]

Question 17.3

Is your weight decreased by the air your body displaces? Would you weight more or less in Denver than in Boston? Why and by how much?

Example 17.2 Suppose that some ice is floating in a container. If the ice melts, will the level of water in the container rise, fall, or stay the same? How much would sea level rise if the ice cap at the North Pole were to melt?

Solution: This is a trick question because to first approximation the answer is "not at all." The ice cap at the North Pole is over water, so the ice is floating in water. In equilibrium (with the ice not moving either up or down), the buoyant force is exactly equal to the weight of the ice. So the ice is displacing a mass of water equal to its own mass. Like an ice cube floating in a glass of water, the volume of ice floating above the water is just the volume by which the ice exceeds the volume of water from which it was frozen. When the ice melts it exactly fills the space of the water that it was displacing! If we look a little farther, the polar ice has considerably less salt content than the sea water in which it is floating and salt water is denser than fresh water. Thus, the less dense water that would result from the melting ice would have a greater volume than the salt water displaced by the ice cap. The density of salt water is only about 2.5% higher than fresh water, so this effect is not great.

This should not lead anyone to be less concerned about global warming, however, because the ice caps on Greenland and Antarctica are over land and not water. If the Greenland ice cap melted it is estimated it would raise the sea level over 20 feet!

17.3.2   Manometers and hydraulic head

If a closed tube is filled with a liquid and then inverted into a container of the same liquid exposed to an ambient pressure as shown in Figure 17.5, the liquid will not completely run out of the tube, but will be supported at some height h by the ambient pressure pushing on the liquid surface compared to the near vacuum[9] left at the top of

---

[8] This principle is only tangentially related to Archimedes' famous "Eureka!" moment when he stepped in a bathtub, observed the overflowing water, and realized that he could find the volume of the irregularly shaped crown that the king suspected was not pure gold. If he collected the displaced water and compared its volume with the volume displaced by an equal mass of pure gold, he would be comparing the volumes of equal masses, i.e., densities.

[9] The space at the top of the tube is not actually a vacuum; it is vapor from the liquid at a pressure equal to the *vapor pressure* of the liquid at the ambient temperature. For water at 20 °C this pressure is about 2.3 kPa, about 2.3% of atmospheric pressure. For mercury, the vapor pressure at 20°C is about 0.17 Pa, about 0.0002% of atmospheric pressure.

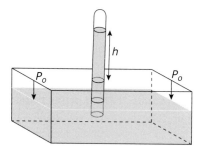

Figure 17.5 A manometer with a column of fluid of height h supported by the atmospheric pressure P_o.

the column. Based on our earlier discussion, it is easy to see that the ambient pressure is equal to the pressure exerted by the column of liquid of height (or depth) h, i.e., $P_o = \rho_{liq}gh$. This principle was extensively used to measure variations in the ambient air pressure due to altitude or weather. The *barometric pressure* of the atmosphere is still commonly quoted in "inches," which refers to the height, in inches, of a column of mercury supported. A *torr* is a pressure unit equal to the pressure that would support 1 millimeter of mercury.

Example 17.3 Standard atmospheric pressure is $P_o = 1$ atmosphere $= 1.013 \times 10^5$ N/ m^2. How many inches of mercury (Hg) does this pressure correspond to? How many inches of water? If the barometric pressure drops from 31 inches to 29 inches of Hg (indicating the approach of a *low pressure area* – usually correlated with storms and precipitation), how much has the pressure decreased in pascals? (1 Pa $= 1$ N/m^2). If you pump up your bicycle tires to 60 lb/in^2, how will the pressure in your tires change due to the barometric pressure change?

Solution: From the relation $P_o = \rho_{liq}gh$ and with a density of mercury $\rho_{Hg} = 13.55 \times 10^3$ kg/m^3, we find $h \cong \frac{1.013 \times 10^5}{(9.8)(13.55 \times 10^3)} = 0.762$ m $\cong 30.03$ inches of Hg. For water with a density $\rho_{H_2O} = 1.00 \times 10^3$ kg/m^3, we find $h \cong 10.3$ m $= 407$ in $\cong 34$ feet!

A decrease in pressure of 2 inches of mercury corresponds to a pressure change $\Delta P = 2$ in$\left(2.54 \times 10^{-2} \frac{m}{in}\right)\left(13.55 \times 10^3 \frac{kg}{m^3}\right)\left(9.8 \frac{m}{s^2}\right) = 6.746 \times 10^3$ Pa. (Note that the percentage change $\frac{\Delta P}{P_o} = \frac{6.746 \times 10^3}{1.013 \times 10^5} = \frac{2}{30.03} = 6.66\%$.) Your tire pressure gage is calibrated in pounds per square inch *above ambient pressure*. Using 1 lb/in$^2 = 6895$ Pa, your tire pressure decreased from $(60)(6895) + (31)(0.0254)(9.8)$ $(13.55 \times 10^3) = 5.18 \times 10^5$ Pa to $(60)(6895) + (29)(0.0254)(9.8)(13.55 \times 10^3)$ $= 5.11 \times 10^5$ Pa.

Hydraulic head is a measure of the pressure expressed in terms of the height (usually in feet) of a column of water that could be supported by that pressure. The fact that one atmosphere of pressure corresponds to a column of water 34 feet high creates an inherent limitation on the vertical height that water can be pumped out of a well by a suction pump located at the top of the well. In a suction pump, the water is pushed up by the pressure difference between the ambient pressure and the reduced pressure created at the surface by the pump. The maximum suction that the pump can generate is to create a vacuum in the line at the top of the well giving a pressure difference equal to the ambient pressure on the water at the bottom of the well – usually air pressure of one atmosphere. As we have seen, one atmosphere of pressure can support a column of water only 34 feet high, so this is a maximum limit on the depth of well that the suction pump can service. In fact, due to inefficiencies in the pump and water flow, surface suction pumps are usually rated for wells less than 25 feet.

So, given this fundamental limit, how is water pumped from wells deeper than 34 feet, and how is water provided to the top stories of high-rise apartment buildings? We will find one answer in the next section, and a more complete solution when we consider fluid flow.

17.3.3   Pascal's Law

A property of incompressible fluids is that, if they are enclosed, they transmit pressure changes uniformly throughout the fluid. This effect is the basis of *Pascal's Law: a change of pressure ΔP applied to an incompressible fluid enclosed in a rigid container is transmitted undiminished to every portion of the fluid and to the walls of its container.* This is easily demonstrated in a container with a free surface. We have already demonstrated that the pressure at any depth in an incompressible fluid with density ρ_{Liq} is given by

$$P(y) = P_o + \rho_{Liq}gy. \tag{17.20}$$

Clearly, if we increase the pressure at the free surface from P_o to $P_o + \Delta P$ the pressure $P(y)$ at any depth will increase by the same ΔP. Since we know that pressure is a scalar, the normal force against any part of the enclosure with area A will be increased by ΔPA.

Pascal's Law is the basis for *hydraulic press* devices, such as the one shown in Figure 17.6. Force F_1 exerted on a piston of area A_1 creates a change in pressure in the fluid of $\Delta P = F_1/A_1$. This ΔP is transmitted through the hydraulic fluid – usually some kind of oil that is light and lubricating for the pistons – through the connecting *hydraulic lines* to a larger piston where it creates a force $F_2 = \Delta PA_2 = F_1\frac{A_2}{A_1}$. By selecting the ratio of the areas of the pistons A_1 and A_2 the force may be increased by any factor desired. Since the volume of the fluid is fixed, the change in volume in the small piston, A_1d_1 (where d_1 is the distance that the force F_1 moves the small cylinder), is the same as the change in the volume A_2d_2 in the large cylinder. Thus

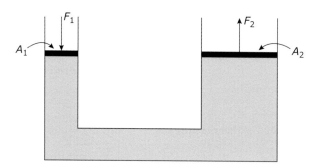

Figure 17.6 A hydraulic press with the force F_1 transmitted through the fluid by Pascal's Law to balance a larger force F_2.

the ratio of the distances that the forces are exerted over is in inverse ratio to the magnitude of the forces

$$\frac{d_1}{d_2} = \frac{A_2}{A_1} = \frac{F_2}{F_1}. \tag{17.21}$$

This implies that $F_1 d_1 = F_2 d_2$, and thus the work done by the small piston is exactly the same as the work done by the large cylinder, as must be the case from conservation of energy. If the pistons are not at the same height, $P_1 = F_1/A_1$ will exceed P_2 by the weight of the column of liquid corresponding to the difference in heights h. In this case, some of the work done by F_1 will go into the change in potential energy of the fluid column; this work will be recovered if the lever is allowed to return to its initial state. Hydraulic pistons are used in hydraulic jacks, hydraulic brakes, power steering, automobile lifts, flight surface controls in airplanes, trash compactors, and many other applications.

Pascal's Law also provides a solution for pumping water from a deep well. A submersible pump is installed at the bottom of the well, which uses electric power transmitted over wires from the surface to operate a piston, which repeatedly pushes water into the pipe leading to the surface. By Pascal's principle, the pressure pulse created by the pump at the bottom of the well will be transmitted by a pipe to the well head at the top. The submersible pump must be capable of exerting a pressure to support the column of water from the well water level to the surface, but the force required can be adjusted by setting the ratio of the piston size to the pipe size. The volume of flow available is limited by the size and speed of the pump piston. Bottom-pumped systems are usually built with elevated water storage tanks which can be pumped full over an extended period and will then provide a steady water pressure based on the gravitation head between the tank and the water faucets. Most large apartment buildings in New York City have water tanks of various types and architectures on top of the building. These tanks are filled by compression pumps on the lower floors, which work continuously in periods of low water demand to provide the required water volume at acceptable pressures during periods of high water use.

17.4   Fluids in motion

At any instant in time, we can characterize the velocity of the fluid at any point by the vector field $\vec{v}\,(x, y, z, t)$. We will begin our study of fluid flow by considering an *ideal fluid* with the following properties:

1.   The fluid is *incompressible and uniform,* $\rho_m =$ constant.
2.   The fluid is assumed to have *negligible viscosity,* so we ignore fluid friction effects, which convert mechanical energy into heat.
3.   The flow is assumed to be *irrotational,* which means we ignore the possibility of eddy currents in the flow. Formally, this means $\vec{\nabla} \times \vec{v} = 0$.
4.   Finally, we will look for solutions that involve steady-state flow conditions, so the velocity field at any point is not changing in time $\frac{\partial \vec{v}}{\partial t} = 0$.

In real fluids, we would usually need to consider all these effects that we are neglecting. An ideal fluid bears approximately the same relation to real fluids as the ideal gas model bears to real gases. As a first approximation, it will yield useful results; we will consider deviations from ideal fluid behavior later in the chapter.

17.4.1   Eulerian flow and streamlines

In steady flow, the path that a particular particle will follow through the fluid flow, as could be made visible by injecting streams of dye into the fluid, is called a *streamline.* The velocity of the particles at any point therefore is tangent to the streamline. The **streamlines never intersect** because the particle at the intersection would have two separate velocity tangents from each of the streamlines, leading to an ambiguity in trajectory. Assuming that the streamlines represent approximately equal mass flow, at points in the fluid where the velocity field is larger, the streamlines are closer together. The streamlines are continuous and, for steady flow, they do not change in time.

Laminar flow is non-fluctuating flow along the streamlines (as opposed to turbulent flow). In Figure 17.7, we see laminar flow around a cylindrical object at rest with the streamlines shown. The flow is clearly faster where the fluid particles have to travel in a longer path to get around the object as indicated by the closer spacing of the streamlines.

In the *Eulerian* description of fluid motion, we concentrate on the time evolution of a control volume of the fluid based on the velocity field in the fluid. The control

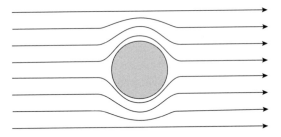

Figure 17.7 Streamlines of a fluid flowing around a cylindrical object.

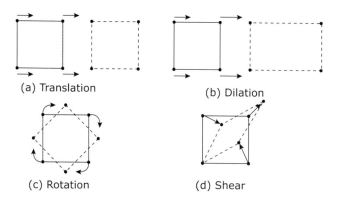

(a) Translation   (b) Dilation

(c) Rotation   (d) Shear

Figure 17.8 Four possible distortions of a control volume in a fluid.

volume moves with the fluid and no fluid crosses the boundaries of the volume. If the velocity field is constant in time and space, as it is far to the left of the obstacle in Figure 17.7, the control volume translates in the direction of the fluid flow without change in shape as shown in Figure 17.8(a). If the velocity is in the x-direction alone ($v_y = 0$) and the velocity increases linearly with x ($v_x = Cx$), then the evolution of the control volume is as shown in Figure 17.8(b). We note that this motion entails a decrease in density since the volume increases but, by definition, no fluid crosses the control volume boundaries. This change in density is shown by the fact that $\vec{\nabla} \cdot \vec{v} = \frac{\partial v_x}{\partial x} = C$. In steady state, this corresponds to a density which decreases as a function of x. For an ideal incompressible fluid with ρ_m constant, on the other hand, we have $\vec{\nabla} \cdot \vec{v} = 0$.

In Figure 17.8(c), we have a rotation of the control volume – v_x is a function of y and v_y is a function of x, so that $\vec{\nabla} \times \vec{v} \neq 0$. Irrotational flow with $\vec{\nabla} \times \vec{v} = 0$ implies that there is no rotation of the Eulerian control volumes.

Finally in Figure 17.8(d), we have both $v_x(x, y)$ and $v_y(x, y)$ changing, resulting in a shear deformation of the control volume. The volume of the new shape is equal to the initial volume, which is consistent with the condition of incompressibility of the fluid, and, in fact, some shear deformation is clearly necessary when the streamlines wrap around an object as in Figure 17.7. **It can be shown that any fluid motion can be constructed from a combination of the fundamental motions shown in Figure 17.8.**

17.4.2 Bernoulli's equation in an ideal fluid

If we apply the vector calculus identity $\vec{\nabla}(\vec{v} \cdot \vec{v}) = 2\vec{v} \times (\vec{\nabla} \times \vec{v}) + 2(\vec{v} \cdot \vec{\nabla})\vec{v}$ to the Euler equation Eq. (17.11), we get the following form, which is helpful for applying the irrotational flow condition

$$\frac{\partial \vec{v}}{\partial t} + \frac{1}{2}\vec{\nabla}(\vec{v} \cdot \vec{v}) - \vec{v} \times (\vec{\nabla} \times \vec{v}) = -\frac{1}{\rho_m}(\vec{\nabla}P - \rho_m g\hat{y}). \tag{17.22}$$

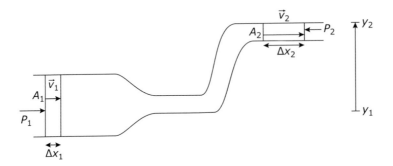

Figure 17.9 Fluid flow in pipe due to a pressure differential.

For our assumptions that $\vec{\nabla} \times \vec{v} = 0$ and the steady-state assumption $\frac{\partial \vec{v}}{\partial t} = 0$, we find

$$\frac{1}{2} \vec{\nabla} \left(\vec{v} \cdot \vec{v} \right) = \vec{\nabla} \frac{v^2}{2} = -\frac{1}{\rho_m} \left(\vec{\nabla} P - \rho_m \, g\hat{y} \right). \tag{17.23}$$

Writing the gravitational force as the gradient of the gravitational potential $\rho_m g\hat{y} = -\vec{\nabla}(\rho_m g y)$ our equation becomes

$$\vec{\nabla} \left[\frac{1}{2}\rho_m v^2 + \rho_m g y + P \right] = 0 \tag{17.24}$$

which implies that the terms inside the brackets, evaluated for any volume element, do not depend on position, and by our steady-state assumption they also do not depend on time. Therefore,

$$\frac{1}{2}\rho_m v^2 + \rho_m \, g y + P = constant. \tag{17.25}$$

This is *Bernoulli's equation* applied to the ideal fluid model.

Bernoulli's equation can be seen to be equivalent to the work–energy theorem: the work done is equal to the change in kinetic energy plus the change in (gravitational) potential energy $W = F\Delta x = \Delta KE + \Delta PE$. Consider the motion of a fluid through a tapered, rising pipe as shown in Figure 17.9.

Equation (17.25) implies that

$$\frac{1}{2}\rho_m v_1^2 + \rho_m \, g y_1 + P_1 = \frac{1}{2}\rho_m v_2^2 + \rho_m \, g y_2 + P_2. \tag{17.26}$$

Because the ideal fluid is incompressible, the volume change $\Delta V_1 = A_1 \Delta x_1$ is the same as $\Delta V_2 = A_2 \, \Delta x_2$. Multiplying the left side of Equation (17.26) by ΔV_1 and the right side by the identical ΔV_2, we have

$$\frac{1}{2}\Delta V_1 \rho_m v_1^2 + \Delta V_1 \rho_m \, g y_1 + \Delta V_1 P_1 = \frac{1}{2}\Delta V_2 \rho_m v_2^2 + \Delta V_2 \rho_m \, g y_2 + \Delta V_2 P_2. \tag{17.27}$$

We identify $\frac{1}{2}\Delta V_1 \rho_m v_1^2 = \frac{1}{2}mv_1^2 = KE_1$ and $\frac{1}{2}\Delta V_2 \rho_m v_2^2 = \frac{1}{2}mv_2^2 = KE_2$ where m is the mass of the fluid in the volumes ΔV_1 and ΔV_2. Similarly, $\Delta V_1 \rho_m \, g \, y_1 = mg \, y_1 = PE_1$ and $\Delta V_2 \rho_m \, g \, y_2 = mg \, y_2 = PE_2$. Rearranging Equation (17.27), we find

$$\Delta V_1 P_1 - \Delta V_2 P_2 = \frac{1}{2}\Delta V_2 \rho_m v_2{}^2 - \frac{1}{2}\Delta V_1 \rho_m v_1{}^2 + \Delta V_2 \rho_m g y_2 - \Delta V_1 \rho_m g y_1 = \Delta KE + \Delta PE.$$

(17.28)

Finally, the terms on the left-hand side can be converted to $\Delta V_1 P_1 = A_1 \Delta x_1 P_1 = F_1 \Delta x_1 = W_1$, which is the work done by the force due to pressure P_1. Similarly, $\Delta V_2 P_2 = F_2 \Delta x_2$ is the work done *on* the fluid due to pressure P_2. So, the left-hand side of the equation is the difference in the work done due to P_1 and the work done due to P_2, which is just the total work done $W_T = W_1 - W_2 = \Delta KE + \Delta PE$ as required by conservation of energy and the work–energy theorem.

Example 17.4 A horizontal circular pipe 8 inches in diameter carries water that flows at a rate of 12 ft³/minute. Find the velocity of the fluid. If the pipe smoothly necks down to 3 inches in diameter, what is the velocity of the fluid in the smaller pipe? If the pressure of the fluid in the larger pipe is 2 psi above atmospheric pressure, what is the pressure in the smaller pipe?

Solution: The flow rate $\Delta V / \Delta t = A \Delta x / \Delta t = Av$. Converting units we find

$$v_1 = \frac{12 \text{ ft}^3 (0.0283 \text{ m}^3/\text{ft}^3)}{(60 \text{ s})\pi[(4 \text{ in})(0.0254 \text{ m/in})]^2} = 0.174 \text{ m/s}.$$

Since the fluid is assumed to incompressible, the flow rate in the smaller pipe is the same with a radius of 1.5 in instead of 4 in. The velocity in the second pipe is thus 1.24 m/s. Applying Bernoulli's equation with no change in height ($\Delta y = 0$), we have

$$\frac{1}{2}\rho_m v_1{}^2 + P_1 = \frac{1}{2}\rho_m v_2{}^2 + P_2$$

$$\Rightarrow P_2 = P_1 - \frac{1}{2}\rho_m(v_2{}^2 - v_1{}^2) = P_1 - \frac{1}{2}(1000 \text{ kg/m}^3)(1.24^2 - 0.174^2) = P_1 - 754 \text{ Pa}.$$

With $P_1 = 1$ atmosphere $+ 2$ lb/in² $= 101300 + 2(6856) = 115090$ Pa, we find $P_2 = 114340$ Pa.

The result that the pressure is less in the smaller pipe with the fluid moving faster than in the larger pipe appears to be counterintuitive. However, the pressure in the larger pipe does the work to accelerate the water, so it must be greater than the pressure in the faster-moving fluid in the smaller pipe.

Since pressure is isotropic, the pressure against the walls in the smaller pipe is also less than the pressure against the walls in the larger pipe. We can use this effect to create a pump. A Venturi tube is a tube with a constriction that increases the velocity of the fluid and lowers the internal pressure. An input tube in the constricted high-velocity, low-pressure section can draw fluid into the pipe as shown in Figure 17.10. A Venturi well pump has two pipes going down into the well. Water is pumped down one pipe where it is forced through a Venturi constriction at the bottom. The reduced pressure in the

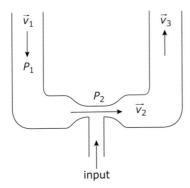

Figure 17.10 A Venturi pump where the pumping pressure is created by the reduction of pressure in regions of higher fluid velocity.

constricted section draws water into the return pipe which returns it to the surface. Venturi pumps can work for wells over 100 feet deep with all the working parts located at the surface. Pumps using the Venturi effect are also used as sump pumps to pump ground water out of the basement of a house. The advantage is that they can be powered with the pressure of water in the mains and thus will continue to work when the electric power fails, a common occurrence during a hurricane when water is likely to flood a basement.

At first glance, it appears that for our ideal fluid the flow through the pipes does not depend on the diameter of the pipe. If the pipes are smaller, the velocity just increases to maintain the continuity of the fluid. In the more usual case where the velocity is limited by the hydraulic head (see Problem 17.9), the volume (or mass) of fluid flow depends on the cross-sectional area of the pipe – πR^2 for a circular pipe. In fact, the dependence on pipe radius is much stronger than that. To understand the dependence of the flow on the pipe diameter we need to remove the condition of ideality – in particular, we need to examine the effect of *fluid viscosity*.

17.4.3   Viscosity in fluids

Thus far, we have assumed that a fluid supports no transverse stresses. This could only be rigorously true if the fluid molecules didn't extend or move in the direction transverse to the fluid motion. Our understanding is that the fluid test volumes that we are considering contain on the order of 10^{10} molecules in random thermal motion at thermal velocities much greater than the fluid velocity. So, when one (virtual) layer of fluid slips over another layer, the molecular exchange between the layers creates a viscous force that reduces the relative velocity between the two layers as the collective fluid motion is degraded, by the collisions, into random thermal motion of the molecules in the fluid. The effect of this viscous fluid force can be modeled in Euler's equation by adding an external force proportional to the vector Laplacian of the velocity field

$$\frac{\partial \vec{v}}{\partial t} + (\vec{v} \cdot \vec{\nabla})\vec{v} = -\frac{1}{\rho_m} \vec{\nabla}P + \frac{1}{\rho_m}\eta \vec{\nabla}^2 \vec{v} = -\frac{1}{\rho_m}\vec{\nabla}P + v\vec{\nabla}^2 \vec{v} \tag{17.29}$$

where η is the *coefficient of viscosity* and $v = \eta/\rho_m$ is the *kinetic viscosity coefficient*.[10] This equation is the *Navier–Stokes equation* for incompressible fluids.

Let's consider the common application of flow of fluid through a cylindrical pipe of radius R, a problem referred to as *Poiseuille flow*. We will assume that the system is in steady state with the velocity everywhere in the z-direction down the pipe $\vec{v} = v_z \hat{z}$ (although v_z can depend on the position r measured from the center of the pipe). The Navier–Stokes equation for this case reduces to

$$\frac{\partial v_z}{\partial t}\hat{z} + v_z \frac{\partial}{\partial z} v_z \hat{z} = -\frac{1}{\rho_m}\vec{\nabla}P + \frac{1}{\rho_m}\eta\nabla^2 v_z \hat{z} \tag{17.30}$$

where $\nabla^2 v_z$ is the (scalar) Laplacian of the scalar value v_z. In steady state, $\frac{\partial v_z}{\partial t} = 0$ and, in addition, since the fluid is incompressible and the flow is parallel to the walls, $\vec{\nabla}\cdot\vec{v}\,\frac{\partial v_z}{\partial z} = 0$ as well, and the equation reduces to

$$\eta\nabla^2 v_z \hat{z} - \vec{\nabla}P = 0. \tag{17.31}$$

The gradient of the pressure in cylindrical coordinates is $\vec{\nabla}P = \frac{\partial P}{\partial r}\hat{r} + \frac{1}{r}\frac{\partial P}{\partial \varphi}\hat{\varphi} + \frac{\partial P}{\partial z}\hat{z}$. The problem has cylindrical symmetry so all derivatives with respect to the azimuthal angle φ are zero, and the $\hat{r}$-component of the equation is just

$$\frac{\partial P}{\partial r}\hat{r} = 0 \tag{17.32}$$

which **implies that the pressure does not depend on the distance from the cylindrical axis r but only on the distance down the pipe z.** The Laplacian in cylindrical coordinates is given by $\nabla^2\Phi = \frac{1}{r}\frac{\partial}{\partial r}\left(r\frac{\partial\Phi}{\partial r}\right) + \frac{1}{r^2}\frac{\partial^2\Phi}{\partial\varphi^2} + \frac{\partial^2\Phi}{\partial z^2}$ and the $\hat{z}$-component of the Navier–Stokes equation is

$$\frac{1}{r}\frac{\partial}{\partial r}\left(r\frac{\partial v_z}{\partial r}\right) + \frac{\partial^2 v_z}{\partial z^2} = -\frac{1}{\eta}\frac{\partial P}{\partial z}. \tag{17.33}$$

By the continuity of the flow $\frac{\partial v_z}{\partial z} = 0$, while the pressure has been shown to not depend on either the radial position r or the angle φ. The pressure is therefore constant in planes of constant z and $\frac{\partial P}{\partial z} = \frac{\Delta P}{\Delta l}$, where ΔP is the total pressure drop across the pipe and Δl is the length of the pipe. Thus

$$\frac{1}{r}\frac{\partial}{\partial r}\left(r\frac{\partial v_z}{\partial r}\right) = -\frac{1}{\eta}\frac{\Delta P}{\Delta l}. \tag{17.34}$$

Multiplying through by r and integrating once, we find

$$\left(r\frac{\partial v_z}{\partial r}\right) = -\frac{1}{\eta}\frac{\Delta P}{\Delta l}\frac{r^2}{2} + C_1. \tag{17.35}$$

---

[10] The symbol v for the *kinetic viscosity coefficient* should not be confused with that of the fluid velocity $\vec{v}$.

Dividing through by r and integrating again, we find the solution

$$v_z = -\frac{1}{4\eta}\frac{\Delta P}{\Delta l}r^2 + C_1 \ln r + C_2 \tag{17.36}$$

where C_1 and C_2 are constants of integration which need to be determined by the boundary conditions. To keep the flow at $r = 0$ from becoming infinite, we set $C_1 = 0$. The other boundary condition is given by the usual assumption that *the fluid velocity adjacent to a solid surface is the same as the velocity of the surface.*

This approximation is easy to justify if we are considering the velocity normal to the surface. If the fluid were to move at a different velocity normal to the surface, the fluid would either penetrate into the surface or the fluid would disengage from the surface leaving a vacuum behind. The phenomenon of *cavitation* where the liquid separates from the solid surface is, in fact, observed in high-speed phenomena in fluids (for example, at the surface of high-speed boat propellers). But in most low-velocity applications, the normal motion of the fluid tracks with the velocity of any solid surface. If we consider the molecular interactions tangent to the surface, the fluid molecules are constantly colliding with the surface molecules of the solid and transferring momentum to the surface. It seems reasonable to assume that the fluid molecules right next to the solid surface would come to an equilibrium situation where their tangential velocity is the same as the tangential velocity of the surface. That is the reasoning that leads to the usual boundary conditions that both the normal and tangential velocity of the fluid at a solid surface will be equal to the velocity of the surface.

In the case of the stationary pipe we are considering, this implies that $v_z(R) = 0$. Thus we have the solution

$$v_z = \frac{1}{4\eta}\frac{\Delta P}{\Delta l}(R^2 - r^2) \tag{17.37}$$

shown in Figure 17.11. This solution, with the velocity entirely in the z-direction and equal to zero at the walls and reaching a maximum velocity in the center is referred to as *Poiseuille flow.*

The total flow is found by integrating the fluid flow $J_m = \rho_m v_z$ over the area of the pipe

$$\frac{\Delta Q}{\Delta t} = \rho_m \int_0^R \frac{1}{4\eta}\frac{\Delta P}{\Delta l}(R^2 - r^2)2\pi r \, dr = \frac{\pi \rho_m}{8\eta}\frac{\Delta P}{\Delta l}R^4. \tag{17.38}$$

This result is called the *Poiseuille formula*. We see that flow in a viscous fluid is extremely sensitive to the pipe size, increasing as the fourth power of the pipe radius.

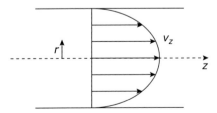

Figure 17.11 Fluid velocity as a function of distance from the center of the pipe for laminar Poiseuille flow in a pipe.

Example 17.5 For a pressure differential of 7 psi, what is the flow rate in gallons per minute in a 100 foot long pipe of 3-inch diameter? How much more water flows in a pipe of diameter 5 inches?

Solution: The viscosity of water is highly temperature dependent. Assuming the problem is at room temperature (20 °C), standard tables give the coefficient of viscosity (*dynamic viscosity*) of water as 1.002×10^{-3} Pa · s. We note that if we simply remove the density ρ_m from Eq. (17.38), we get the flow in m^3/s (instead of kg/s). Applying unit conversions we find for the flow rate

$$\frac{\pi}{8\eta}\frac{\Delta P}{\Delta l}R^4 = \frac{\pi 7\left(6895\,\frac{\text{Pa}}{\text{psi}}\right)}{8(1.002 \times 10^{-3}\,\text{Pa·s})}\frac{\left(\frac{3}{2}\,\text{in}\right)^4(0.0254\,\frac{\text{m}}{\text{in}})^3}{100(12)}\left(264\,\frac{\text{gallons}}{\text{m}^3}\right) = 345\,\frac{\text{gallons}}{\text{s}}$$

Increasing the diameter to 5 inches increases the flow rate by a ratio of $\left(\frac{5}{3}\right)^4 = 7.71$ giving a flow rate of 2663 gallons/s.

17.4.4 Reynolds number and turbulent flow

If we express the Navier–Stokes equation

$$\frac{\partial \vec{v}}{\partial t} + (\vec{v} \cdot \vec{\nabla})\vec{v} = -\frac{1}{\rho_m}\vec{\nabla}P + v\vec{\nabla}^2\vec{v} \tag{17.39}$$

in terms of normalized variables for velocity, length, and pressure,

$$\vec{\omega} = \frac{\vec{v}}{V} \qquad \vec{x}' = \frac{\vec{x}}{L} \qquad P_n = \frac{P}{P_1} \tag{17.40}$$

where V, L, and P_1 are "typical values" of velocity, length, and pressure for the problem we are considering,[11] we obtain a dimensionless equation

$$\frac{\partial \vec{\omega}}{\partial t} + (\vec{\omega} \cdot \vec{\nabla})\vec{\omega} = -\frac{P_1}{\rho_m V^2}\vec{\nabla}P_n + \frac{v}{VL}\vec{\nabla}^2\vec{\omega} = -\frac{P_1}{\rho_m V^2}\vec{\nabla}P_n + \frac{1}{Re}\vec{\nabla}^2\vec{\omega}. \tag{17.41}$$

The dimensionless constant $Re = \frac{VL}{v} = \frac{VL\rho_m}{\eta}$ is called the *Reynolds number* and it represents the ratio of the *inertial* fluid forces from the left-hand side of the equation, $\frac{\partial \vec{\omega}}{\partial t} + (\vec{\omega} \cdot \vec{\nabla})\,\vec{\omega}$, to the viscous forces given by the $\frac{1}{Re}\vec{\nabla}^2\vec{\omega}$ term. When the inertial forces dominate and the ideal liquid restrictions on vortices ($\vec{\nabla} \times \vec{\omega} = 0$ or irrotational flow) are removed, the non-linear term $(\vec{\omega} \cdot \vec{\nabla})\,\vec{\omega}$ can make small changes in initial conditions result in large changes in the final solution. This is referred to as *turbulent flow*, the opposite of the *laminar flow* of ideal fluids that we have been considering so far. For small Reynolds numbers, the viscous forces damp out the extreme fluid motions and the fluid dynamics problems are more tractable.

---

[11] For example, for flow in a pipe we might select V to be the velocity at the center of the pipe, L to be the diameter of the pipe, and P_1 to be the pressure at $z = 0$.

Nobel Laureate Richard Feynman described fluid flow with turbulence as "the most important unsolved problem of classical physics." The problem is that dynamics at high Reynolds number are *chaotic* and, essentially, incalculable. These are the problems where infinitesimal changes in the initial conditions make enormous changes in the solution: "a butterfly's wing-beat in Brazil causes a tornado in Texas." Problems approaching this limit include calculating the fluid motion around a high-speed propeller, shed vortices of air currents behind a landing airplane that buffet the following planes, the gas/plasma flow around reentry vehicles entering the atmosphere, or even falling sheets of water in a waterfall. The fluid flow in any of these situations is too complex and too sensitive to small changes in the initial conditions for detailed calculation to be made.

While flow in the turbulent limit is not calculable, this does not imply that turbulence is necessarily undesirable, even when it could be avoided. The dimples on golf balls, for example are designed to promote a transition to turbulence in which the drag is reduced (even though the surface fluid friction is greater) by disturbing the boundary layer formation. It has also been found that there are predictable patterns in systems approaching the chaotic limit, including fluids during the approach to turbulent flow.

Example 17.6 Calculate the Reynolds number for the flow in the two diameters of pipe in Example 17.5.

Solution: The velocity of water in the center of the 3-inch pipe ($r = 0$) is

$$v_z = \frac{1}{4\eta} \frac{\Delta P}{\Delta l}(R^2) = \frac{1}{4(1.002 \times 10^{-3})} \frac{7(6895)}{100(0.3048)}(3 \times 0.0254)^2 = 550 \text{ m/s}.$$

Using the *diameter* of the pipe as the characteristic length, we find

$$Re = \frac{VL\rho_m}{\eta} = \frac{(550)(3 * 0.0254) * (1000)}{1.002 \times 10^{-3}} = 4.18 \times 10^7.$$

(Note that the Reynolds number is dimensionless.) The 5-inch pipe has a maximum velocity $(5/3)^2$ larger and a characteristic length $(5/3)$ larger, which implies a Reynolds number $Re = (2.09 \times 10^7)(5/3)^3 = 1.16 \times 10^8$. These are high values, characteristic of turbulent flow, as you might guess thinking about flow from a fire hose. Turbulence in Poiseuille flow in pipes typically begins in the range of Reynolds number from about 2000 to 4000.

Question 17.4

Estimate the Reynolds number of a tractor trailer moving at 60 miles per hour on a highway. Do you think it is possible to calculate the air currents that shake your car when you pass the truck?

Chapter 17 problems

1. Suppose that a ship is floating on closed lock. (a) If the crew drops the metallic cargo into the water, what happens to the level of the water? (b) What will happen to the water level if the cargo it dumps in it has a density less than that of water and it floats? Justify your answers.

2. Consider an incompressible fluid flowing steadily horizontally through a pipe of diminishing diameter. If the speed of the fluid is 2 m/s when the diameter is d_1, and 3 m/s when the diameter is d_2 ($d_1 > d_2$) and the horizontal distance between the two points is 12 cm, estimate the acceleration of the fluid.

3. If the velocity vector is given by $v_x = Kx$; $v_y = -Ky$; $v_z = 0$, where K is a constant greater than zero, calculate and plot the streamlines. [*Hint:* First show that $\frac{dr}{V} = \frac{dx}{v_x} = \frac{dy}{v_y} = \frac{dz}{v_z}$, where dr is the arc length along the velocity vector field V and dx, dy, dz are the corresponding arc lengths along the x-, y-, and z-components of the velocity vector.]

4. For the velocity field $V = Kyt\hat{x} - Kxt\hat{y} + 0k$, calculate (a) the *curl* $\vec{V}$ and (b) $\vec{\nabla}^2 \vec{V}$.

5. Consider a cylindrical and a conical container of the same base and height both filled with water. Which one exerts a larger force on the base? Justify your answer.

6. A helium balloon is tied inside an accelerating car moving in a straight line. Which way will the balloon move? What if the car is moving on a curved path with constant speed? Justify your answers.

7. Consider an incompressible fluid flowing steadily horizontally through a pipe from point 1 to point 2. The cross-sectional area and the velocity at points 1 and 2 are A, V and a, v, correspondingly, with $A > a$. Using Bernoulli's equation and the equation of continuity, show that the speed of the fluid at point 1 is given by:

$$V = \sqrt{\frac{2a\Delta p}{\rho(a^2 - A^2)}}$$

    where Δp ($= p_2 - p_1$) is the pressure difference between points 1 and 2 and ρ is the density of the fluid. This equation allows us to calculate the fluid speed at particular points in a pipe, as Δp can be measured using a Venturi meter.

8. Using the result in the previous problem, calculate the rate of water flow (in m³/s), if the inner diameter of the large pipe is 8 cm, of the small pipe $\frac{8}{\sqrt{2}}$ cm and the pressure difference between the two points is 14 kPa.

9. Show by energy conservation that the velocity of water flow from a horizontal pipe in the bottom of a large tank is given by $V = \sqrt{2gh}$, where h is the height of water in the tank.

Bibliography

John Allison, *Electronic Engineering Semiconductors and Devices*, McGraw-Hill, 1990.

Neil W. Ashcroft and N. David Mermin, *Solid State Physics*, Cengage Learning, 1976.

Henry A. Bent, *The Second Law: An Introduction to Classical and Statistical Thermodynamics*, Oxford University Press, 1965.

Max Born and Emil Wolf, *Principles of Optics, 7th Edition*, Cambridge University Press, 1999.

Audrey L. Companion, *Chemical Bonding*, McGraw-Hill, 1964.

Charles A. Coulson, *Waves*, Oliver and Boyd, 1955.

Robert Eisberg and Robert Resnick, *Quantum Physics of Atoms, Molecules, Solids, Nuclei, and Particles, 2nd Edition*, Wiley & Sons, 1985.

Richard P. Feynman, Robert B. Leighton, and Matthew Sands, *The Feynman Lectures on Physics*, Addison-Wesley, 1963.

Grant R. Fowles, *Introduction to Modern Optics, 2nd Edition*, Dover Publication, 1975.

Robert A. Granger, *Fluid Mechanics*, Dover Books, 1995.

David Halliday, Robert Resnick, and Jearl Walker, *Fundamentals of Physics, 9th Edition*, Wiley & Sons, 2011.

Magdy Iskander, *Electromagnetic Fields and Waves*, Prentice-Hall, 1992.

Harold Jeffreys, *Cartesian Tensors*, Cambridge University Press, 1963.

Linus Pauling, *College Chemistry*, W. H. Freeman and Co., 1964.

William Prager, *Introduction to Mechanics of Continua*, Dover Phoenix Editions, 2004.

Frederick Reif, *Fundamentals of Statistical and Thermal Physics*, McGraw-Hill, 1965.

John R. Reitz, Frederick J. Milford, and Robert W. Christy, *Foundations of Electromagnetic Theory, 4th Edition*, Addison Wesley, 1993.

Rolf H. Sabersky, Allen J. Acosta, Edward G. Hauptmann, and E. M. Gates, *Fluid Flow: A First Course in Fluid Mechanics, 4th Edition*, Prentice-Hall, 1998.

Liang Chi Shen and Jin Au Kong, *Applied Electromagnetism, 3rd Edition*, PWS Publishers, 1995.

Ronald F. Soohoo, *Microwave Magnetics*, Harper and Row, 1985.

M. F. Spotts, *Mechanical Design Analysis*, Prentice-Hall, 1964.

Ben G. Streetman and Sanjay Banerjee, *Solid State Electronic Devices, 5th Edition*, Prentice-Hall, 2000.

D. J. Tritton, *Physical Fluid Dynamics*, Oxford Science Publications, 1988.

Carmine Vittoria, *Microwave Properties of Magnetic Films*, World Scientific, 1993.

Jearl Walker, *The Flying Circus of Physics, 2nd Edition*, Wiley & Sons, 2007.

Charles M. Wolfe, Nick Holonyak, Jr., and Gregory E. Stillman, *Physical Properties of Semiconductors*, Prentice-Hall, 1989.

Index